CAMBRIDGE STUDIES IN MODERN OPTICS: 9

The elements of nonlinear optics

The elements of nonlinear optics

PAUL N. BUTCHER

Professor of Theoretical Physics
University of Warwick
Coventry, Warwickshire, England

DAVID COTTER

Scientific Advisor
British Telecom Research Laboratories
Martlesham Heath, Ipswich, Suffolk, England

The right of the University of Cambridge to print and sell all manner of books was granted by Henry VIII in 1534. The University has printed and published continuously since 1584.

CAMBRIDGE UNIVERSITY PRESS

CAMBRIDGE
NEW YORK PORT CHESTER
MELBOURNE SYDNEY

Published by the Press Syndicate of the University of Cambridge
The Pitt Building, Trumpington Street, Cambridge CB2 1RP
40 West 20th Street, New York NY 10011, USA
10 Stamford Road, Oakleigh, Melbourne 3166, Australia

First published 1990

Printed in Great Britain at The Bath Press, Avon

British Library cataloguing in publication data

Butcher, Paul N.
The elements of nonlinear optics.
1. Nonlinear optics
I. Title II. Cotter, David
535′.2

Library of Congress cataloguing in publication data

Butcher, Paul N.
The elements of nonlinear optics / Paul N. Butcher, David Cotter.
p. cm.
Includes bibliographical references.
ISBN 0-521-34183-3
1. Nonlinear optics. I. Cotter, D. (David), 1950– .
II. Title. III. Title: Nonlinear optics.
QC446.2.B88 1990
535.2--dc20 89-22187 CIP

ISBN 0 521 34183 3

dc

Contents

Preface

In recent years there has been a rapid expansion of activity in the field of nonlinear optics. Judging by the proliferation of published papers, conferences, international collaborations and enterprises, more people than ever before are now involved in research and applications of nonlinear optics. This intense activity has been stimulated largely by the increasing interest in applying optics and laser technology in telecommunications and information processing, and has been propelled by significant advances in nonlinear-optical materials.

The origins of these recent developments can be traced through three decades of work since the invention of the laser and the first observations of nonlinear-optical phenomena by Franken *et al* (1961). From the earliest days it was recognised that such phenomena can have useful practical applications; for example, effects such as optical-frequency doubling allow the generation of coherent radiation at wavelengths different from those of the available lasers. In the 1960s many of the most fundamental discoveries and investigations were made. Work was then mainly concerned with the interaction of high-power lasers with inorganic dielectric crystals, gases and liquids. Effects such as parametric wave-mixing, stimulated Raman scattering and self-focusing of laser beams were investigated intensively. The invention of the wavelength-tunable dye laser paved the way for the development during the 1970s of many tunable sources utilising nonlinear effects, such as harmonic generation, sum- and difference-frequency mixing, and stimulated Raman scattering. In this way tunable coherent radiation became available in a wide spectral range from the far infrared to vacuum ultraviolet. Some of these sources used dielectric crystals as the nonlinear medium, whilst others used atomic and molecular gases; in the latter cases the efficiency was increased by tuning the input radiation close to the resonant frequencies of the medium. These new tunable light sources were used increasingly for spectroscopy, and numerous spectroscopic techniques were developed which themselves relied on nonlinear-optical processes, such as coherent anti-Stokes Raman scattering, and multi-photon absorption and ionisation. Mode-locked lasers capable of generating pulses of sub-nanosecond duration became more widely available during the 1970s, allowing time-resolved nonlinear spectroscopy and coherent transient effects to be studied.

The mid- to late-1970s also saw the advent of telecommunications using optical fibres, and this has been the most important stimulus for the explosion of effort in nonlinear optics during the last decade. By the late 1980s and early 1990s it is common to find information carried on a laser beam in the process of communication, storage, retrieval, printing or sensing, and there is increasing effort to achieve ever greater data-processing capabilities, by using ultrafast optoelectronic interactions and by harnessing the properties of laser light for highly-parallel manipulation and interconnection of signals. Nonlinear optics provides the key to many of these future developments, and exploits the properties of new materials (such as multiple-quantum-well semiconductors, fibre wave-guides, organic polymers and photorefractive materials). New effects (such as optical bistability, phase conjugation and optical solitons) have been discovered which can make use of these nonlinearities for signal processing. In summary, the field of nonlinear optics has become enormously diverse, with applications in many branches of science and engineering.

In writing this book our objective was *not* to review this vast range of activity. Several excellent books have already appeared which cover specialised aspects and, most notably, the recent book by Shen (1984) provides a broad review. However, the diversity of the field has meant that, especially for the newcomer, the underlying principles often lie hidden in the technical literature amongst a wealth of detail. Therefore, this book is intended to fulfil the need for a self-contained account of the most important principles and theory of nonlinear optics, fully developed from basic concepts. The book is written for graduate students of physics and electrical engineering, as well as for the increasing numbers of scientists and engineers entering the field. The book is also directed to established researchers who need a source reference for the derivation of fundamental formulae, with confidence that the notation, vexing numerical factors and units are treated consistently. No prior knowledge of quantum mechanics, apart from the most simple notions, is required of the reader. All the essential quantum-mechanical apparatus is derived fully and explained. Only a familiarity with simple calculus, vectors and electromagnetic theory is assumed.

In 1965 one of us (PNB) published a monograph *Nonlinear Optical Phenomena*, based on a series of lectures at Ohio State University during 1963–4; this work has been out of print for many years. Those readers familiar with the monograph may recognise some sections and passages in the earlier chapters of this book. The original material has been revised extensively to bring it into line with modern conventions and practice, and represents not quite one-quarter of the present book.

Chapter 1 is a brief general introduction to nonlinear optics. The starting point for the main development in Chapter 2 is the constitutive relations between the polarisation and electric field. The chapter also contains a representative catalogue of nonlinear-optical phenomena, and particular attention is paid to conventions and the various numerical factors. Chapter 3 on quantum mechanics serves two purposes: with the unfamiliar reader in mind, we provide a self-contained tutorial review starting from basic concepts; we also derive various formulae for later use. Throughout the book, with a few simple exceptions, we use the semiclassical approach in which the light is treated classically but the nonlinear medium is treated quantum-mechanically. Chapter 4 is the core of the first half of the book, in which the formulae for the linear and nonlinear susceptibilities are derived. Two different approaches are fully developed: that using electric-dipole operators, and the alternative in terms of momentum operators. The latter are of greatest value in the discussion of systems, such as semiconductors, which contain mobile charged particles. Topics such as resonance enhancement and local-field factors are also discussed here. The various symmetry properties of the susceptibilities are discussed in Chapter 5. In Chapter 6 we develop the descriptions appropriate to resonant processes, which are of increasing importance today. We place particular emphasis on the links between the simple model for a two-level system and the susceptibility formalism, and show how the two approaches complement each other. Chapter 7 is concerned with propagation and wave-coupling processes in nonlinear media. The object here is to provide practical descriptions of processes such as harmonic generation, frequency mixing, parametric amplification and stimulated scattering. More specialised topics, including phase conjugation and soliton effects in optical fibres, are also discussed.

There are very many kinds of nonlinear-optical media of interest. However, because of the need to keep the book to a reasonable size, we have selected only one material system for detailed discussion (in Chapter 8). We chose semiconductors for a number of reasons: first, semiconductors provide useful illustrations of many aspects from earlier chapters; second, in recent years there have been very significant advances in semiconductor materials and these are likely to remain of the greatest technological importance; and lastly, this reflects the current research interests of the authors. Finally, in Chapter 9 we describe the background to the present intense interest in new artificial materials for nonlinear optics. Examples of applications of the general theory are scattered throughout the book. We use SI units, and the relations between alternative systems of units and definitions are given in appendices.

Because of the vast scope of the subject and the limits of space, the contents of the book are inevitably a compromise. We hope, however, that the choice of topics will provide an understanding of the broad principles. We consider that the material presented here represents the minimum knowledge necessary for an appreciation of key aspects of the current literature.

We are grateful to our many past and present colleagues who have influenced the writing of this book; we are especially indebted to T.P. McLean, R. Loudon, D.C. Hanna and M.A. Yuratich. Nonlinear optics has developed through the efforts of countless researchers. For brevity, however, references are limited to those which supplement the text. We sincerely apologise to the many authors of important works whom we have failed to cite by name. We acknowledge with thanks the permission of those authors and publishers whose figures are reproduced or adapted here.

This book was set in type by the authors using the *eroff* and *eqn* text-formatting codes. Permission to use a *UNIX* system at British Telecom Research Laboratories is gratefully acknowledged. We are grateful to C.J. Todd for his support, and to the system administrators for much invaluable help. Finally, we thank sincerely Freda Butcher and Sarah Cotter, Colin, Kirsty and Simon, for their tolerance and encouragement.

Coventry, Warwickshire *Paul Butcher*

Woodbridge, Suffolk *David Cotter*

January 1990

1

Introduction

'Nonlinear' optical phenomena are not part of our everyday experience. Their discovery and development were possible only after the invention of the laser.

In optics we are concerned with the interaction of light with matter. At the relatively low light intensities that normally occur in nature, the optical properties of materials are quite independent of the intensity of illumination. If light waves are able to penetrate and pass through a medium, this occurs without any interaction between the waves. These are the optical properties of matter that are familiar to us through our visual sense. However, if the illumination is made sufficiently intense, the optical properties begin to depend on the intensity and other characteristics of the light. The light waves may then interact with each other as well as with the medium. This is the realm of nonlinear optics. The intensities necessary to observe these effects can be obtained by using the output from a coherent light source such as a laser. Such behaviour provides insight into the structure and properties of matter. It is also utilised to great effect in nonlinear-optical devices and techniques which have important applications in many branches of science and engineering.

Another effect of light on matter can sometimes be to induce changes in the chemical composition; such 'photochemical' processes lie outside the subject of this book.

1.1 Origins of optical nonlinearity

We now consider in a simple way how nonlinear-optical behaviour might arise. The materials which concern us in optics can be thought of as a collection of charged particles: electrons and ion cores. When an electric field is applied the charges move; the positive charges tend to move in the direction of the field, whilst negative ones move the opposite way. In conductors, some of the charged particles are free to move through the material for as long as the electric field is applied, giving rise to a flow of electric current. In dielectric materials, on the other hand, the charged

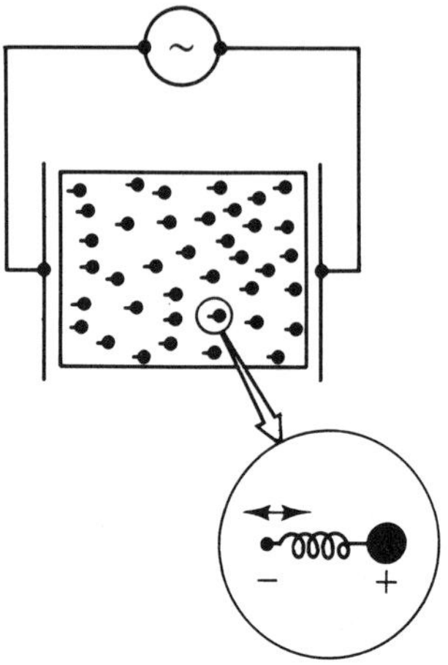

Fig. 1.1 Schematic representation of the motion of bound charges in a dielectric medium which is subject to an electric field alternating at the frequency of visible light. The motion of the ions is insignificant compared to that of the much lighter electrons.

particles are bound together, although the bonds do have a certain 'elasticity'. Therefore, the motion of the charges is *transitory* when the field is first applied; they are displaced slightly from their usual positions. This small movement – positive charges in one direction and negative ones in the other – results in a collection of induced electric-dipole moments. In other words, the effect of the field on a dielectric medium is to induce a polarisation.

A light wave consists of electric and magnetic fields which vary sinusoidally at 'optical' frequencies ($\sim 10^{13} - 10^{17}$ Hz). The motion of the charged particles in a dielectric medium in response to an optical electric field is therefore oscillatory; they form oscillating dipoles. The effect of the optical magnetic field on the particles is much weaker and we neglect it here. The positively-charged particles – the ion cores – have much greater mass than electrons and so, for high optical frequencies (in the ultraviolet and visible regions of the spectrum), it is the motion of the electrons that is most significant. The response of an electron to the optical electric field is that of a particle in an anharmonic potential well. We can think of this in terms of a simple mechanical analogy. Suppose the electron, of mass m and charge $-e$, is attached to the mother ion by a spring, as shown in Fig. 1.1. For simplicity we consider the case shown where the electric dipoles are all oriented in the same way, in the direction of the field. The position of the electron varies in response to the electric field $E(t)$ in a manner governed by the equation of motion for an oscillator:

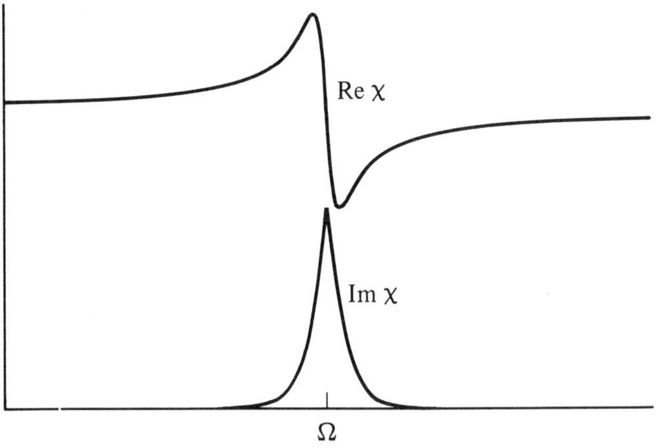

Fig. 1.2 Variation of the real and imaginary parts of the susceptibility χ with frequency in the region of a resonant frequency Ω.

$$m\left[\frac{d^2x}{dt^2} + 2\Gamma\frac{dx}{dt} + \Omega^2 x - (\xi^{(2)}x^2 + \xi^{(3)}x^3 + \cdots)\right] = -eE(t), \quad (1.1)$$

where x is the displacement from the mean position, Ω is the resonance frequency, and Γ is a damping constant. The term on the right-hand side of (1.1) represents the force exerted on the electron by the applied field which drives the oscillations. We ignore the anharmonic terms $\xi^{(2)}x^2 + \xi^{(3)}x^3 + \cdots$ for the moment, and consider the harmonic response to an applied electric field of the form:

$$E(t) = E_0\cos(\omega t) = \tfrac{1}{2}E_0[\exp(-i\omega t)+\exp(i\omega t)], \quad (1.2)$$

where ω is the optical frequency. Substituting (1.2) into (1.1) gives a linear equation whose solution is

$$x = \frac{-eE_0}{2m}\,\frac{\exp(-i\omega t)}{\Omega^2-2i\Gamma\omega-\omega^2} + \text{c.c.}, \quad (1.3)$$

where c.c. denotes the complex conjugate. If there are N electric dipoles per unit volume, the polarisation induced in the medium is $P = -Nex$. We can express this linear dependence of the polarisation P on the field E in terms of the susceptibility χ as

$$P = \tfrac{1}{2}\,\varepsilon_0\chi E_0\exp(-i\omega t) + \text{c.c.}, \quad (1.4)$$

where

$$\chi = \frac{Ne^2}{\varepsilon_0 m}\,\frac{1}{\Omega^2-2i\Gamma\omega-\omega^2}, \quad (1.5)$$

and ε_0 is the free-space permittivity. The electric dipoles (and the

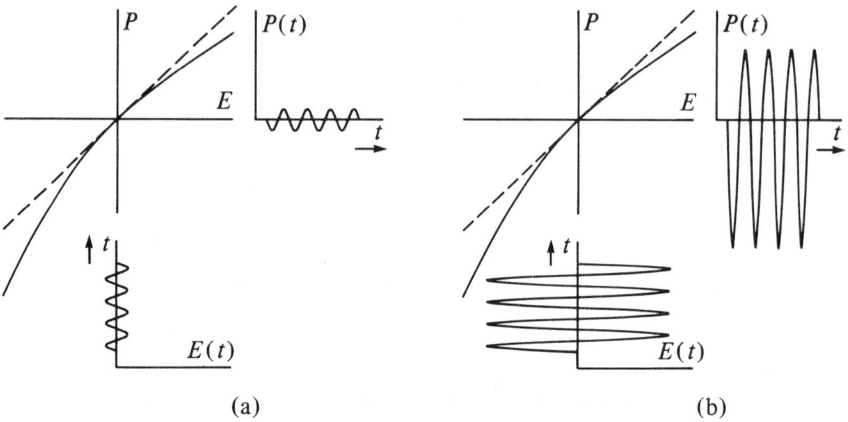

Fig. 1.3 The effect of a nonlinear dependence of the polarisation P on the electric field E is shown. For small input fields (a), P does not depart significantly from a linear dependence (dashed line). At larger fields (b), the polarisation has a distorted waveform which contains significant components at harmonic frequencies.

polarisation) therefore oscillate at the same frequency as the incident optical field. They radiate into the medium and modify the way in which the wave propagates. Since the electric displacement $D=\varepsilon_0 E+P$, we see that the dielectric constant is $1+\chi$ and the refractive index is $\mathrm{Re}\sqrt{1+\chi}$. Losses in the medium are allowed for by the imaginary part of χ, which takes into account the component of P in quadrature with the field. The real and imaginary parts of (1.5) are sketched in Fig. 1.2. These describe the familiar 'linear' optical properties of a medium.

In physics the linear dependence of one physical quantity on another is almost always an approximation, which is valid over only a limited range of values. In our case, the motion of the charged particles in a dielectric medium can be considered to be linear with the applied field only if the displacement x is small. For larger displacements the restoring force is significantly nonlinear in x; in terms of the mechanical analogy, the spring becomes distorted when the extension or compression is large. This nonlinearity is accounted for in (1.1) by including the terms which represent an additional anharmonic restoring force $m(\xi^{(2)}x^2+\xi^{(3)}x^3+\cdots)$, where $\xi^{(2)}$, $\xi^{(3)}$,... are constants. In Fig. 1.3 we show how an anharmonic response gives rise to an induced polarisation which can be considered to be either linear (to a good approximation) or significantly nonlinear, depending on the magnitude of the applied field. Spectral analysis of the polarisation wave in case (b) shows that, in addition to the major component oscillating at the input frequency ω, it contains significant components oscillating at the harmonic frequencies 2ω, 3ω,..., and a d.c. component (at zero frequency). This is analogous to the well-known harmonic distortion of signals in an electrical circuit whose response is not perfectly linear. Now, an important fact in electromagnetic theory is that an oscillating electric dipole emits a radiation field at the frequency of oscillation. This is also true of a collection of dipoles. (We made use of this fact in the previous discussion of linear-optical properties.) Therefore the component of the polarisation that oscillates at the second-harmonic frequency 2ω can radiate a field at 2ω. This is the process of second-harmonic generation.

When the anharmonic terms are included, there is no longer an exact solution for the equation of motion (1.1). However, provided the anharmonic terms are small compared with the harmonic one, we can solve (1.1) to successive orders of approximation by expressing x as a power series in E. Equivalently, we can expand the polarisation P in the form:

$$P = \varepsilon_0 (\chi^{(1)}E + \chi^{(2)}E^2 + \chi^{(3)}E^3 + \cdots). \tag{1.6}$$

Here $\chi^{(1)}$ denotes the linear susceptibility discussed previously, and the quantities $\chi^{(2)}$, $\chi^{(3)}$,... are called the nonlinear susceptibilities of the medium.

Shortly we shall review some of the many nonlinear-optical phenomena that can occur. First, however, we consider how large the incident optical field must be to allow atoms and molecules to reveal their nonlinear properties. From the above discussion it is apparent that, for significant nonlinearity arising from the anharmonic motion of electrons, we require an incident field which is not entirely negligible in comparison with the internal field E_a which binds together the electrons and ions; typically $E_a \sim 3\times10^{10}$ V m^{-1}. To obtain an optical field of such a magnitude, an incident intensity of $\sim 10^{14}$ W cm^{-2} is required. Intensities of this magnitude can be achieved by focusing the powerful picosecond-duration pulses that are obtainable from mode-locked lasers. However, such high intensities are not in fact necessary for the observation of many nonlinear-optical effects. One reason is that, provided the assembly of induced dipoles oscillates coherently (*i.e.*, with a definite phase relationship between them), the field that they radiate individually can, in certain circumstances, add together constructively to produce a much larger total intensity. In nonlinear optics this condition of constructive interference is known as 'phase matching'. For example, the characteristic length for significant second-harmonic generation under phase-matched conditions is $L \sim \lambda E_a/E$, where E is the incident optical field and λ is its wavelength.

For $\lambda = 1\ \mu$m, we see that $L \sim 1$ cm for a power density of 1 MW cm^{-2} (3×10^6 V m^{-1} incident field). This intensity can be readily achieved using a pulsed laser of modest size. Indeed, the first observation of second-harmonic generation by Franken *et al* (1961) (which marked the beginning of nonlinear optics) used one of the very earliest lasers. It was found subsequently that, in favourable circumstances, the incident laser beam can be converted to a beam at double the frequency with an efficiency exceeding 50%. This provides a valuable method for obtaining powerful coherent radiation at a wavelength shorter than the available laser.

The intensity required to observe some nonlinear processes can be reduced further by many orders of magnitude by choosing one or more of the optical frequencies so that they lie close to a resonant frequency of the oscillating dipoles; this is termed 'resonance enhancement'. In nonlinear optics, resonance enhancement is utilised in two ways: First, it allows nonlinear processes and devices to operate effectively at lower power levels, thus increasing their range of use and efficiency. Second, resonant nonlinear phenomena provide the basis for 'nonlinear spectroscopy'; the observation of these effects can provide information about the structure of matter that is not accessible using conventional 'linear' optical spectroscopy.

The anharmonic-oscillator model was used by Bloembergen (1965) and Garrett and Robinson (1966) to estimate the second- and third-order nonlinear susceptibilities of dielectric materials. Here we have considered the case when the optical frequency lies in the ultraviolet and visible spectral region, so that the oscillatory motion of the electrons is particularly significant. At lower frequencies, the motion of the ions becomes more important (molecular vibration and rotation in gases and liquids, and ionic-lattice vibration in the case of solids). In some liquids, the effect of the incident light beam can be to reorientate the molecules, and this can play a significant rôle in the nonlinear optical response. In later chapters the more rigorous methods of quantum mechanics are used to derive formulae for the nonlinear susceptibilities in the general case. Also, in writing the power-series expansion (1.6) for a realistic medium, we should take into account the vector character of the electric field and polarisation, and the tensor character and frequency dependence of the susceptibilities. These details are filled in later.

1.2 Effects arising from the quadratic polarisation

The quadratic polarisation $P^{(2)} = \varepsilon_0 \chi^{(2)} E^2$ gives rise to effects which are basically all mixing phenomena, involving the generation of sum and difference frequencies, but they take a variety of forms. When the applied optical field contains just one frequency, the quadratic polarisation contains a static term and a term oscillating at twice the applied frequency. The static polarisation produces a d.c. electric field in the medium, *i.e.,* we have an optical rectification effect. As we have already seen, the polarisation oscillating at twice the applied frequency radiates into the medium, giving rise to second-harmonic generation. More effects arise when the applied optical field contains two frequencies. The simplest case occurs when one of the frequencies is zero, *i.e.,* when we send an optical wave through the medium in the presence of a d.c. electric field. The quadratic polarisation then contains a term proportional to the product of the optical and d.c. fields. There is therefore an extra term in the polarisation which is linear in the optical field and whose magnitude is proportional to the d.c. field. In its effect on the optical wave, this extra term is equivalent to changing $\chi^{(1)}$ by an amount proportional to the d.c. field. Hence, the refractive index at the optical frequency depends on the d.c. field. This is the linear electrooptic (or Pockels) effect, which is widely used in optical modulators. Electro-optic effects, which are well described within the framework of nonlinear optics, are nevertheless exceptions to our opening remark that nonlinear effects require intense illumination for their observation. Indeed, the discovery of the Pockels effect in 1893 predated the invention of the laser by nearly 70 years.

Perhaps the most interesting phenomenon arising from the quadratic polarisation is 'parametric amplification'. This effect occurs when we send a small optical signal, at frequency ω_S, through the medium in the presence of a powerful optical field, called the pump, at a higher frequency ω_P. The pump and signal beat together to produce a field at the difference frequency $\omega_I = \omega_P - \omega_S$ which is called the idler field; it is proportional to the product of the pump and signal fields. Then a second beating action takes place; the idler beats with the pump to produce a term in the polarisation at the difference frequency $\omega_P - \omega_I = \omega_S$ which is proportional to the product of the signal field and the pump intensity (*i.e.,* the pump field squared). Thus, as a result of this double beating action, there is an extra term in the total polarisation which is linear in the signal field. In its effect on the signal field, this extra term is equivalent to changing $\chi^{(1)}$ by an amount proportional to the pump intensity.

This effect is called a 'parametric' interaction because the pump field may be regarded as modulating the parameter $\chi^{(1)}$ at the pump frequency. The parametric interaction is particularly strong when it is phase matched; as explained later, this is equivalent to ensuring that the propagation constants k_S, k_I and k_P for the waves at the frequencies ω_S, ω_I and ω_P, respectively, satisfy the condition $k_S + k_I = k_P$. In that case, power is removed from the pump wave to the benefit of both the signal and idler waves, which are amplified. This effect is put to good use in optical devices such as certain tunable sources, parametric amplifiers and oscillators.

It should be pointed out that $\chi^{(2)}$ vanishes in media with inversion symmetry. In these media, opposite directions are completely equivalent and so the polarisation must change sign when the optical electric field is reversed. Hence there can be no even powers of the field in the expansion of the polarisation. This fact is the simplest result which follows from a consideration of the restrictions imposed on the susceptibilities by the point-group symmetry of the medium they describe. These restrictions are discussed more fully later. In a medium with inversion symmetry, the first nonlinear term in the polarisation is the cubic term which we now consider.

1.3 Effects arising from the cubic polarisation

The cubic polarisation $P^{(3)} = \varepsilon_0 \chi^{(3)} E^3$ gives rise to third-harmonic generation and related mixing phenomena. Again, the simplest situation occurs when an optical wave propagates through the medium in the presence of a d.c. field; the cubic polarisation leads to a change in the refractive index proportional to the square of the d.c. field. This is the quadratic electrooptic (or d.c. Kerr) effect which is used in many fast-acting optical shutters. Perhaps the most important of all third-order processes is the intensity-dependent refractive index. An optical field passing through the nonlinear medium induces a cubic polarisation which is proportional to the third power of the field. In its effect on the wave, this term is equivalent to changing the effective value of $\chi^{(1)}$ to $\chi^{(1)} + \chi^{(3)} E^2$; in other words, the refractive index is changed by an amount proportional to the optical intensity. As described later, this effect is involved in a wide variety of important processes, including self-focusing of a laser beam, self-phase and frequency modulation, 'soliton' pulse propagation, and 'phase conjugate' reflection.

The intensity-dependent refractive index is, in many cases, the key nonlinear effect used in optical switching and signal-processing devices. A schematic representation of a switching device is shown in Fig. 1.4. In linear optics, a Fabry-Perot étalon has the property that it blocks an incident monochromatic beam, unless the wavelength of the light satisfies the resonance condition $2nd = p\lambda$ so that a high transmission is obtained. Here n is the refractive index of the material between the two mirrors,

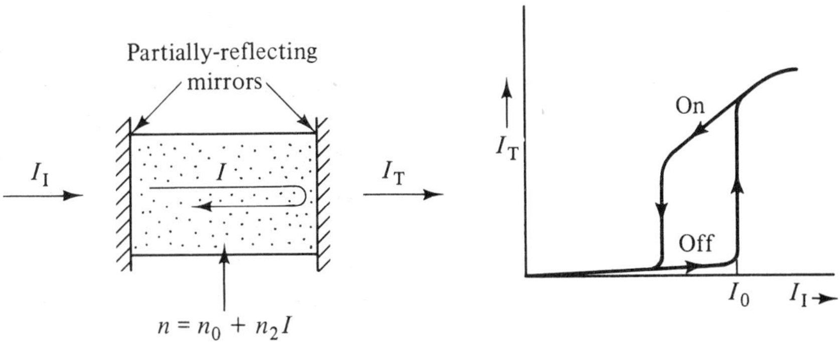

Fig. 1.4 Optical switching behaviour is exhibited by a Fabry-Perot étalon that contains a material having a refractive index n which depends on the optical intensity I. The hysteresis loop in the curve of transmitted intensity I_T versus input intensity I_I implies bistable 'on' and 'off' states for some values of I_I.

d is the mirror separation, λ is the optical wavelength and p is an integer number. Now let us assume that the space between the mirrors of a Fabry-Perot étalon is filled with a material which exhibits a nonlinear refractive index. We suppose also that, at low incident intensity, the wavelength λ and other parameters fail to satisfy the resonance condition. Therefore, the transmitted light beam is blocked; this is the 'off' state. If the intensity is now increased gradually, the refractive index n changes. Eventually a value of n is reached which satisfies the resonance condition ($I_I = I_0$ in Fig. 1.4), and so the transmitted beam is switched 'on'. A more detailed analysis of the operation of the nonlinear Fabry-Perot étalon shows that, in certain circumstances, it can exhibit a 'latching' action, or bistability, rather like an electronic bistable logic circuit (Gibbs, 1985). The important point is that the device operation is controlled by the light beam itself; this is an example of 'all-optical' switching.

All of the nonlinear effects which we have described so far are essentially classical. However, the cubic polarisation also gives rise to some effects which are essentially quantal. These arise when we propagate a small signal wave at a frequency ω_S through the medium in the presence of a strong pump wave at frequency ω_P, the frequencies being chosen so that $\omega_P \pm \omega_S = \Omega$, where Ω is some transition frequency of the medium. When $\omega_P + \omega_S = \Omega$ the transition can take place through the simultaneous absorption of a pump photon and a signal photon. Thus we have a two-photon absorption process which produces signal attenuation even though the signal frequency itself is not equal to a transition frequency of the medium. (The imaginary part of $\chi^{(1)}$, which describes linear absorption, arises from one-photon absorption processes and, as shown in Fig. 1.2, vanishes unless the signal frequency is close to a transition frequency of the medium.) Two-photon absorption was one of the very first nonlinear-optical effects to be discovered, at about the same time as second-harmonic generation (Kaiser and Garrett, 1961). In the other case, when $\omega_P - \omega_S = \Omega$, the transition can take place by the simultaneous absorption of a pump photon and *emission* of a signal photon. This is the stimulated Raman effect which results in amplification of the input signal. Both the two-photon absorption and stimulated Raman effects arise from the term in the cubic polarisation which is proportional to the product of the signal field and the pump intensity. In its effect on the signal wave, this term is equivalent to changing $\chi^{(1)}$ by an amount proportional to the pump intensity.

Whilst second- and third-order nonlinear effects are the most important and widely studied, processes involving fifth-, seventh- and yet higher-order interactions can be observed (Reintjes, 1984).

1.4 Summary of the book

In this book we develop the theoretical description of nonlinear-optical phenomena starting from first principles. The plan is as follows. In Chapter 2 the 'constitutive relation' between the induced polarisation and the driving field is written down in a more precise form than has been used in this introduction; this is done by invoking some simple and fundamental physical principles. The susceptibilities, which are frequency-dependent tensors, are defined by the constitutive relation. We discuss some of the key properties of the susceptibility tensors which are inherent in their definition. The next stage is to obtain explicit formulae for the susceptibility tensors by solving the equations of motion for the charged particles in the medium under the influence of an electric field. This is done in Chapter 4 using a quantum-mechanical approach which is more rigorous and general than the classical model of an anharmonic oscillator discussed here. Various approaches to the problem of calculating nonlinear coefficients for practical media are discussed. The calculations are preceded (in Chapter 3) by a tutorial review of the relevant aspects of quantum mechanics and quantum statistics. Chapter 5 is devoted to a discussion of the important symmetry properties of the susceptibility tensors.

As mentioned previously, the effectiveness of nonlinear processes can be increased greatly by exploiting resonances of the medium. In the case of strong resonance, a particular process may receive a significant contribution from many orders of nonlinearity acting simultaneously – in other words, the analysis based on small perturbations breaks down. The susceptibility formalism is then less practical and alternative ways of describing resonant nonlinear processes are required; this is the subject of Chapter 6.

Once the polarisation induced in the medium has been determined, the next step is to examine how this couples back to the radiation field. This is done in Chapter 7 by considering the solution of Maxwell's equations when the nonlinear contributions to the polarisation are taken into account. Aspects such as phase matching become important at this stage. This study leads us back to a more detailed treatment of some of the nonlinear-optical phenomena outline above.

The effects which we have discussed up to now involve the response of *bound* charges to an applied field. However, nonlinear-optical effects also arise from the motion of *free* charge carriers, and we devote Chapter 8 to a discussion of these in the case of semiconductors. Such optical nonlinearity in semiconductors is particularly important for practical applications because of the high degree of control that we have over the free-carrier densities and therefore on the performance of devices which make use of them. In the last chapter we review some of the recent work aimed towards the fabrication of artificial semiconductor materials having enhanced nonlinear properties.

2

The constitutive relation

At the heart of the conventional description of nonlinear optics is the constitutive relation between the polarisation $\mathbf{P}(t)$ at time t and the driving field $\mathbf{E}(t)$. In later chapters, quantum mechanics are used to derive explicit formulae for the induced polarisation, and it is also shown how the polarisation acts as the driving source in nonlinear processes. The purpose of this chapter is first to consider the general form of the constitutive relation; it is formulated more precisely than in Chapter 1, and we show that by applying a few simple physical principles it is possible to deduce important fundamental and universal properties of nonlinear-optical phenomena. The constitutive relation is considered from two closely-linked approaches: one based on time-domain response functions, and the other – perhaps more familiar – in terms of susceptibilities. Both approaches, and also combinations of the two, are useful in different circumstances, and the choice depends on factors such as the bandwidth or pulse duration of the applied light and the speed of response of the nonlinear medium. Towards the close of the chapter, alternative constitutive relations are considered briefly.

2.1 Time-domain response functions

In Chapter 1 the polarisation of a medium under the influence of an applied electric field is described in terms of a power series in the field. It is convenient here to express the power series as

$$\mathbf{P}(t) = \mathbf{P}^{(0)}(t) + \mathbf{P}^{(1)}(t) + \mathbf{P}^{(2)}(t) + \cdots + \mathbf{P}^{(n)}(t) + \cdots \qquad (2.1)$$

where $\mathbf{P}^{(1)}(t)$ is linear in the field, $\mathbf{P}^{(2)}(t)$ is quadratic, and so on. (In this book, vector and tensor quantities appear in bold typeface.) The term $\mathbf{P}^{(0)}(t)$, which is independent of the field, would represent, for example, the static polarisation found in some crystals. Here we consider the *local* response, in which the polarisation at a point in the nonlinear medium is determined by the electric field at that point (this is a restriction which is lifted later). The general form of the various terms in the series (2.1) is found by invoking a fundamental physical principle: time-invariance.

This should not be misunderstood to mean that any time dependence in the physical system is to be excluded; indeed many of the most interesting problems in nonlinear optics are those in which the applied electric field and the resulting nonlinear-optical response are time-dependent. A major proportion of this book is devoted to just such problems. What is meant by time-invariance is that the dynamical properties of the system are assumed to be unchanged by a translation of the time origin; in this case, a time-displacement of the driving electric field merely results in a corresponding time-displacement of the induced polarisation.

2.1.1 Linear response

We apply this principle first to determine the form of the linear polarisation. Since $\mathbf{P}^{(1)}(t)$ is linear in $\mathbf{E}(t)$, it may be expressed in the form:

$$\mathbf{P}^{(1)}(t) = \varepsilon_0 \int_{-\infty}^{+\infty} \mathrm{d}\tau \, \mathbf{T}^{(1)}(t;\tau)\cdot\mathbf{E}(\tau)\,, \tag{2.2}$$

where $\mathbf{T}^{(1)}(t;\tau)$ is a second-rank tensor which is a function of the two times t and τ. This equation is simply the most general possible linear relation between the time-dependent vectors $\mathbf{P}(t)$ and $\mathbf{E}(t)$. In suffix notation it reads:

$$P_\mu^{(1)}(t) = \varepsilon_0 \int_{-\infty}^{+\infty} \mathrm{d}\tau \, T_{\mu\alpha}^{(1)}(t;\tau)E_\alpha(\tau) \tag{2.3}$$

where the Greek subscripts take the values x, y and z which label the cartesian coordinate axes, and we invoke the 'repeated-index summation convention'; that is to say, repeated subscripts on the right-hand side of (2.3) are understood to be summed over x, y and z. This implies that (2.3) may be expanded fully as:

$$P_x^{(1)}(t) = \varepsilon_0 \int_{-\infty}^{+\infty} \mathrm{d}\tau \left[T_{xx}^{(1)}(t;\tau)E_x(\tau) + T_{xy}^{(1)}(t;\tau)E_y(\tau) + T_{xz}^{(1)}(t;\tau)E_z(\tau)\right] \tag{2.4}$$

with two further equations for $P_y^{(1)}(t)$ and $P_z^{(1)}(t)$. Merely by replacing t in (2.2) by $t+t_0$, where t_0 is an arbitrary time, we see that

$$\mathbf{P}^{(1)}(t+t_0) = \varepsilon_0 \int_{-\infty}^{+\infty} \mathrm{d}\tau \, \mathbf{T}^{(1)}(t+t_0,\tau)\cdot\mathbf{E}(\tau)\,. \tag{2.5}$$

Now, from the principle of time-invariance stated above, $\mathbf{P}^{(1)}(t+t_0)$ must be identical to the linear polarisation induced by the time-displaced field $\mathbf{E}(t+t_0)$. Hence by replacing $\mathbf{E}(\tau)$ by $\mathbf{E}(\tau+t_0)$ in (2.2) we obtain:

$$\mathbf{P}^{(1)}(t+t_0) = \varepsilon_0 \int_{-\infty}^{+\infty} \mathrm{d}\tau \, \mathbf{T}^{(1)}(t;\tau)\cdot\mathbf{E}(\tau+t_0)\,,$$

or with a simple change of variable $\tau \rightarrow \tau' - t_0$:

$$\mathbf{P}^{(1)}(t + t_0) = \varepsilon_0 \int_{-\infty}^{+\infty} d\tau' \, \mathbf{T}^{(1)}(t; \tau' - t_0) \cdot \mathbf{E}(\tau') . \tag{2.6}$$

By comparing equations (2.5) and (2.6) we see that

$$\mathbf{T}^{(1)}(t + t_0; \tau) = \mathbf{T}^{(1)}(t; \tau - t_0) \tag{2.7}$$

for all t, τ, and t_0. This equation determines the way in which the time variables t and τ appear in $\mathbf{T}^{(1)}(t;\tau)$. Thus, by setting $t = 0$ and then replacing the arbitrary time t_0 by t we have

$$\mathbf{T}^{(1)}(t; \tau) = \mathbf{T}^{(1)}(0; \tau - t) . \tag{2.8}$$

Hence $\mathbf{T}^{(1)}(t;\tau)$ in fact depends only on the difference between the times t and τ and not on their individual values. To make this point explicit in our formulae we write:

$$\mathbf{T}^{(1)}(t; \tau) \equiv \mathbf{R}^{(1)}(t - \tau) , \tag{2.9}$$

where $\mathbf{R}^{(1)}(t-\tau)$ is a second-rank tensor depending only on the time difference $t - \tau$.

By substituting (2.9) into (2.2) we obtain the canonical form for the linear polarisation which is dictated by the principle of time-invariance:

$$\begin{aligned} \mathbf{P}^{(1)}(t) &= \varepsilon_0 \int_{-\infty}^{+\infty} d\tau \, \mathbf{R}^{(1)}(t - \tau) \cdot \mathbf{E}(\tau) \\ &= \varepsilon_0 \int_{-\infty}^{+\infty} d\tau' \, \mathbf{R}^{(1)}(\tau') \cdot \mathbf{E}(t - \tau') \end{aligned} \tag{2.10}$$

where, in the second line, we have made the change of variable $\tau \rightarrow t - \tau'$. The tensor $\mathbf{R}^{(1)}(\tau)$ is called the linear polarisation response function of the medium; it is the tensorial analogue of the linear impulse-response function that is familiar in electrical circuit theory. Its form is subject to two restrictions: First, $\mathbf{R}^{(1)}(\tau)$ must vanish when τ is negative to ensure that $\mathbf{P}^{(1)}(t)$ depends only on values of the field for times before t; this is the causality condition. The second requirement is the reality condition; since $\mathbf{E}(t)$ and $\mathbf{P}^{(1)}(t)$ are both real, the response function must also be real.

2.1.2 Quadratic nonlinear response

We may now apply essentially the same arguments to determine the form of the lowest-order nonlinear polarisation, the term $\mathbf{P}^{(2)}(t)$ which is quadratic in $\mathbf{E}(t)$. It may be expressed in the form:

$$\mathbf{P}^{(2)}(t) = \varepsilon_0 \int_{-\infty}^{+\infty} d\tau_1 \int_{-\infty}^{+\infty} d\tau_2 \, \mathbf{T}^{(2)}(t; \tau_1, \tau_2) \vdots \mathbf{E}(\tau_1) \mathbf{E}(\tau_2) , \tag{2.11}$$

where $\mathbf{T}^{(2)}(t;\tau_1,\tau_2)$ is a third-rank tensor which is a function of the three times t, τ_1 and τ_2. Equation (2.11) is simply the most general possible expression for $\mathbf{P}^{(2)}(t)$ which is quadratic in $\mathbf{E}(t)$. (The convention for the placing of the free-space permittivity ε_0 is detailed in §2.2.) In suffix notation (2.11) reads:

$$P_\mu^{(2)}(t) = \varepsilon_0 \int_{-\infty}^{+\infty} d\tau_1 \int_{-\infty}^{+\infty} d\tau_2\, T_{\mu\alpha\beta}^{(2)}(t;\tau_1,\tau_2) E_\alpha(\tau_1) E_\beta(\tau_2). \qquad (2.11a)$$

The tensor $T_{\mu\alpha\beta}^{(2)}(t;\tau_1,\tau_2)$ uniquely determines the quadratic polarisation in the medium. However, because the right-hand side of (2.11) is a quadratic form, the tensor $T_{\mu\alpha\beta}^{(2)}(t;\tau_1,\tau_2)$ is not itself unique. To see this we express $T_{\mu\alpha\beta}^{(2)}(t;\tau_1,\tau_2)$ as a sum of a symmetric part and an anti-symmetric part, as follows:

$$T_{\mu\alpha\beta}^{(2)}(t;\tau_1,\tau_2) = S_{\mu\alpha\beta}^{(2)}(t;\tau_1,\tau_2) + A_{\mu\alpha\beta}^{(2)}(t;\tau_1,\tau_2) \qquad (2.12)$$

where

$$S_{\mu\alpha\beta}^{(2)} = \tfrac{1}{2}\left[T_{\mu\alpha\beta}^{(2)}(t;\tau_1,\tau_2) + T_{\mu\beta\alpha}^{(2)}(t;\tau_2,\tau_1)\right] \qquad (2.13)$$

and

$$A_{\mu\alpha\beta}^{(2)} = \tfrac{1}{2}\left[T_{\mu\alpha\beta}^{(2)}(t;\tau_1,\tau_2) - T_{\mu\beta\alpha}^{(2)}(t;\tau_2,\tau_1)\right], \qquad (2.14)$$

which are respectively symmetric and antisymmetric under the interchange of the pairs of dummy variables $\alpha\tau_1$ and $\beta\tau_2$. However it can be seen that such an interchange in (2.11) leaves that expression unchanged. Thus it follows that $A_{\mu\alpha\beta}^{(2)}(t;\tau_1,\tau_2)$ makes no contribution to $P_\mu(t)$ and is therefore arbitrary. To remove this arbitrariness in $T_{\mu\alpha\beta}^{(2)}(t;\tau_1,\tau_2)$ we set its antisymmetric part to zero. Then $T_{\mu\alpha\beta}^{(2)}(t;\tau_1,\tau_2)$ is unique and symmetric, *i.e.*,

$$T_{\mu\alpha\beta}^{(2)}(t;\tau_1,\tau_2) = T_{\mu\beta\alpha}^{(2)}(t;\tau_2,\tau_1). \qquad (2.15)$$

By applying the principle of time-invariance as we did previously for the linear response function, we find that

$$\mathbf{T}^{(2)}(t+t_0;\tau_1,\tau_2) = \mathbf{T}^{(2)}(t;\tau_1-t_0,\tau_2-t_0) \qquad (2.16)$$

for all t, t_0, τ_1 and τ_2. Hence, by setting $t=0$ and then replacing the arbitrary time t_0 by t, we find that $\mathbf{T}^{(2)}(t;\tau_1,\tau_2)$ depends only on the two time differences $t-\tau_1$ and $t-\tau_2$. To make this fact explicit in our formulae we write:

$$\mathbf{T}^{(2)}(t;\tau_1,\tau_2) \equiv \mathbf{R}^{(2)}(t-\tau_1,t-\tau_2). \qquad (2.17)$$

By substituting (2.17) into (2.11) we obtain the canonical form for the quadratic polarisation:

$$\mathbf{P}^{(2)}(t) = \varepsilon_0 \int_{-\infty}^{+\infty} d\tau_1 \int_{-\infty}^{+\infty} d\tau_2\, \mathbf{R}^{(2)}(t-\tau_1, t-\tau_2) : \mathbf{E}(\tau_1)\mathbf{E}(\tau_2)$$
$$= \varepsilon_0 \int_{-\infty}^{+\infty} d\tau_1' \int_{-\infty}^{+\infty} d\tau_2'\, \mathbf{R}^{(2)}(\tau_1', \tau_2') : \mathbf{E}(t-\tau_1')\mathbf{E}(t-\tau_2') \quad (2.18)$$

where, in the second line, we have made the change of variables $\tau_1 \to t-\tau_1'$, $\tau_2 \to t-\tau_2'$.

The tensor $\mathbf{R}^{(2)}(\tau_1, \tau_2)$ may be called the quadratic polarisation response function of the medium. The causality requirement dictates that $\mathbf{R}^{(2)}(\tau_1, \tau_2)$ is zero when either τ_1 or τ_2 is negative, while the reality condition dictates that $\mathbf{R}^{(2)}(\tau_1, \tau_2)$ is real. Moreover we have shown in (2.15) that $R^{(2)}_{\mu\alpha\beta}(\tau_1, \tau_2)$ is invariant under the interchange of the pairs (α, τ_1) and (β, τ_2). This property, and its generalisation to higher orders, is known as 'intrinsic permutation symmetry'. Permutation symmetry and its physical consequences are discussed in more detail in §§2.3, 4.3 and 5.1.

2.1.3 Higher-order nonlinearity

These arguments may obviously be extended to determine the form of the nth-order polarisation $\mathbf{P}^{(n)}(t)$ which is proportional to the nth power of the field $\mathbf{E}(t)$. First $\mathbf{P}^{(n)}(t)$ is expressed in terms of the $(n+1)$-rank tensor $\mathbf{T}^{(n)}(t; \tau_1, \tau_2, ..., \tau_n)$ as:

$$\mathbf{P}^{(n)}(t) = \varepsilon_0 \int_{-\infty}^{+\infty} d\tau_1 \cdots \int_{-\infty}^{+\infty} d\tau_n\, \mathbf{T}^{(n)}(t; \tau_1, ..., \tau_n) \,|\, \mathbf{E}(\tau_1) \cdots \mathbf{E}(\tau_n) \quad (2.19)$$

where a vertical bar has been used to replace the column of n dots conventionally required to indicate a contraction. The polarisation $\mathbf{P}^{(n)}(t)$ is determined uniquely by (2.19), but the tensor $\mathbf{T}^{(n)}$ is not unique because of the $n!$ possible different orders in which the terms $\mathbf{E}(\tau_1) \cdots \mathbf{E}(\tau_n)$ may be written. Similar to the second-order case described above, this arbitrariness in the definition of $\mathbf{T}^{(n)}$ is removed by specifying that

$$T^{(n)}_{\mu\alpha_1 \cdots \alpha_n}(t; \tau_1, ..., \tau_n) = \frac{1}{n!} \mathbf{S}\, T^{(n)}_{\mu\alpha_1 \cdots \alpha_n}(t; \tau_1, ..., \tau_n), \quad (2.20)$$

where the symmetrising operation $\mathbf{S}$ indicates a summation over all the tensors obtained by making the $n!$ permutations of the n pairs (α_1, τ_1), (α_2, τ_2), ..., (α_n, τ_n). This is the intrinsic permutation-symmetry property in nth order.

From the principle of time-invariance we obtain:

$$\mathbf{T}^{(n)}(t; \tau_1 - t_0, ..., \tau_n - t_0) = \mathbf{T}^{(n)}(t + t_0; \tau_1, ..., \tau_n), \quad (2.21)$$

and hence our definition for the nth-order polarisation response function $\mathbf{R}^{(n)}(\tau_1,...,\tau_n)$ is determined by

$$\mathbf{T}^{(n)}(t;\tau_1,...,\tau_n) \equiv \mathbf{R}^{(n)}(t-\tau_1,...,t-\tau_n)\,. \tag{2.22}$$

From whence we obtain the canonical form:

$$\begin{aligned}\mathbf{P}^{(n)}(t) &= \varepsilon_0 \int_{-\infty}^{+\infty} \mathrm{d}\tau_1 \cdots \int_{-\infty}^{+\infty} \mathrm{d}\tau_n\, \mathbf{R}^{(n)}(t-\tau_1,...,t-\tau_n) \,|\, \mathbf{E}(\tau_1)\cdots\mathbf{E}(\tau_n) \\ &= \varepsilon_0 \int_{-\infty}^{+\infty} \mathrm{d}\tau_1 \cdots \int_{-\infty}^{+\infty} \mathrm{d}\tau_n\, \mathbf{R}^{(n)}(\tau_1,...,\tau_n) \,|\, \mathbf{E}(t-\tau_1)\cdots\mathbf{E}(t-\tau_n)\,.\end{aligned} \tag{2.23}$$

In suffix notation (2.23) reads:

$$P_\mu^{(n)}(t) = \varepsilon_0 \int_{-\infty}^{+\infty} \mathrm{d}\tau_1 \cdots \int_{-\infty}^{+\infty} \mathrm{d}\tau_n\, R_{\mu\alpha_1\cdots\alpha_n}(\tau_1,...,\tau_n) \times E_{\alpha_1}(t-\tau_1)\cdots E_{\alpha_n}(t-\tau_n)\,. \tag{2.24}$$

The nth-order tensor $\mathbf{R}^{(n)}(\tau_1,...,\tau_n)$ is of rank $n+1$, and is a real function of the n time variables $\tau_1,...,\tau_n$. It vanishes when any one of the τ_i is negative, and is invariant under any of the $n!$ permutations of the n pairs (α_1,τ_1), (α_2,τ_2), ..., (α_n,τ_n).

2.2 Frequency domain – the susceptibility tensors

As we have just seen, the time-domain polarisation response tensors in all orders provide a complete description of the optical properties of the medium. However, an alternative – and more widely used – description is provided by frequency-domain response functions, known as the susceptibility tensors. Both the time-domain response functions and susceptibility tensors provide useful descriptions of nonlinear-optical properties, and the choice of which is the most appropriate in given circumstances is discussed in §2.4.

2.2.1 The complex frequency plane

The susceptibility tensors arise when the electric field $\mathbf{E}(t)$ is expressed in terms of its Fourier transform $\mathbf{E}(\omega)$ by means of the Fourier integral identity:

$$\mathbf{E}(t) = \int_{-\infty}^{+\infty} \mathrm{d}\omega\, \mathbf{E}(\omega)\exp(-i\omega t)\,, \tag{2.25}$$

where

$$\mathbf{E}(\omega) = \frac{1}{2\pi}\int_{-\infty}^{+\infty} \mathrm{d}\tau\, \mathbf{E}(\tau)\exp(i\omega\tau)\,. \tag{2.26}$$

These relations require a few comments before we proceed. The conventions adopted in (2.25) and (2.26) regarding the sign in the exponent and the factor 2π are those common in quantum-mechanical calculations. In most applications of Fourier transforms the frequency ω is taken to be real, and certainly for all practical purposes only real frequencies are meaningful. Nevertheless, in the treatment of susceptibilities that follows we shall allow ω to lie in the upper half of the complex frequency plane. The reason for this is a matter of mathematical convenience, which will become clear shortly. First we must ensure that this does not prejudice the validity and convergence of the Fourier transform pair. By substituting $\omega = x + iy$ in the right-hand side of (2.26) we obtain

$$\frac{1}{2\pi}\int_{-\infty}^{+\infty} d\tau\, \mathbf{E}(\tau)\exp[(ix - y)\tau], \tag{2.27}$$

which converges provided that $y > 0$ (*i.e.*, ω lies in the upper half-plane), and provided also that we impose the constraint that $\mathbf{E}(t)$ vanishes in the remote past; this is not a significant restriction in practice. We shall now take the integration path in (2.25) to be in the upper half-plane and parallel to the real axis, so that $d\omega \to dx$. Then, by substituting (2.26) into the right-hand side of (2.25), we obtain

$$\frac{1}{2\pi}\int_{-\infty}^{+\infty} dx \int_{-\infty}^{+\infty} d\tau\, \mathbf{E}(\tau)\exp[y(t-\tau)]\exp[ix(\tau - t)]. \tag{2.28}$$

We now note that

$$\frac{1}{2\pi}\lim_{X\to\infty}\int_{-X}^{+X} dx\, \exp[ix(\tau - t)] = \lim_{X\to\infty}\left\{\frac{\sin[X(\tau - t)]}{\pi(\tau - t)}\right\}, \tag{2.29}$$

as can be shown by direct evaluation of the integral. The function in braces in (2.29) displays very singular behaviour as $X\to\infty$. Its value tends to infinity as $\tau - t \to 0$. Moreover, the oscillations (with respect to τ) in the sine function become more rapid, and consequently the singular peak in the vicinity of $\tau = t$ becomes narrower. At the same time we observe that the integral over τ (from $-\infty$ to $+\infty$) of the function in braces in (2.29) yields unity. Thus we obtain the useful identity:

$$\frac{1}{2\pi}\int_{-\infty}^{+\infty} dx\, \exp[ix(\tau - t)] = \delta(\tau - t) \tag{2.30}$$

where $\delta(\theta)$ is the Dirac delta-function: $\delta(\theta) = 0$ for $\theta \neq 0$, $\delta(\theta) \to \infty$ for $\theta = 0$, such that $\int_{-\infty}^{+\infty} d\theta\, \delta(\theta - a) f(\theta) = f(a)$. By substituting (2.30) into (2.28), we thus obtain:

$$\int_{-\infty}^{+\infty} d\tau\, \mathbf{E}(\tau) \exp[y(t-\tau)]\, \delta(\tau - t) = \mathbf{E}(t), \tag{2.31}$$

and so we see that the right-hand side of (2.25) is indeed identical to $\mathbf{E}(t)$. Finally, since $\mathbf{E}(t)$ is real, the Fourier coefficients must satisfy

$$[\mathbf{E}(\omega)]^* = \mathbf{E}(-\omega^*), \tag{2.32}$$

where the symbol * denotes the complex conjugate. The relation (2.32) can be verified readily by comparing the complex conjugate of (2.27) with a corresponding expression of (2.26) for $\mathbf{E}(-\omega^*)$.

2.2.2 Linear susceptibility

When (2.25) is substituted into (2.10), we obtain for the first-order polarisation:

$$\mathbf{P}^{(1)}(t) = \varepsilon_0 \int_{-\infty}^{+\infty} d\omega \int_{-\infty}^{+\infty} d\tau\, \mathbf{R}^{(1)}(\tau) \cdot \mathbf{E}(\omega) \exp[-i\omega(t-\tau)]$$

$$= \varepsilon_0 \int_{-\infty}^{+\infty} d\omega\, \boldsymbol{\chi}^{(1)}(-\omega_\sigma;\omega) \cdot \mathbf{E}(\omega) \exp(-i\omega_\sigma t) \tag{2.33}$$

where

$$\boldsymbol{\chi}^{(1)}(-\omega_\sigma;\omega) = \int_{-\infty}^{+\infty} d\tau\, \mathbf{R}^{(1)}(\tau) \exp(i\omega\tau), \tag{2.34}$$

which is, by definition, the linear susceptibility tensor. In these equations $\omega_\sigma = \omega$. The broader meaning of ω_σ will become apparent shortly when we consider the nonlinear susceptibilities. (The additional argument $-\omega_\sigma$ before the semicolon in our definition of $\boldsymbol{\chi}^{(1)}$ may seem superfluous; however it is included for consistency with the later notation for higher-order susceptibilities.)

The reason for putting ω in the upper half of the complex plane should now become clear. We have already seen that the causality condition ensures that $\mathbf{R}^{(1)}(\tau)$ vanishes when $\tau<0$. Also, if ω lies in the upper half-plane, then $\exp(i\omega_\sigma\tau)$ tends to zero as $\tau \rightarrow +\infty$. Thus the integral in (2.34) converges for ω in the upper half-plane (*i.e.*, $\boldsymbol{\chi}^{(1)}(-\omega_\sigma;\omega)$ is analytic there). Any realistic physical medium is subject to relaxation processes, so that the response function $\mathbf{R}^{(1)}(\tau)$ will tend to zero as $\tau \rightarrow +\infty$, and the frequency ω can be taken to be real. However, it will greatly simplify our analysis if we choose to neglect relaxation processes for the time being (relaxation will be introduced in a phenomenological fashion in Chapters 4 and 6). If we do this then $\mathbf{R}^{(1)}(\tau)$ remains finite in the remote future and so ω must be taken to lie in the upper half-plane. This is a mathematical device which ensures the convergence of integrals such as in (2.34) and in our quantum-mechanical calculations later.

Both the electric field $\mathbf{E}(t)$ and polarisation $\mathbf{P}(t)$ are real, and thus from (2.10) it follows that the response function $\mathbf{R}^{(1)}(t)$ is also real. It therefore also follows, by taking the complex conjugate of (2.34), that

$$[\boldsymbol{\chi}^{(1)}(-\omega_\sigma;\omega)]^* = \boldsymbol{\chi}^{(1)}(\omega_\sigma^*;-\omega^*). \tag{2.35}$$

This is the reality condition for the linear susceptibility tensor.

2.2.3 Second-order susceptibility

Turning now to the second-order polarisation, we substitute (2.25) into (2.18) to obtain†

$$\mathbf{P}^{(2)}(t) = \varepsilon_0 \int_{-\infty}^{+\infty} \mathrm{d}\omega_1 \int_{-\infty}^{+\infty} \mathrm{d}\omega_2 \int_{-\infty}^{+\infty} \mathrm{d}\tau_1 \int_{-\infty}^{+\infty} \mathrm{d}\tau_2\, \mathbf{R}^{(2)}(\tau_1,\tau_2) : \mathbf{E}(\omega_1)\mathbf{E}(\omega_2)\exp[-i\{\omega_1(t-\tau_1)+\omega_2(t-\tau_2)\}]$$

$$= \varepsilon_0 \int_{-\infty}^{+\infty} \mathrm{d}\omega_1 \int_{-\infty}^{+\infty} \mathrm{d}\omega_2\, \boldsymbol{\chi}^{(2)}(-\omega_\sigma;\omega_1,\omega_2) : \mathbf{E}(\omega_1)\mathbf{E}(\omega_2)\exp(-i\omega_\sigma t), \tag{2.36}$$

where $\omega_\sigma = \omega_1 + \omega_2$, and the quadratic susceptibility tensor is defined by

$$\boldsymbol{\chi}^{(2)}(-\omega_\sigma;\omega_1,\omega_2) = \int_{-\infty}^{+\infty} \mathrm{d}\tau_1 \int_{-\infty}^{+\infty} \mathrm{d}\tau_2\, \mathbf{R}^{(2)}(\tau_1,\tau_2)\exp[i(\omega_1\tau_1+\omega_2\tau_2)]. \tag{2.37}$$

The meaning of ω_σ now becomes clearer; it is the sum of the optical driving frequencies in the susceptibility tensor. This interpretation is valid for all orders of nonlinearity.

Causality implies that $\boldsymbol{\chi}^{(2)}(-\omega_\sigma;\omega_1,\omega_2)$ is analytic when both ω_1 and ω_2 lie in the upper half-plane, while the reality condition generalises to

$$[\boldsymbol{\chi}^{(2)}(-\omega_\sigma;\omega_1,\omega_2)]^* = \boldsymbol{\chi}^{(2)}(\omega_\sigma^*;-\omega_1^*,-\omega_2^*). \tag{2.38}$$

The intrinsic permutation symmetry of $\mathbf{R}^{(2)}(\tau_1,\tau_2)$ carries over to

† In SI units, the placing of ε_0 is conventional in the expression (2.33) for the linear polarisation. Retaining it for the nonlinear polarisation, (2.36) and (2.40), is logical and means furthermore that the nonlinear susceptibilities $\boldsymbol{\chi}^{(n)}$ take the simple dimensions of $(\mathrm{V\,m^{-1}})^{1-n}$; *i.e.,* (electric field)$^{1-n}$. The use of esu was commonplace in the early nonlinear-optics literature, but thankfully is now superseded to a great extent by SI. The relations between the esu and SI systems of units of relevance in nonlinear optics are detailed in Appendix 2. The notation in this book is mainly consistent with that used by Hanna *et al* (1979).

$\boldsymbol{\chi}^{(2)}(-\omega_\sigma;\omega_1,\omega_2)$ in the sense that we can write:

$$\chi^{(2)}_{\mu\alpha\beta}(-\omega_\sigma;\omega_1,\omega_2) = \chi^{(2)}_{\mu\beta\alpha}(-\omega_\sigma;\omega_2,\omega_1), \tag{2.39}$$

i.e., $\chi^{(2)}_{\mu\alpha\beta}(-\omega_\sigma;\omega_1,\omega_2)$ is invariant under the interchange of the pairs (α,ω_1) and (β,ω_2). The property (2.39) can be verified by inserting the subscripts $\mu\alpha\beta$ into (2.37) and then making the interchanges $(\alpha,\omega_1,\tau_1) \leftrightarrow (\beta,\omega_2,\tau_2)$.

2.2.4 *n*th-order susceptibility

Finally, in *n*th order, by substituting (2.25) into (2.23) we obtain:

$$\mathbf{P}^{(n)}(t) = \varepsilon_0 \int_{-\infty}^{+\infty} \mathrm{d}\omega_1 \cdots \int_{-\infty}^{+\infty} \mathrm{d}\omega_n\, \boldsymbol{\chi}^{(n)}(-\omega_\sigma;\omega_1,\ldots,\omega_n) |\, \mathbf{E}(\omega_1)\cdots\mathbf{E}(\omega_n)\exp(-i\omega_\sigma t) \tag{2.40}$$

where

$$\boldsymbol{\chi}^{(n)}(-\omega_\sigma;\omega_1,\ldots,\omega_n) = \int_{-\infty}^{+\infty} \mathrm{d}\tau_1 \cdots \int_{-\infty}^{+\infty} \mathrm{d}\tau_n\, \mathbf{R}^{(n)}(\tau_1,\ldots,\tau_n)\exp\left[i\sum_{j=1}^{n}\omega_j\tau_j\right] \tag{2.41}$$

and

$$\omega_\sigma = \omega_1 + \omega_2 + \cdots + \omega_n . \tag{2.42}$$

Equation (2.41) defines the *n*th-order susceptibility tensor. Causality implies that $\boldsymbol{\chi}^{(n)}(-\omega_\sigma;\omega_1,\ldots,\omega_n)$ is analytic when all the frequencies lie in the upper half-plane. The reality condition generalises to

$$[\boldsymbol{\chi}^{(n)}(-\omega_\sigma;\omega_1,\ldots,\omega_n)]^* = \boldsymbol{\chi}^{(n)}(\omega_\sigma{}^*;-\omega_1{}^*,\ldots,-\omega_n{}^*). \tag{2.43}$$

Intrinsic permutation symmetry implies that $\chi^{(n)}_{\mu\alpha_1\cdots\alpha_n}(-\omega_\sigma;\omega_1,\ldots,\omega_n)$ is invariant under all $n!$ permutations of the n pairs (α_1,ω_1), (α_2,ω_2), ..., (α_n,ω_n).

There are several further important symmetry properties of the nonlinear susceptibilities and response functions; the discussion of these (in Chapter 5) is deferred until after the derivation of explicit formulae for the susceptibility tensors in Chapter 4. In the section which follows, we examine an important application of intrinsic permutation symmetry.

2.3 Superposition of monochromatic waves

2.3.1 Fourier components of the polarisation

The usefulness of the susceptibility tensors arises from the fact that we are often concerned with fields that are superpositions of monochromatic waves. The Fourier transform of the field, (2.25), taken for real ω then involves delta-functions. In a relaxing medium the integration paths in (2.33), (2.36) and (2.40) may be identified with the real axis. The evaluation of the integrals is then straightforward and the polarisation is completely determined by the values of the susceptibility tensors at the various frequencies involved. Thus, by expanding the polarisation $\mathbf{P}(t)$ in the frequency domain,

$$\mathbf{P}^{(n)}(t) = \int_{-\infty}^{+\infty} d\omega\, \mathbf{P}^{(n)}(\omega) \exp(-i\omega t) \tag{2.44}$$

where

$$\mathbf{P}^{(n)}(\omega) = \frac{1}{2\pi} \int_{-\infty}^{+\infty} d\tau\, \mathbf{P}^{(n)}(\tau) \exp(i\omega\tau)\,, \tag{2.45}$$

we obtain, from (2.30) and (2.40),

$$\mathbf{P}^{(n)}(\omega) = \varepsilon_0 \int_{-\infty}^{+\infty} d\omega_1 \cdots \int_{-\infty}^{+\infty} d\omega_n\, \boldsymbol{\chi}^{(n)}(-\omega_\sigma;\omega_1,\ldots,\omega_n) \,|\, \mathbf{E}(\omega_1) \cdots \mathbf{E}(\omega_n)\, \delta(\omega-\omega_\sigma). \tag{2.46}$$

Similar to the expansion of $\mathbf{P}(t)$ as a power series in the fields (2.1), we have here expanded the Fourier component of the polarisation at the frequency ω_σ as a power series, so that

$$\mathbf{P}(\omega) = \sum_{r=0}^{\infty} \mathbf{P}^{(r)}(\omega)\,. \tag{2.47}$$

For completeness we have again included the $r=0$ term (as in (2.1)), representing any static polarisation which may be present. Notice that in (2.46), for any given ω, there may be several sets of frequencies $\omega_1, \ldots, \omega_n$ (each of which may be positive, negative or zero) which satisfy $\omega - \omega_\sigma = 0$.

It is convenient, for later reference, to write (2.46) in suffix notation:

$$\left[P^{(n)}(\omega)\right]_\mu = \varepsilon_0 \sum_{\alpha_1 \cdots \alpha_n} \int_{-\infty}^{+\infty} d\omega_1 \cdots \int_{-\infty}^{+\infty} d\omega_n\, \chi^{(n)}_{\mu\alpha_1\cdots\alpha_n}(-\omega_\sigma;\omega_1,\ldots,\omega_n) \times \left[E(\omega_1)\right]_{\alpha_1} \cdots \left[E(\omega_n)\right]_{\alpha_n} \delta(\omega-\omega_\sigma). \tag{2.48}$$

The summation sign is included as a reminder that repeated cartesian-coordinate subscripts $\alpha_1, \ldots, \alpha_n$ are to be summed over x, y and z.

The usefulness of the contracted notation ω_σ, introduced in the previous section, for identifying the various Fourier components of the nonlinear polarisation (2.46) can now be more fully appreciated. Furthermore, the ω_σ notation is especially useful in §§4.3 and 5.1 when we describe the *overall* permutation-symmetry property.

2.3.2 Polarisation induced by monochromatic waves

We now consider in more detail the evaluation of the integrals in (2.46) for specific nonlinear processes, in the important practical case when the applied field consists of a superposition of monochromatic waves. In this case a useful simplification can be made by invoking the intrinsic permutation-symmetry property.

We write ω' for any one of the frequencies involved, which may be positive or zero. Then (2.25) may be written as

$$\mathbf{E}(t) = \tfrac{1}{2}\sum_{\omega'\geq 0}\Big[\mathbf{E}_{\omega'}\exp(-i\omega' t) + \mathbf{E}_{-\omega'}\exp(i\omega' t)\Big], \qquad (2.49)$$

where, since $\mathbf{E}(t)$ is real, $\mathbf{E}_{-\omega'} = \mathbf{E}_{\omega'}{}^*$. It follows immediately from (2.25) and (2.49) that the Fourier transform $\mathbf{E}(\omega)$ of $\mathbf{E}(t)$ is given by

$$\mathbf{E}(\omega) = \tfrac{1}{2}\sum_{\omega'\geq 0}\Big[\mathbf{E}_{\omega'}\,\delta(\omega-\omega') + \mathbf{E}_{-\omega'}\,\delta(\omega+\omega')\Big]. \qquad (2.50)$$

The Fourier transform $\mathbf{E}(\omega)$ has the units of $\mathrm{V\,s\,m^{-1}}$, whereas the monochromatic-wave amplitude $\mathbf{E}_{\omega'}$ has units of $\mathrm{V\,m^{-1}}$. It will be noticed that we follow the convention of including a factor of $\frac{1}{2}$ in (2.49). Consequently the intensity for a running wave with frequency ω' is

$$\begin{aligned} I_{\omega'} &= \varepsilon_0\, c\, n(\omega') < \tfrac{1}{2}[\mathbf{E}_{\omega'}\exp(-i\omega' t) + \mathbf{E}_{\omega'}{}^*\exp(i\omega' t)] \\ &\qquad \cdot \tfrac{1}{2}[\mathbf{E}_{\omega'}\exp(-i\omega' t) + \mathbf{E}_{\omega'}{}^*\exp(i\omega' t)] > \\ &= \tfrac{1}{2}\varepsilon_0\, c\, n(\omega')\,|\mathbf{E}_{\omega'}|^2, \qquad \text{(units: W m}^{-2}\text{)} \end{aligned} \qquad (2.51)$$

where the angle brackets here denote a cycle average, and $n(\omega')$ is the refractive index at frequency ω'. When $\mathbf{E}(t)$ is given by (2.49), we may rewrite (2.40) for the polarisation $\mathbf{P}^{(n)}(t)$ in the form:

$$\mathbf{P}^{(n)}(t) = \tfrac{1}{2}\sum_{\omega\geq 0}\Big[\mathbf{P}^{(n)}_{\omega}\exp(-i\omega t) + \mathbf{P}^{(n)}_{-\omega}\exp(i\omega t)\Big], \qquad (2.52)$$

where, since $\mathbf{P}^{(n)}(t)$ is real, $\mathbf{P}^{(n)}_{-\omega} = (\mathbf{P}^{(n)}_{\omega})^*$. By substituting (2.50) into (2.46), we can obtain an expression for $\mathbf{P}^{(n)}_{\omega}$. The cartesian μ-component [*cf.* (2.48)] is given by

$$\begin{aligned} (P^{(n)}_{\omega_\sigma})_\mu = 2\varepsilon_o \sum_{\alpha_1\cdots\alpha_n} \Big[\\ &\chi^{(n)}_{\mu\alpha_1\cdots\alpha_n}(-\omega_\sigma;\omega_1,\omega_2,\dots,\omega_n)\tfrac{1}{2}(E_{\omega_1})_{\alpha_1}\tfrac{1}{2}(E_{\omega_2})_{\alpha_2}\cdots\tfrac{1}{2}(E_{\omega_n})_{\alpha_n} \quad \text{(a)} \\ &+\chi^{(n)}_{\mu\alpha_1\cdots\alpha_n}(-\omega_\sigma;\omega_2,\omega_1,\dots,\omega_n)\tfrac{1}{2}(E_{\omega_2})_{\alpha_1}\tfrac{1}{2}(E_{\omega_1})_{\alpha_2}\cdots\tfrac{1}{2}(E_{\omega_n})_{\alpha_n} \quad \text{(b)} \\ &+ \text{further distinguishable terms}\Big], \end{aligned} \qquad (2.53)$$

where $\omega_\sigma = \omega_1 + \omega_2 + \cdots + \omega_n$ as before. Notice, however, that we are now considering a specific frequency ω_σ (≥ 0), and that $\omega_1, \dots, \omega_n$ each denote any of the frequencies (positive, negative or zero) which appear inside the brackets in the expression (2.49) for $\mathbf{E}(t)$, and which together satisfy $\omega_\sigma = \omega_1 + \omega_2 + \cdots + \omega_n$.

The right-hand side of (2.53) contains only *distinguishable* terms, by which we mean terms corresponding to each of the *distinct* arrangements of the set $\omega_1, ..., \omega_n$ whose sum is ω_σ. The number of distinguishable arrangements is reduced if two or more of the frequencies $\omega_1, ..., \omega_n$ are 'degenerate', *i.e.*, the same. In identifying the distinct combinations for a particular nonlinear process, it should be noted that a frequency and its negative must be considered distinguishable.

Let us consider an example: For clarity, the first two terms in the summation in (2.53) have been labelled (a) and (b) respectively. If $\omega_1 = \omega_2$, then it can be seen that the term (b) is indistinguishable from term (a), and does not therefore occur. Consider, on the other hand, the situation $\omega_1 \neq \omega_2$; the terms (a) and (b) are then distinguishable and both must be included in the expansion of (2.53). In this case, however, by relabelling the dummy subscripts (α_1, α_2) in term (b) as (α_2, α_1), and by recognising that because of the intrinsic permutation symmetry we can freely interchange $(\alpha_1, \omega_1) \leftrightarrow (\alpha_2, \omega_2)$, it can be seen that the two terms (a) and (b) are in fact the same. By this example, we have shown that although two or more terms occurring in the summation in (2.53) may be distinguishable in the strict sense we have defined above and therefore must be included, they nevertheless each have the same value because of permutation symmetry.

The general conclusion is that in place of each of the distinguishable terms in the summation in (2.53), only one term need be written. The number of times this term occurs is equal to the number of distinguishable permutations of the set of frequencies $\omega_1, ..., \omega_n$.

2.3.3 Conventions and catalogue of nonlinear phenomena

Because we have chosen to include a factor $\frac{1}{2}$ in the definition (2.49), a factor of $\frac{1}{2}$ is associated with each $\mathbf{E}_{\omega_j}$ appearing in (2.53), but only provided that $\omega_j \neq 0$. Similarly, the factor 2 in the right-hand side of (2.53) arises from the $\frac{1}{2}$ in the definition (2.52). In general it is rather tiresome to keep track of the various numerical factors, and this is a frequent source of error. For this reason it is convenient to follow a definitive notation (Ward and New, 1969); for a superposition of monochromatic waves, and by invoking intrinsic permutation symmetry as explained above, we can write (2.53) in the form:

$$(P_{\omega_\sigma}^{(n)})_\mu = \varepsilon_0 \sum_{\alpha_1 \cdots \alpha_n} \sum_{\omega} K(-\omega_\sigma; \omega_1, ..., \omega_n)$$
$$\times \chi^{(n)}_{\mu\alpha_1 \cdots \alpha_n}(-\omega_\sigma; \omega_1, ..., \omega_n)(E_{\omega_1})_{\alpha_1} \cdots (E_{\omega_n})_{\alpha_n}, \quad (2.54)$$

or alternatively, in vector notation,

$$\mathbf{P}^{(n)}_{\omega_\sigma} = \varepsilon_0 \sum_\omega K(-\omega_\sigma;\omega_1,\ldots,\omega_n)\boldsymbol{\chi}^{(n)}(-\omega_\sigma;\omega_1,\ldots,\omega_n)|\mathbf{E}_{\omega_1}\cdots\mathbf{E}_{\omega_n}. \tag{2.55}$$

The first summation in (2.54) has the same significance as in (2.48). The second summation $\sum_\omega$ serves as a reminder to sum over all of the distinct sets of $\omega_1,\ldots,\omega_n$. (The spectrum of monochromatic waves in (2.49) could contain several distinct sets of frequencies $\omega_1,\ldots,\omega_n$ which satisfy $\omega_\sigma = \omega_1+\omega_2+\cdots+\omega_n$. Most often, however, there is only one set – generally experiments are designed to avoid this ambiguity – and therefore the symbol $\sum_\omega$ is usually omitted henceforth, although it remains implied.) Because of the intrinsic permutation symmetry, the frequencies $\omega_1,\ldots,\omega_n$ can be written in any arbitrary sequence in (2.54) and (2.55); for monochromatic waves this is certainly of no consequence. Later, however, we consider resonant and time-dependent nonlinear processes, and in that case the ordering may have physical significance.

In (2.54) and (2.55), K is the numerical factor just described, and it can be defined formally as

$$K(-\omega_\sigma;\omega_1,\ldots,\omega_n) = 2^{l+m-n}p \tag{2.56}$$

where p is the number of distinct permutations of $\omega_1,\ldots,\omega_n$, n is the order of nonlinearity, m of the set of n frequencies $\omega_1,\ldots,\omega_n$ are zero (*i.e.*, d.c. fields), and $l=1$ if $\omega_\sigma\neq 0$, otherwise $l=0$. If, for example, $\omega_\sigma,\omega_1,\ldots,\omega_n$ are all non-zero, and $\omega_1,\ldots,\omega_n$ are all different, then $K=2^{1-n}n!$. Table 2.1 contains a representative list of nonlinear processes with values of K given; some of the processes have already been mentioned in the introductory chapter, and some are discussed in more detail later (see Subject index).

Unfortunately the nonlinear-optics literature contains a tortuous jumble of inconsistent definitions. The inclusion of the factor $\frac{1}{2}$ in the definition (2.49) for the wave amplitudes is not a universal convention. Yet more confusing, a discussion of a particular nonlinear process may include an expression for the nonlinear polarisation, similar to (2.54) or (2.55), except with the numerical factor K absorbed into a newly-defined susceptibility. This has a number of disadvantages: the E-field amplitudes are often ill-defined (and so the vexing factors of 2 cannot be traced); it becomes confusing to compare the magnitudes of susceptibilities for different processes; and also the redefined susceptibility undergoes discontinuous jumps in magnitude when any of the frequency arguments are allowed to become equal or tend to zero.

Table 2.1: Representative catalogue of nonlinear phenomena listing K factors and frequency arguments of $\chi^{(n)}$

Process	Order n	$-\omega_\sigma;\omega_1,\dots,\omega_n$	K
Linear absorption/emission and refractive index	1	$-\omega;\omega$	1
Optical rectification (optically-induced d.c. field)	2	$0;\omega,-\omega$	$\frac{1}{2}$
Pockels effect (linear electrooptic effect)	2	$-\omega;0,\omega$	2
Second-harmonic generation	2	$-2\omega;\omega,\omega$	$\frac{1}{2}$
Sum- and difference-frequency mixing, parametric amplification and oscillation	2	$-\omega_3;\omega_1,\pm\omega_2$	1
d.c. Kerr effect (quadratic electrooptic effect)	3	$-\omega;0,0,\omega$	3
d.c.-induced second-harmonic generation	3	$-2\omega;0,\omega,\omega$	$\frac{3}{2}$
Third-harmonic generation	3	$-3\omega;\omega,\omega,\omega$	$\frac{1}{4}$
General four-wave mixing	3	$-\omega_4;\omega_1,\omega_2,\omega_3$	$\frac{3}{2}$
Third-order sum- and difference-frequency mixing	3	$-\omega_3;\pm\omega_1,\omega_2,\omega_2$	$\frac{3}{4}$
Coherent anti-Stokes Raman scattering	3	$-\omega_{AS};\omega_P,\omega_P,-\omega_S$	$\frac{3}{4}$
Optical Kerr effect (optically-induced birefringence), cross-phase modulation, stimulated Raman scattering, stimulated Brillouin scattering	3	$-\omega_S;\omega_P,-\omega_P,\omega_S$	$\frac{3}{2}$
Intensity-dependent refractive index, optical Kerr effect (self-induced and cross-induced birefringence), self-focusing, self-phase and cross-phase modulation, degenerate four-wave mixing	3	$-\omega;\omega,-\omega,\omega$	$\frac{3}{4}$
Two-photon absorption/ionisation /emission	3	$-\omega_1;-\omega_2,\omega_2,\omega_1$ or $-\omega;-\omega,\omega,\omega$	$\frac{3}{2}$ $\frac{3}{4}$

Table 2.1 (*continued*)

Process	Order n	$-\omega_\sigma;\omega_1,...,\omega_n$	K
Fifth-harmonic generation	5	$-5\omega;\omega,\omega,\omega,\omega,\omega$	$\frac{1}{16}$
Stimulated hyper-Raman scattering	5	$-\omega_S;\omega_{P1},\omega_{P2},-\omega_{P2},-\omega_{P1},\omega_S$ or $-\omega_S;\omega_P,\omega_P,-\omega_P,-\omega_P,\omega_S$	$\frac{15}{2}$ $\frac{15}{8}$
m-photon absorption /ionisation /emission (single frequency)	$2m-1$	$-\omega;-\omega,...,-\omega,\omega,...,\omega$	‡
nth harmonic generation	n	$-n\omega;\omega,...,\omega$	2^{1-n}

‡ $= 2^{2(1-m)}(2m-1)!/(m-1)!m!$

The K values given apply if differently-labelled frequencies are different in value. As can be seen from this table, a susceptibility with a particular set of frequency arguments can describe a number of distinct (although related) processes. For example, the form of the susceptibilities given for two-photon absorption are the same as those for stimulated Raman scattering and the intensity-dependent refractive index, except that the frequency arguments have been written in a different sequence. The ordering of arguments only assumes some significance when considering resonant processes (see §4.5).
The real and imaginary parts of a susceptibility may describe physically distinct phenomena; for example, the real and imaginary parts of $\chi^{(1)}(-\omega;\omega)$ describe the refractive index and linear absorption/emission respectively. The real and imaginary parts of $\chi^{(3)}(-\omega_S;\omega_P,-\omega_P,\omega_S)$ can describe processes as diverse as the optical Kerr effect and stimulated Raman scattering, respectively.
Lettered subscripts refer to the conventional naming of frequencies in Raman spectroscopy: S = Stokes, AS = anti-Stokes, P = pump.

In this book we follow the convention of including the factor K explicitly in the expressions (2.54) and (2.55) for a superposition of monochromatic waves. As already mentioned, its consistency is an aid to accuracy. With this convention, the susceptibilities $\chi^{(n)}(-\omega_\sigma;\omega_1,...,\omega_n)$ vary smoothly without discontinuities as the various frequency arguments tend to zero or become degenerate (for example, as $\omega_1 \rightarrow \omega_2$).

2.3.4 Scalar form of the susceptibilities

It is often useful to lighten the notation by expressing the vector equation (2.55) in scalar form, by writing:

$$\mathbf{E}_{\omega_j} = \mathbf{e}_j E_j , \tag{2.57}$$

where the amplitude E_j is a complex scalar, and $\mathbf{e}_j$ is a unit vector in the direction of polarisation of the field, with the general property $\mathbf{e}_j \cdot \mathbf{e}_j{}^* = 1$. (For an elliptically-polarised field, $\mathbf{e}_j$ is complex; for linear polarisation $\mathbf{e}_j$ is real.) We can then define a scalar nonlinear susceptibility by

$$\begin{aligned}\chi^{(n)}(-\omega_\sigma;\omega_1,\ldots,\omega_n) \\ &= \mathbf{e}_\sigma{}^* \cdot \boldsymbol{\chi}^{(n)}(-\omega_\sigma;\omega_1,\ldots,\omega_n) \,|\, \mathbf{e}_1 \cdots \mathbf{e}_n \\ &= \sum_{\mu\alpha_1\cdots\alpha_n} \chi^{(n)}_{\mu\alpha_1\cdots\alpha_n}(-\omega_\sigma;\omega_1,\ldots,\omega_n)\,(e_\sigma{}^*)_\mu\,(e_1)_{\alpha_1}\cdots(e_n)_{\alpha_n},\end{aligned} \tag{2.58}$$

and thus the polarisation $\mathbf{P}^{(n)}_{\omega_\sigma}$, with unit polarisation vector $\mathbf{e}_\sigma$, becomes:

$$P_\sigma^{(n)} = \varepsilon_0 \sum_\omega K(-\omega_\sigma;\omega_1,\ldots,\omega_n)\,\chi^{(n)}(-\omega_\sigma;\omega_1,\ldots,\omega_n)\,E_1\cdots E_n, \tag{2.59}$$

where the amplitude $P_\sigma^{(n)}$ is a complex scalar defined similarly to (2.57); *i.e.*, $\mathbf{P}^{(n)}_{\omega_\sigma} = \mathbf{e}_\sigma P_\sigma^{(n)}$. In (2.58) the polarisation vector $\mathbf{e}_i{}^*$ is to be associated with a negative frequency $-\omega_i$ which appears as an argument of $\chi^{(n)}$. The scalar quantity $\chi^{(n)}$ so defined is sometimes called the *effective* susceptibility for a particular process. The factors $(e_j)_{\alpha_k}$ are direction cosines with respect to the α_k axis. The definition (2.58) involves 3^{n+1} direction cosines in general, although frequently some of the waves are copolarised so that fewer direction cosines need to be evaluated. (The polarisation-vector direction cosines should not, however, be confused with the direction cosines of the wave vectors $\mathbf{k}$ introduced in §2.5 and Chapter 7.)

2.4 Time-dependent problems

The beautiful symmetry of the time- and frequency-domain descriptions of the constitutive relations can be seen by comparing (2.23) and (2.40). The frequency-domain approach in terms of susceptibilities is the one most frequently used and, as described in the previous section, it is certainly the more appropriate description when considering monochromatic or nearly-monochromatic fields (as may be obtained, for example, from continuous-wave single-mode lasers). On the other hand, in recent years there have been dramatic advances in the development of short-pulse lasers; by using a variety of mode-locking and pulse-compression techniques it is now possible to generate intense optical pulses as short as a few femtoseconds (equivalent to only a few cycles of the optical carrier wave). Such pulses may be shorter in duration than the fundamental ultrafast relaxation processes of the nonlinear medium; such as, for example, the orientational relaxation and thermalisation of photoexcited free carriers in semiconductors. These ultrafast processes can be probed by observing the nonlinear optical response which is induced by short-pulse lasers. In this case the time-domain description is more appropriate. In fact, the approach in terms of time-domain response functions could be used for pulsed fields of less than one period of the optical carrier wave, since the function $\mathbf{R}^{(n)}(\tau_1, \ldots, \tau_n)$ in (2.23) describes the material response to a sequence of delta-function fields at times $\tau_1, \ldots, \tau_n$.

There is a third approach to the description of the nonlinear optical response, and it is widely used. This approach does not fall completely within the categories of time- or frequency-domain descriptions, but instead is a hybrid of the two; it may be called the 'quasi-monochromatic' description.

2.4.1 Quasi-monochromatic waves

Suppose the applied field consists of a superposition of quasi-monochromatic waves (such as the outputs from pulsed or modulated laser sources) each of which has some characteristic centre frequency ω'. By making the time-dependence explicit, the superposition can be written, by analogy with (2.49), as

$$\mathbf{E}(t) = \tfrac{1}{2} \sum_{\omega' \geq 0} \left[\mathbf{E}_{\omega'}(t) \exp(-i\omega' t) + \mathbf{E}_{-\omega'}(t) \exp(i\omega' t)\right] \tag{2.60}$$

where, since $\mathbf{E}(t)$ is real, $\mathbf{E}_{-\omega'}(t) = \mathbf{E}_{\omega'}{}^{*}(t)$, and the envelope functions $\mathbf{E}_{\omega'}(t)$ are assumed to be slowly-varying on the time scale of $(\omega')^{-1}$. Because the envelopes $\mathbf{E}_{\omega'}(t)$ are time-dependent, we are no longer considering waves each having a δ-function spectrum, as in (2.50); instead each wave occupies some spectral bandwidth centred at ω'. (The envelope function contains both amplitude and phase information; therefore a more precise definition of the centre frequency ω' is unnecessary.) Similar to (2.60), the polarisation $\mathbf{P}^{(n)}(t)$ can be written as

$$\mathbf{P}^{(n)}(t) = \tfrac{1}{2} \sum_{\omega'' \geq 0} \left[\mathbf{P}^{(n)}_{\omega''}(t) \exp(-i\omega'' t) + \mathbf{P}^{(n)}_{-\omega''}(t) \exp(i\omega'' t)\right], \tag{2.61}$$

where $\mathbf{P}^{(n)}_{-\omega''}(t) = \mathbf{P}^{(n)}_{\omega''}{}^{*}(t)$, and the envelope functions $\mathbf{P}^{(n)}_{\omega''}(t)$ are slowly varying on the time scale of $(\omega'')^{-1}$. Let us assume that we are interested in a particular nonlinear process that gives rise to an nth-order polarisation at a frequency ω_σ having the envelope function $\mathbf{P}^{(n)}_{\omega_\sigma}(t)$. We recall the expression (2.23) for the total polarisation $\mathbf{P}^{(n)}(t)$ in the time-domain:

$$\mathbf{P}^{(n)}(t) = \varepsilon_0 \int_{-\infty}^{+\infty} \mathrm{d}\tau_1 \cdots \int_{-\infty}^{+\infty} \mathrm{d}\tau_n \, \mathbf{R}^{(n)}(t-\tau_1, \ldots, t-\tau_n) | \mathbf{E}(\tau_1) \cdots \mathbf{E}(\tau_n). \tag{2.62}$$

By substituting (2.60) in (2.62), assembling the various frequency components (similar to the procedure leading to (2.55)), and picking out the particular component at ω_σ, we obtain the expression:

$$\mathbf{P}^{(n)}_{\omega_\sigma}(t) = \varepsilon_0 K(-\omega_\sigma;\omega_1,\ldots,\omega_n) \int_{-\infty}^{+\infty} \mathrm{d}\tau_1 \cdots \int_{-\infty}^{+\infty} \mathrm{d}\tau_n \, \mathbf{R}^{(n)}(t-\tau_1,\ldots,t-\tau_n) \,|\, \mathbf{E}_{\omega_1}(\tau_1) \cdots \mathbf{E}_{\omega_n}(\tau_n) \exp\left[i \sum_{r=1}^{n} \omega_r (t-\tau_r) \right]. \quad (2.63)$$

It is convenient to introduce a tensor $\mathbf{\Phi}^{(n)}$ of rank $n+1$, defined as follows:

$$\mathbf{\Phi}^{(n)}_{\omega_\sigma;\omega_1,\ldots,\omega_n}(\tau_1,\ldots,\tau_n) = \mathbf{R}^{(n)}(\tau_1,\ldots,\tau_n) \exp\left[i \sum_{r=1}^{n} \omega_r \tau_r \right]. \quad (2.64)$$

Then (2.63) can be rewritten in the simpler form

$$\mathbf{P}^{(n)}_{\omega_\sigma}(t) = \varepsilon_0 K(-\omega_\sigma;\omega_1,\ldots,\omega_n) \int_{-\infty}^{+\infty} \mathrm{d}\tau_1 \cdots \int_{-\infty}^{+\infty} \mathrm{d}\tau_n \times \mathbf{\Phi}^{(n)}_{\omega_\sigma;\omega_1,\ldots,\omega_n}(t-\tau_1,\ldots,t-\tau_n) \,|\, \mathbf{E}_{\omega_1}(\tau_1) \cdots \mathbf{E}_{\omega_n}(\tau_n). \quad (2.65)$$

For want of a better name, we call $\mathbf{\Phi}^{(n)}$ the 'envelope response function'. By comparing (2.62) and (2.65), it can be seen that $\mathbf{\Phi}^{(n)}$ has some of the appearance of a time-domain response function, yet we notice in (2.65) that $\mathbf{\Phi}^{(n)}$ relates the wave *envelopes* associated with the polarisation and electric field. Thus, for many time-dependent problems in nonlinear optics, the envelope response function $\mathbf{\Phi}^{(n)}$ provides a useful practical description – more so than either the susceptibility or the true time-domain response function $\mathbf{R}^{(n)}$. (In the nonlinear-optics literature, the response function that we denote by $\mathbf{\Phi}^{(n)}$ is sometimes called – incorrectly – the susceptibility.)

In common with $\boldsymbol{\chi}^{(n)}$ and $\mathbf{R}^{(n)}$, the response function $\mathbf{\Phi}^{(n)}$ is a property of the medium. The causality requirement dictates that $\mathbf{\Phi}^{(n)}(\tau_1,\ldots,\tau_n)$ vanishes when any of the arguments $\tau_1,\ldots,\tau_n$ is negative. From (2.41) we obtain the relation between $\mathbf{\Phi}^{(n)}$ and the susceptibility:

$$\boldsymbol{\chi}^{(n)}(-\omega_\sigma;\omega_1,\ldots,\omega_n) = \int_{-\infty}^{+\infty} \mathrm{d}\tau_1 \cdots \int_{-\infty}^{+\infty} \mathrm{d}\tau_n \, \mathbf{\Phi}^{(n)}_{\omega_\sigma;\omega_1,\ldots,\omega_n}(\tau_1,\ldots,\tau_n). \quad (2.66)$$

To lighten the notation, the frequency subscripts on $\mathbf{\Phi}^{(n)}$ will be omitted from now on, but it should be realised that $\mathbf{\Phi}^{(n)}$ is specific to a particular set of frequencies $\omega_\sigma, \omega_1, \ldots, \omega_n$ (as is usually clear from the context).

As shown below, there are two important limiting cases of (2.65). These correspond to the cases when the fluctuations of the incident fields occur on a time scale which is either much shorter or longer than the time constants associated with the response function $\boldsymbol{\Phi}^{(n)}$. The characteristic times for field fluctuations may be determined by the spectral intensity of the light source, the optical pulse duration, or both.

2.4.2 The adiabatic limit

The first important limiting case of (2.65) is the adiabatic limit, in which the amplitude fluctuations of the applied field are much slower than the relaxation time for the polarisation induced in the medium (this is explained more fully in §6.1). In this limit, the response depends only on the instantaneous values of the field envelopes, and this can be expressed as

$$\boldsymbol{\Phi}^{(n)}(t-\tau_1,\ldots,t-\tau_n) \rightarrow \mathbf{S}^{(n)}\,\delta(t-\tau_1)\cdots\delta(t-\tau_n), \tag{2.67}$$

where $\mathbf{S}^{(n)}$ is a time-independent tensor of rank $n+1$. Then (2.65) becomes

$$\mathbf{P}_{\omega_\sigma}^{(n)}(t) = \varepsilon_0 K(-\omega_\sigma;\omega_1,\ldots,\omega_n)\,\mathbf{S}^{(n)}\,|\,\mathbf{E}_{\omega_1}(t)\cdots\mathbf{E}_{\omega_n}(t). \tag{2.68}$$

In the limit we are considering, (2.68) must also be valid for continuous-wave fields (time-independent envelopes). So, by comparing with (2.55), we see that $\mathbf{S}^{(n)} = \boldsymbol{\chi}^{(n)}(-\omega_\sigma;\omega_1,\ldots,\omega_n)$. We thus obtain the adiabatic limit for (2.65):

$$\mathbf{P}_{\omega_\sigma}^{(n)}(t) = \varepsilon_0 K(-\omega_\sigma;\omega_1,\ldots,\omega_n)\,\boldsymbol{\chi}^{(n)}(-\omega_\sigma;\omega_1,\ldots,\omega_n)|\mathbf{E}_{\omega_1}(t)\cdots\mathbf{E}_{\omega_n}(t), \tag{2.69}$$

which can also be written in the convenient scalar form (see §2.3.4):

$$P_\sigma^{(n)}(t) = \varepsilon_0 K(-\omega_\sigma;\omega_1,\ldots,\omega_n)\,\chi^{(n)}(-\omega_\sigma;\omega_1,\ldots,\omega_n)\,E_1(t)\cdots E_n(t). \tag{2.70}$$

These adiabatic equations are obviously the slowly-time-dependent analogues of the monochromatic-wave description, (2.55) and (2.59) respectively.

2.4.3 The transient limit

For applied fields consisting of short pulses, it may be appropriate to take an opposite limit (the 'impulse' or 'transient limit') in which a single pulse can be considered to be very much shorter than the time for the induced polarisation to decay; this relaxation time is described in §6.1. In the transient limit the pulses can be used to probe directly the response of the medium to a sequence of optical-frequency impulses at times $\tau_1,\ldots,\tau_n$. An experimental arrangement often used (usually called 'pump-and-probe') is to provide variable time delays between the incident pulses and observe some effect resulting from the induced polarisation as a function of the delay times. (A situation of this type is described in §6.5.) Let us consider, for example, a third-order nonlinear process involving the incident quasi-monochromatic waves with envelopes $\mathbf{E}_{\omega_1}(t)$, $\mathbf{E}_{\omega_2}(t)$ and $\mathbf{E}_{\omega_3}(t)$. Suppose that we can represent the fields as

$$\mathbf{E}_{\omega_j}(t) = \mathbf{A}_{\omega_j} f_i(t), \tag{2.71}$$

where the $f_j(t)$ are normalised functions describing the pulse envelopes. The normalisation is such that $|\int_{-\infty}^{+\infty} \mathrm{d}t\, f_j(t)| = 1$, and thus $\mathbf{A}_{\omega_j}$ is the complex time-integral of the pulse $\mathbf{E}_{\omega_j}(t)$. (The pulse 'area' is a key parameter in the theory of coherent-resonant interactions, described in §6.2.) The incident fields are distinguishable by having distinct values of one or more of the properties: mean frequency ω_j, time dependence $f_j(t)$, E-field polarisation vector $\mathbf{e}_j$, and propagation wave vector (denoted by $\mathbf{k}_j$ in Chapter 7). Since, in the case we are considering, the pulses are much shorter than the relaxation times, they appear to the medium as impulses, and therefore the pulse envelopes can each be approximated by delta-functions: $f_j(t) \rightarrow \delta(t - t_j)$, where the t_j are the times of incidence. Then, evaluating the integrals in (2.65), we obtain

$$\mathbf{P}^{(3)}_{\omega_\sigma}(t) = \varepsilon_0 K(-\omega_\sigma; \omega_1, \omega_2, \omega_3)\, \boldsymbol{\Phi}^{(3)}(t - t_1, t - t_2, t - t_3) \,|\, \mathbf{A}_{\omega_1} \mathbf{A}_{\omega_2} \mathbf{A}_{\omega_3} . \tag{2.72}$$

Thus an observation of $\mathbf{P}^{(3)}_{\omega_\sigma}(t)$ as a function of the variable time delays $t - t_j$ provides a direct measure of the envelope response function $\boldsymbol{\Phi}^{(3)}(\tau_1, \tau_2, \tau_3)$. Experiments of this kind can provide information about the magnitude of the optical nonlinearity and the dominant relaxation rates. It is also a common practice to quote the magnitude of the nonlinearity in terms of an equivalent susceptibility, as though the incident fields were truly continuous-wave and monochromatic. An illustrative example is given in §6.1.

A further discussion of the notions of adiabatic and transient fields and material response is given in Chapter 6, where resonant nonlinear processes are considered in detail.

2.5 Spatial dispersion

We now recall the statement at the beginning of this chapter that our discussion of the constitutive relation is confined to the case where the nonlinear medium displays only a *local* response, *i.e.* the polarisation at a point in the medium is assumed to be determined completely by the electric field at that point. This restriction was made for simplicity.

However, a similar analysis of the constitutive relation can be based on the less restrictive assumption that the polarisation at a point is determined by the electric field in the *neighbourhood* of that point. The principle of time-invariance, which we considered in §2.1, must then be augmented by a corresponding principle of spatial invariance, *i.e.*, homogeneity. It is then still possible to introduce time-domain response functions for the polarisation, but they now involve position coordinates as well as time variables. For example, in place of (2.23), the polarisation $\mathbf{P}^{(n)}(t,\mathbf{r})$ at a position $\mathbf{r}$ can be expressed in terms of the field $\mathbf{E}(t,\mathbf{r})$ as:

$$\mathbf{P}^{(n)}(t,\mathbf{r}) = \varepsilon_0 \int_{-\infty}^{+\infty} d\tau_1 \cdots \int_{-\infty}^{+\infty} d\tau_n \int_{-\infty}^{+\infty} d\mathbf{r}_1 \cdots \int_{-\infty}^{+\infty} d\mathbf{r}_n$$

$$\times \mathbf{R}^{(n)}(\tau_1,\mathbf{r}_1,\ldots,\tau_n,\mathbf{r}_n) \,|\, \mathbf{E}(t-\tau_1,\mathbf{r}_1) \cdots \mathbf{E}(t-\tau_n,\mathbf{r}_n). \quad (2.73)$$

The intrinsic permutation-symmetry property now dictates that the response function $R^{(n)}_{\mu\alpha_1\cdots\alpha_n}(\tau_1,\mathbf{r}_1,\ldots,\tau_n,\mathbf{r}_n)$ is invariant under any of the $n!$ permutations of the n triplets $(\alpha_1,\tau_1,\mathbf{r}_1),\ldots,(\alpha_n,\tau_n,\mathbf{r}_n)$. Moreover, susceptibility tensors may still be introduced by Fourier analysis of the electric field with respect to both time *and* space:

$$\mathbf{E}(t,\mathbf{r}) = \int_{-\infty}^{+\infty} d\omega \int_{-\infty}^{+\infty} d\mathbf{k}\, \mathbf{E}(\omega,\mathbf{k}) \exp[-i(\omega t - \mathbf{k}\cdot\mathbf{r})]. \quad (2.74)$$

However, the susceptibility tensors then depend on the optical wave vectors $\mathbf{k}$ as well as frequencies ω, *i.e.*, we have spatial dispersion in addition to temporal dispersion (Ginsburg, 1958; Flytzanis, 1975). Thus in place of (2.40) and (2.41) we have the more general relation:

$$\mathbf{P}^{(n)}(t,\mathbf{r}) = \varepsilon_0 \int_{-\infty}^{+\infty} d\omega_1 \cdots \int_{-\infty}^{+\infty} d\omega_n \int_{-\infty}^{+\infty} d\mathbf{k}_1 \cdots \int_{-\infty}^{+\infty} d\mathbf{k}_n$$

$$\times \boldsymbol{\chi}^{(n)}(-\omega_\sigma;\omega_1,\mathbf{k}_1,\ldots,\omega_n,\mathbf{k}_n) \,|\, \mathbf{E}(\omega_1,\mathbf{k}_1) \cdots \mathbf{E}(\omega_n,\mathbf{k}_n)$$

$$\times \exp[-i(\omega_\sigma t - \mathbf{k}_P\cdot\mathbf{r})] \quad (2.75)$$

where

$$\boldsymbol{\chi}^{(n)}(-\omega_\sigma;\omega_1,\mathbf{k}_1,\ldots,\omega_n,\mathbf{k}_n) = \int_{-\infty}^{+\infty} d\tau_1 \cdots \int_{-\infty}^{+\infty} d\tau_n \int_{-\infty}^{+\infty} d\mathbf{r}_1 \cdots \int_{-\infty}^{+\infty} d\mathbf{r}_n$$

$$\times \mathbf{R}^{(n)}(\tau_1,\mathbf{r}_1\ldots,\tau_n,\mathbf{r}_n) \exp\left[i\sum_{j=1}^{n}(\omega_j\tau_j - \mathbf{k}_j\cdot\mathbf{r}_j)\right], \quad (2.76)$$

with $\omega_\sigma = \sum_{j=1}^{n}\omega_j$ and $\mathbf{k}_P = \sum_{j=1}^{n}\mathbf{k}_j$. Once again, intrinsic permutation symmetry implies that the susceptibility $\chi^{(n)}_{\mu\alpha_1\cdots\alpha_n}(-\omega_\sigma;\omega_1,\mathbf{k}_1,\ldots,\omega_n,\mathbf{k}_n)$ is invariant under all $n!$ permutations of the n triplets $(\alpha_1,\omega_1,\mathbf{k}_1),\ldots,(\alpha_n,\omega_n,\mathbf{k}_n)$.

The constitutive relations which include spatial dispersion find an application in the theory of the microscopic response of condensed matter, in which allowance is made for interactions between the microscopic polarisable units (*e.g.*, atoms) which make up the structure (Flytzanis, 1975). If the units are strongly coupled, the field at one point $\mathbf{r}$ can induce large changes of polarisation at an adjacent point $\mathbf{r}'$ and this can give rise to cooperative effects such as polaritons (Kittel, 1986; Hayes and Loudon, 1978). However, this is not usually the case. Also, optical wavelengths are usually large compared with the dimensions of the polarisable units, so that the local field may be considered uniform. In that case, the spatial dispersion (*i.e.*, the $\mathbf{k}$-dependence) of $\boldsymbol{\chi}^{(n)}$ may be neglected. This is the electric-dipole approximation, and it is followed throughout this book. We also neglect effects due to the optical magnetic field, which are generally very weak. (An exception occurs in Appendix 9 where we indicate how the explicit formulae for the susceptibilities can sometimes be modified in a simple way to describe optical processes which involve electric-quadrupolar and magnetic-dipolar interactions. In the latter case we must consider the constitutive relation for the magnetisation $\mathbf{M}$ in terms of the magnetic field $\mathbf{H}$, in addition to that for the polarisation $\mathbf{P}$ in terms of $\mathbf{E}$.)

2.6 Alternative approaches

Finally, we consider briefly the possibility of alternative constitutive relations. In the conventional formulation, as presented in previous sections, a particular material parameter (the polarisation) is expressed as a function of the electric field. The usefulness of this is that the polarisation is a natural source term in Maxwell's equations (see Chapter 7). Although this conventional constitutive relation is found to be perfectly adequate for the description of the great majority of nonlinear optical phenomena, it is not essential or unique. Alternative relations can be written in terms of other material parameters, such as the absorption coefficient (Miller, 1984) or the molecular orientation in a liquid crystal (Shen, 1984a). Yet another approach (Miller, 1964; Goldstone and Garmire, 1984) is to use an inverted form of the usual constitutive relation (*i.e.*, the electric field is expressed in terms of the polarisation). Miller (1964) used this inverted form in his definition of the $\boldsymbol{\Delta}$-tensor (§4.7.3). The motive for considering these alternative forms of the constitutive relation is that new physics is sometimes revealed.

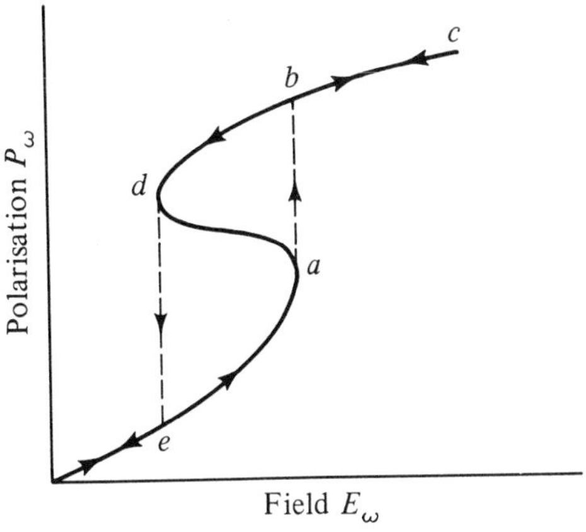

Fig. 2.1 Form of the constitutive relation resulting in bistability in the polarisation. The field is single-valued in the polarisation, but not *vice versa*.

For example, suppose the relation between the polarisation and the optical field were as shown in Fig. 2.1; this is the form that is calculated by taking the classical 'Duffing' oscillator as a model for the oscillation of electrons (see Goldstone and Garmire (1984) for a detailed description). This leads to bistability since, for a given value of the field, there can be more than one possible value of the polarisation. Referring to Fig. 2.1, if the field amplitude E_ω is increased gradually from zero, a situation is reached at point *a* where the polarisation undergoes a discontinuous jump to point *b*. Whilst increasing E_ω further to *c*, the polarisation increases smoothly without discontinuities. However, on decreasing E_ω from *c*, the polarisation undergoes a further discontinuous jump from point *d* to point *e*. There exists a range of values of the polarisation, corresponding to the curve between *a* and *d*, which is not observed. Bistability (or multistability) in the polarisation can be seen in several macroscopic effects (Goldstone and Garmire, 1984; Gibbs, 1985). This class of optical bistability is an intrinsic property of the medium, by virtue of the particular form of the nonlinear constitutive relation, and no mirrors or other external optical feedback elements are required for its observation (in contrast with the device shown in Fig. 1.4). The object of this brief discussion is to point out that such optical bistable behaviour could not be predicted from the usual phenomenological description of the polarisation as a power series in the electric field because the electric field is treated as the independent parameter and the multivalued nature of the polarisation cannot be described. In these circumstances it is sometimes convenient to consider the electric field as a function of the polarisation because, as indicated in Fig. 2.1, this may remain single-valued. Similar bistable behaviour can be seen in particular forms of nonlinearity in optical absorption (Miller, 1984), and again no susceptibility rigorously expandable in a power series only of the electric field (with constant coefficients) can predict this bistability.

As already stated, the conventional constitutive relation is appropriate for the great majority of problems in nonlinear optics. It is not our purpose to review all the possible alternatives and their applications. However, the underlying fundamental principles which have been described earlier in this chapter remain valid whichever relation is used, and could be more widely applied.

3

Review of quantum mechanics

The preceding chapter describes some general properties of the constitutive relation and the susceptibility tensors – properties which follow from the causality and reality conditions and the principle of time-invariance. Some further important symmetry properties are considered in Chapter 5, but before doing so it is necessary to derive explicit formulae for the susceptibility tensors. This is done in Chapter 4, by studying the dynamical behaviour of the charged particles in the medium under the influence of an electric field. For this purpose the techniques of quantum mechanics and quantum statistics are used, and those techniques are first reviewed briefly here. For those readers less familiar with the principles of quantum mechanics, this chapter will serve as an introduction and summary, focusing attention on the particular aspects of the general theory that are required later. The main aim of this tutorial review is to introduce the density operator and its governing equation of motion. The final objective, reached in the last section of the chapter, is to derive the terms in a perturbation series for the density operator; this will be employed to calculate the susceptibilities.

More detailed developments of the theory of quantum mechanics may be found in the standard texts (*e.g.*, Schiff, 1968; Dirac, 1958; Tolman, 1938; Merzbacher, 1970).

3.1 Basic concepts

The fundamental concept of quantum mechanics is that the state, *i.e.*, the instantaneous physical condition, of a dynamical system can be described by a normalised wave function ψ of the position and spin coordinates of the system. By 'normalised' we mean that

$$\int d\varsigma\, \psi^*\psi = 1 \tag{3.1}$$

where the integration is over all the coordinates of the system (assuming that it includes a summation over spin coordinates when this is appropriate). With the normalisation condition (3.1), the wave function which describes a particular state is completely determined, apart from a trivial phase factor, and all the physical properties of the state can be derived from the wave function.

The general procedure for calculating physical properties is as follows. Each of the dynamical variables of the system (*i.e.*, physical quantities such as the position coordinates, momenta, energy, *etc.*) is represented in the formalism by a linear operator. The process of measuring a physical quantity then corresponds to operating on the wave function by the corresponding operator. Consider an arbitrary dynamical variable which is represented by the operator O. Then the 'expectation value' of O, when the system is in the state ψ, is given by

$$<O> = \int \mathrm{d}\varsigma\, \psi^* O\, \psi. \tag{3.2}$$

This is the mean value of O which will be obtained by making repeated measurements when the system is known to be in the state ψ prior to each measurement.

In order to determine uniquely the state of a system with F degrees of freedom, it is necessary to measure simultaneously F dynamical variables for which simultaneous measurement is possible, *e.g.*, the positions or the momenta of the particles in the system but not a mixture of the two. When this has been done at some instant of time, the subsequent time-development of the state is determined by the equation of motion of the wave function, *i.e.*, by the time-dependent Schrödinger equation:

$$i\hbar\, \mathrm{d}\psi/\mathrm{d}t = H\psi \tag{3.3}$$

where $\hbar = h/2\pi$ is the Dirac constant (Appendix 1) and H is the Hamiltonian operator. The Hamiltonian is the operator which represents the energy of the system. The form of H can usually be derived from the classical Hamiltonian by replacing the cartesian position coordinates and their conjugate momenta by the corresponding operators. The cartesian coordinate x is replaced by the multiplicative operator x, and the conjugate momentum p_x is replaced by the differential operator $-i\hbar\,\partial/\partial x$. This particular way of describing the evolution of the state of a system, according to (3.3), is known as the 'Schrödinger picture'.

3.2 Representations and matrix operators

The above remarks contain, in briefest outline, most of the aspects of the basic theory of quantum mechanics which will be used in the subsequent analysis. We turn now to a discussion of representations and operators which provide a convenient technique for carrying out particular calculations.

3.2.1 Representations

A representation is a set of wave functions $\{u_i\}$ which is both orthonormal and complete; the label i serves to distinguish the various members of the set from one another. By 'orthonormal' is meant that

$$\int d\varsigma\, u_i^* u_j = \delta_{ij}\,, \tag{3.4}$$

where the Kronecker delta is defined as $\delta_{ij} = 1$ when $i = j$, and is zero otherwise. By 'complete' is meant that an arbitrary wave function ψ can be expanded in terms of $\{u_i\}$ as follows:

$$\psi = \sum_i a_i u_i \tag{3.5}$$

where

$$a_i = \int d\varsigma\, u_i^* \psi\,. \tag{3.6}$$

The latter expression for a_i can be easily verified by substituting for ψ in the right-hand side of (3.6) and applying the condition (3.4). Apart from the two conditions of orthonormality and completeness, the representation is at our disposal. Which representation one should choose is dictated by the particular characteristics of the calculation in hand.

3.2.2 Matrix representation of operators

The coefficients a_i in (3.5) may be conveniently regarded as the elements of a column matrix. Thus, in any particular representation, the wave function ψ may be specified by means of a column matrix $[a_i]$. It is also convenient to specify an operator O by means of a square matrix $[O_{ij}]$ whose (ij)th element is defined by

$$O_{ij} = \int d\varsigma\, u_i^* O\, u_j\,. \tag{3.7}$$

An alternative notation for O_{ij} is the 'bra–ket' form introduced by Dirac (Schiff 1968, p.164):

$$O_{ij} = \langle i|O|j\rangle\,. \tag{3.7a}$$

This notation is particularly useful when i and j involve several different quantum numbers because they may be written out more clearly on the right-hand side of (3.7a). We use it in Chapter 8.

There are two basic rules governing the manipulation of operator matrices: first, the linear combination rule, which states that the matrix of a linear combination of several operators is equal to the same linear combination of the matrices of the individual operators. To show the validity of this rule, it is sufficient to consider the operator O which is a linear combination of two other operators O' and O''; $O = lO' + mO''$, where l and m are arbitrary scalars. Then, from (3.7), we see that:

$$O_{ij} = \int d\varsigma\, u_i^*(l\, O' + m\, O'')\, u_{ij}$$
$$= l\, O'_{ij} + m\, O''_{ij}. \tag{3.8}$$

The extension of this simple demonstration to larger linear combinations is straightforward. The second basic rule is the matrix product rule, which states that the matrix of a product of several operators is equal to the product of the matrices of the individual operators. Again, to prove this rule it is sufficient to consider the product of two operators O and O'. Then, from (3.7) ,

$$(O\, O')_{ij} = \int d\varsigma\, u_i^*\, O\, O' u_j\,. \tag{3.9}$$

We now apply the expansion theorem (3.5) and (3.6) to $O'u_j$. Thus we have

$$O'u_j = \sum_k u_k \int d\varsigma\, u_k^*\, O' u_j$$
$$= \sum_k u_k\, O'_{kj}. \tag{3.10}$$

By substituting (3.9) into (3.8) we obtain

$$(O\, O')_{ij} = \int d\varsigma\, u_i^* O \sum_k u_k\, O'_{kj}$$
$$= \sum_k O_{ik}\, O'_{kj}\,, \tag{3.11}$$

which is the matrix product rule for two operators, expressed in terms of the elements of the matrices. The extension of the rule to products of more than two operators is also straightforward. Thus, for example,

$$(O\, O' O'')_{ij} = \sum_k (O\, O')_{ik}\, O''_{kj}$$
$$= \sum_k \sum_l O_{il}\, O'_{lk}\, O''_{kj}. \tag{3.12}$$

3.2.3 Hermitian operators

The expectation-value formula (3.2) can be expressed in the representation $\{u_i\}$ by substituting for ψ from (3.5). Thus we have

$$<O> = \int d\varsigma \sum_i a_i^* u_i^* O \sum_j a_j\, u_j$$
$$= \sum_i \sum_j a_i^* a_j\, O_{ij}. \tag{3.13}$$

If the dynamical variable O is real, then it can be seen from (3.2) that $<O>$ must be real for arbitrary ψ, *i.e.*, for arbitrary $\{a_i\}$. Then, if $<O>$ is real, by equating (3.13) with its complex conjugate and interchanging the dummy subscripts i and j, it can be shown that $[O_{ij}]$ is an Hermitian matrix, *i.e.*,

$$O_{ij}{}^{*} = O_{ji}\,. \tag{3.14}$$

In this case the operator is said to be Hermitian.

We thus arrive at the important result that real dynamical variables are represented in the formalism by Hermitian operators which have Hermitian matrices. Notice the property, directly evident from (3.14), that the diagonal elements O_{ii} of an Hermitian matrix must be real.

In the subsequent analysis we will make use of a generalised form of the property (3.14) of an Hermitian operator; namely,

$$\left(\int d\varsigma\, \psi^* O \phi\right)^* = \int d\varsigma\, \phi^* O \psi, \tag{3.15}$$

where ψ and ϕ are arbitrary wave functions, which need not belong to the same representation. The property (3.15) can be proved by expanding both ψ and ϕ in terms of $\{u_i\}$ and making use of (3.14).

3.3 Traces

We now come to a discussion of a numerical characteristic of an operator which plays a central rôle in the further development of the theory; namely, the trace. The trace of an operator O is defined to be the sum of its diagonal matrix elements and is denoted by $\mathrm{Tr}(O)$, thus

$$\mathrm{Tr}(O) = \sum_i O_{ii}\,. \tag{3.16}$$

As implied by the simplicity of the statement (3.16), the trace of an operator is independent of the representation used for its evaluation. To verify this fact we consider two distinct representations $\{u_i\}$ and $\{v_i\}$. Then, in the representation $\{u_i\}$ the trace of O is given by (3.16), where O_{ii} is understood to be evaluated by means of (3.7). On the other hand, in the representation $\{v_i\}$, the trace of O is given by

$$\mathrm{Tr}(O) = \sum_i \int d\varsigma\, v_i{}^* O\, v_i\,. \tag{3.17}$$

We now show that (3.17) is indeed identical to (3.16).

We first make use of the expansion theorem (3.5) and (3.6) to expand v_i in terms of $\{u_i\}$; thus we have

$$v_i = \sum_j a_{ij}\, u_j \tag{3.18}$$

where

$$a_{ij} = \int d\varsigma\, u_j{}^* v_i\,. \tag{3.19}$$

When (3.18) is substituted into (3.17) we obtain:

$$\begin{aligned}\mathrm{Tr}(O) &= \sum_i \int \mathrm{d}\varsigma \sum_j a_{ij}{}^* u_j{}^* O \sum_k a_{ik} u_k \\ &= \sum_i \sum_j \sum_k a_{ij}{}^* a_{ik} O_{jk} ,\end{aligned} \tag{3.20}$$

where O_{jk} is understood to be evaluated in the representation $\{u_i\}$. To complete the proof we observe that, from (3.19),

$$\begin{aligned}\sum_i a_{ij}{}^* a_{ik} &= \sum_i \int \mathrm{d}\varsigma\, u_k{}^* v_i \int \mathrm{d}\varsigma\, u_j v_i{}^* \\ &= \int \mathrm{d}\varsigma\, u_k{}^* \sum_i v_i \int \mathrm{d}\varsigma\, u_j v_i{}^* \\ &= \int \mathrm{d}\varsigma\, u_k{}^* u_j \\ &= \delta_{kj} .\end{aligned} \tag{3.21}$$

In the second line we have interchanged the summatio n and integration; the sum over i is then recognised as the expansion of u_i in terms of $\{v_i\}$; the final result follows from (3.4). With the aid of (3.21) the sum over i in (3.20) can be carried out immediately, to yield

$$\begin{aligned}\mathrm{Tr}(O) &= \sum_j \sum_k \delta_{kj} O_{jk} \\ &= \sum_j O_{jj} ,\end{aligned} \tag{3.22}$$

which is identical to (3.16).

There are two basic rules governing the manipulation of traces which we shall frequently employ: first, the linear combination rule, which states that the trace of a linear combination of several operators is equal to the same linear combination of the traces of the individual operators. This rule follows immediately from the linear combination rule (3.8) for the matrices of the operators. The second rule is the cyclic permutation rule, which states that the trace of the product of several operators is invariant under cyclic permutations of the factors. To prove this property it is sufficient to consider the product of two operators O and O'. Then, making use of the matrix product rule (3.11), we have

$$\begin{aligned}\mathrm{Tr}(O\,O') &= \sum_i (O\,O')_{ii} = \sum_i \sum_j O_{ij} O'_{ji} \\ &= \sum_i \sum_j O'_{ji} O_{ij} = \sum_j (O'\,O)_{jj} \\ &= \mathrm{Tr}(O'\,O) .\end{aligned} \tag{3.23}$$

In making the step between the first and second lines, we recognise that we are free to rearrange the order of the scalar matrix elements. The extension of the cyclic permutation rule to products of more than two operators is straightforward. Thus, for example,

$$\mathrm{Tr}(O\,O'O'') = \mathrm{Tr}(O''O\,O') = \mathrm{Tr}(O'O''O) . \tag{3.24}$$

3.4 The projection operator

The real significance of traces lies in the fact that the physical content of quantum mechanics can always be expressed in terms of them. In particular, the expectation-value formula (3.13) can be expressed as a trace. For this purpose we introduce the *projection operator* $P(\psi)$ for the state ψ by means of the defining relation

$$P(\psi)\theta = \psi \int d\varsigma\, \psi^* \theta, \tag{3.25}$$

where θ is an arbitrary wave function. The projection operator derives its name from the fact that, if ψ is a member of a complete orthonormal set, then $P(\psi)$ picks out the contribution to the expansion of θ from the state ψ. By making use of the defining relation (3.25), we obtain for the (ji)th matrix element of $P(\psi)$:

$$\begin{aligned} P(\psi)_{ji} &= \int d\varsigma\, u_j^* P(\psi) u_i \\ &= \int d\varsigma\, u_j^* \psi \int d\varsigma\, \psi^* u_i \\ &= a_j a_i^*, \end{aligned} \tag{3.26}$$

where the coefficients a_i and a_j are defined by (3.6). By substituting this result in (3.13) and invoking the matrix product rule (3.11), we obtain the desired formula

$$\begin{aligned} \langle O \rangle &= \sum_i \sum_j P(\psi)_{ji} O_{ij} \\ &= \sum_j (P(\psi) O)_{jj} \\ &= \mathrm{Tr}(P(\psi) O), \end{aligned} \tag{3.27}$$

which, we note, is independent of the representation.

Since it can be used to calculate expectation values, the projection operator provides a description of the physical state of the system which is completely equivalent to that provided by the wave function. For our purpose the description in terms of the projection operator is more suitable. The time-development of the system, as well as its instantaneous condition, can also be discussed in terms of the projection operator; that is to say, it is possible to derive an equation of motion for $P(\psi)$ which may be used instead of the equation of motion (3.3) for the wave function ψ (the time-dependent Schrödinger equation). It is sufficient for this purpose to suppose that the arbitrary wave function θ in (3.25) is independent of time. Then, by differentiating (3.25) with respect to time t and making use of (3.3), we obtain

$$\begin{aligned} i\hbar\, d\{P(\psi)\theta\}/dt &= i\hbar\,(d\psi/dt)\int d\varsigma\, \psi^*\theta + \psi \int d\varsigma\, i\hbar\,(d\psi^*/dt)\theta \\ &= H\psi \int d\varsigma\, \psi^*\theta - \psi \int d\varsigma\, (H\psi)^*\theta. \end{aligned} \tag{3.28}$$

From the defining relation (3.25), it is evident that the first term on the second line of (3.28) is $HP(\psi)\theta$. Moreover, since energy is a real dynamical variable and therefore the corresponding operator (the Hamiltonian H) is Hermitian, we see with the aid of (3.15) that the second term on the second line of (3.28) is

$$\begin{aligned} \psi \int \mathrm{d}\varsigma\, (H\psi)^* \theta &= \psi\left(\int \mathrm{d}\varsigma\, \theta^* H \psi\right)^* \\ &= \psi \int \mathrm{d}\varsigma\, \psi^* H \theta \\ &= P(\psi) H \theta . \end{aligned} \tag{3.29}$$

Hence (3.28) becomes

$$i\hbar\, \mathrm{d}\{P(\psi)\theta\}/\mathrm{d}t = HP(\psi)\theta - P(\psi)H\theta . \tag{3.30}$$

Finally, since θ is arbitrary, it can be cancelled out of (3.30) to yield the equation of motion of the projection operator $P(\psi)$:

$$\begin{aligned} i\hbar\, \mathrm{d}\{P(\psi)\}/\mathrm{d}t &= HP(\psi) - P(\psi)H \\ &= [H, P(\psi)] \end{aligned} \tag{3.31}$$

where we have introduced the conventional bracket notation for the commutator of two operators.

3.5 The density operator

We are now in a position to introduce the density operator and determine its equation of motion. The expectation-value formula (3.2) is the basic rule for the physical interpretation of the formalism of quantum mechanics. However, in order to use it we must know the state of the system precisely so that the wave function ψ is uniquely determined (apart from a trivial phase factor). Such precise knowledge is seldom at our disposal. Let us consider, as an example, a system containing M spin-free particles. Then a precise determination of the state of the system involves the measurement of $3M$ quantities (since the system has $3M$ degrees of freedom); the quantities to be measured might, for example, be the particle momenta. For a macroscopic system comprising, for example, a 1 mm^3 cube of solid material, $M \sim 10^{20}$ typically, and the task must be abandoned as hopeless. Even in small volumes of very dilute gases, the number of entities is usually so large that one cannot expect to know the state of the system precisely. The most that can really be hoped for is statistical information. We can use whatever limited knowledge is available about the state of the system to determine the probability p_n that the system is in the state ψ_n, where n labels the different possible states in which the system might actually be. Setting $p_n = 0$ for all values of n except one, for which $p_n = 1$, gives us the exceptional case in which the system is known with certainty (such a very exceptional case might occur when, for example, isolated atoms are prepared and maintained in predetermined states). We must now generalise the mean-value formula (3.2) so as to cope with the most generally-occurring situation in which p_n is nonzero for several values of n; we must pass from the realm of quantum mechanics to that of quantum statistics.

The required generalisation can be written down immediately in a concise form when the mean-value formula is expressed as in (3.27). (Indeed, it was for this reason that (3.27) was derived.) When the system is in the state ψ_n, then the mean value of an operator O is obtained by replacing ψ by ψ_n in (3.27). Hence, by weighting this expression with the probability p_n that the system is actually in the state ψ_n, we obtain an expression for the mean value of O which is appropriate when we have only statistical information about the system:

$$\begin{aligned} \langle O \rangle &= \sum_n p_n \operatorname{Tr}[P(\psi_n)\, O] \\ &= \operatorname{Tr}(\rho\, O), \end{aligned} \tag{3.32}$$

where

$$\rho = \sum_n p_n\, P(\psi_n), \tag{3.33}$$

which is known as the density operator of the system. The second equality in (3.32) follows immediately from the linear combination rule for traces. The density operator derives its name from the fact that it is the quantum-mechanical analogue of the probability density in phase space which is used in classical statistical mechanics.

3.5.1 Equation of motion

The important advantage of the density operator, apparent from (3.33), is that it describes the observable properties of an *ensemble* of quantum-mechanical entities. Since it can be used to calculate expectation values, the density operator provides a physical description of the ensemble system, which is appropriate when we have only statistical information about its state. To obtain the equation of motion of ρ, we observe that the probabilities p_n are determined by our initial information about the system and do not change with time. Hence, by differentiating (3.33) with respect to time t and using (3.31), we obtain

$$\begin{aligned} i\hbar\, \mathrm{d}\rho/\mathrm{d}t &= \sum_n p_n\, [H, P(\psi_n)] \\ &= [H, \sum_n p_n\, P(\psi_n)] \\ &= [H, \rho]. \end{aligned} \tag{3.34}$$

This is the equation of motion for the density operator. In the second line we have made use of a commutator identity, which will be frequently employed in the subsequent analysis, namely:

$$\left[\sum_i a_i O_i , \sum_j b_j O'_j\right] = \sum_i \sum_j a_i b_j [O_i , O'_j]. \tag{3.35}$$

Here O_i and O'_j are arbitrary operators and a_i and b_j are arbitrary constants. The validity of (3.35) can be shown by writing out in full the commutators (as defined in the second line of (3.31)) and performing a simple algebraic rearrangement.

3.5.2 Thermal equilibrium

The expectation-value formula in terms of the density operator (3.32) and the equation of motion (3.34) form the basis of our subsequent calculations. In the applications of these equations which we shall make, the system is in thermal equilibrium initially and is subsequently perturbed by an electromagnetic field. A knowledge of the form of the density operator in the thermal equilibrium situation provides the boundary condition necessary to solve (3.34). It is straightforward to obtain. As mentioned previously, the energy of the system is represented in the formalism by the Hamiltonian operator. In the absence of the field, the Hamiltonian will be denoted by H_0 (the Hamiltonian H in the presence of the field will differ from H_0 by terms representing the coupling of the system to the field). Then the allowed energy levels of the unperturbed system E_n and the corresponding wave functions ψ_n are determined by the eigenvalue equation

$$H_0 \psi_n = \mathrm{E}_n \psi_n . \tag{3.36}$$

It can be shown that the set of energy eigenstates $\{\psi_n\}$ can be chosen to be orthonormal (see *e.g.*, Merzbacher (1970) pp. 149–50); this follows from the fact that H_0 is Hermitian. Also, it is a fundamental assumption of quantum mechanics that the eigenfunctions of a real measurable dynamical variable form a complete set. Hence the set of energy eigenstates $\{\psi_n\}$ forms a representation; it is usually known as the 'energy representation' (see also §3.6).

As is implied by the notation, the energy eigenstates ψ_n are precisely the states for which the probabilities p_n are known in the thermal-equilibrium situation. For any macroscopic system, the probability that the system is in the particular energy eigenstate ψ_n, with energy E_n, is given by the familiar Boltzmann distribution:

$$p_n = \eta \exp(-\mathrm{E}_n/kT), \tag{3.37}$$

where η is a normalisation constant chosen so that $\sum_n p_n = 1$, k is the

Boltzmann constant, and T is the absolute temperature of the system. We shall accept the probability distribution (3.37) as being axiomatic (its derivation from more primitive postulates would be too great a diversion here; an excellent account of the derivation of (3.37) by the technique of maximum entropy inference has been given by Jaynes (1957)).

The thermal-equilibrium density operator ρ_0 is obtained by substituting (3.37) into (3.33), to give the formula

$$\rho_0 = \eta \sum_n \exp(-E_n/kT) P(\psi_n). \tag{3.38}$$

This formula can be given a much simpler appearance by introducing the exponential Hamiltonian operator $\exp(-H_0/kT)$, which can be regarded as defined by the power-series expansion:

$$\exp(-H_0/kT) = \sum_{r=0}^{\infty} \frac{1}{r!}(-H_0/kT)^r. \tag{3.39}$$

Thus we have the simple result:

$$\rho_0 = \eta \exp(-H_0/kT). \tag{3.40}$$

To verify this equation we observe that, using (3.36),

$$\begin{aligned} \exp(-H_0/kT)\psi_n &= \sum_{r=0}^{\infty} \frac{1}{r!}(-H_0/kT)^r \psi_n \\ &= \sum_{r=0}^{\infty} \frac{1}{r!}(-E_n/kT)^r \psi_n \\ &= \exp(-E_n/kT)\psi_n. \end{aligned} \tag{3.41}$$

We now apply ρ_0, as given by (3.38), to an arbitrary wave function

$$\theta = \sum_n b_n \psi_n \tag{3.42}$$

and make use of (3.41) and the definition of the projection operator $P(\psi_n)$. Thus we obtain

$$\begin{aligned} \rho_0 \theta &= \eta \sum_n \exp(-E_n/kT) b_n \psi_n \\ &= \eta \sum_n \exp(-H_0/kT) b_n \psi_n \\ &= \eta \exp(-H_0/kT)\theta. \end{aligned} \tag{3.43}$$

Since θ is an arbitrary wave function, it may be cancelled out of this equation to yield (3.40).

The value of the normalisation constant η in (3.40) may be obtained by setting the operator $O = 1$ in the expectation-value formula (3.32). Since $\langle 1 \rangle = 1$, we see that we must have $\mathrm{Tr}(\rho) = 1$ for any density operator. Hence, by setting $\rho = \rho_0$, we arrive at the result:

$$\eta = \{\mathrm{Tr}[\exp(-H_0/kT)]\}^{-1}. \tag{3.44}$$

3.6 The energy representation

Before moving on to consider the evolution of the density operator in more detail, we now make a few remarks about the energy representation introduced in §3.5.2. As we have seen, the evaluation of a trace may be carried out in any representation. However, because of the fundamental rôle played by the Hamiltonian in determining the dynamical behaviour of a system, the energy representation is often the most convenient one to use. We can show that the matrices H_0 and ρ_0 take a simple diagonal form in the energy representation; since the set $\{\psi_n\}$ is orthonormal, we see from (3.36) and the definition (3.7), that

$$\begin{aligned}(H_0)_{mn} &= \int d\varsigma\, \psi_m^* H_0 \psi_n \\ &= \int d\varsigma\, \psi_m^* \mathrm{E}_n \psi_n \\ &= \mathrm{E}_n \delta_{mn}.\end{aligned} \tag{3.45}$$

Moreover, from (3.40) and (3.41), we have

$$\begin{aligned}(\rho_0)_{mn} &= \int d\varsigma\, \psi_m^* \eta \exp(-H_0/kT)\psi_n \\ &= \int d\varsigma\, \psi_m^* \eta \exp(-\mathrm{E}_n/kT)\psi_n \\ &= \eta \exp(-\mathrm{E}_n/kT)\delta_{mn}.\end{aligned} \tag{3.46}$$

Indeed, it is clear from the previous discussion of $\exp(-H_0/kT)$ that, if $f(H_0)$ is any function of H_0 which can be expanded as a power series, then

$$f(H_0)\psi_n = f(\mathrm{E}_n)\psi_n. \tag{3.47}$$

Hence, the matrix of $f(H_0)$ is diagonal in the energy representation:

$$\begin{aligned}f(H_0)_{mn} &= \int d\varsigma\, \psi_m^* f(H_0)\psi_n \\ &= \int d\varsigma\, \psi_m^* f(\mathrm{E}_n)\psi_n \\ &= f(\mathrm{E}_n)\delta_{mn}.\end{aligned} \tag{3.48}$$

3.7 A perturbation series for the density operator

In earlier sections we have seen that for an ensemble system, for which only statistical information about its state is known, the physical characteristics at any time t are determined by the density operator $\rho(t)$ at that time. In this, the last and longest section of the present chapter, we calculate the density operator for a system which is initially in thermal equilibrium and is subsequently perturbed by external forces. In the next chapter we shall consider the particular case where the system consists of the charged particles in a medium and the external forces are produced by electromagnetic waves. However, there is nothing to be gained by making this specialisation at this stage; to do so would merely obscure the generality of the following calculation, without simplifying it at all.

3.7.1 Outline

Consider a system which has the Hamiltonian:

$$H = H_0 + H_{\mathrm{I}}(t), \tag{3.49}$$

where H_0 is the Hamiltonian in the equilibrium situation when external forces are absent, and $H_{\mathrm{I}}(t)$ represents the perturbation introduced by the external forces. Suppose that the external forces vanish in the remote past ($t \to -\infty$) and that the system is then in thermal equilibrium at a temperature T. Thus, $H_{\mathrm{I}}(-\infty)=0$ and $\rho(-\infty)=\rho_0$, where the thermal-equilibrium density operator ρ_0 is given by (3.40). The time-development of the density operator is governed by the equation of motion (3.34), which may be rewritten in a more convenient form by substituting for H from (3.49) and expressing the commutator in (3.34) as a sum of two commutators with the aid of the commutator identity (3.35). Thus we have

$$\begin{aligned} i\hbar\, \mathrm{d}\{\rho(t)\}/\mathrm{d}t &= [H, \rho(t)] \\ &= [H_0 + H_{\mathrm{I}}(t), \rho(t)] \\ &= [H_0, \rho(t)] + [H_{\mathrm{I}}(t), \rho(t)]. \end{aligned} \tag{3.50}$$

We observe that ρ_0 satisfies this equation when $H_{\mathrm{I}}(t)=0$; this can be shown by noting that ρ_0 is independent of t, and also that ρ_0 commutes with H_0 since, as evident from (3.40) and (3.41), ρ_0 can be expanded in a power series in H_0.

Our task is to solve (3.50) when $H_{\mathrm{I}}(t) \neq 0$ and subject to the boundary condition $\rho(-\infty) = \rho_0$.

The solution may be obtained by expressing $\rho(t)$ in the form of a perturbation series (*i.e.*, a power series in $H_{\mathrm{I}}(t)$) by writing

$$\rho(t) = \rho_0 + \rho_1(t) + \rho_2(t) + \cdots + \rho_n(t) + \cdots, \tag{3.51}$$

where ρ_0 is given by (3.40), $\rho_1(t)$ is linear in $H_{\mathrm{I}}(t)$, $\rho_2(t)$ is quadratic in $H_1(t)$, and so on. In view of the boundary condition $\rho(-\infty) = \rho_0$, we see that each term in (3.51) after ρ_0 must vanish in the remote past. Thus we have the boundary conditions:

$$\rho_r(-\infty) = 0; \qquad r = 1,2,\ldots,\infty. \tag{3.52}$$

The differential equations satisfied by the various terms in (3.51) are derived by substituting (3.51) into (3.50). By equating terms involving the same power of $H_{\mathrm{I}}(t)$, the following chain of equations is obtained:

$$i\hbar\, \mathrm{d}\rho_0/\mathrm{d}t = [H_0, \rho_0] \tag{3.53}$$
$$i\hbar\, \mathrm{d}\{\rho_1(t)\}/\mathrm{d}t = [H_0, \rho_1(t)] + [H_{\mathrm{I}}(t), \rho_0] \tag{3.54}$$
$$i\hbar\, \mathrm{d}\{\rho_2(t)\}/\mathrm{d}t = [H_0, \rho_2(t)] + [H_{\mathrm{I}}(t), \rho_1(t)] \tag{3.55}$$
$$\cdots\cdots$$
$$i\hbar\, \mathrm{d}\{\rho_n(t)\}/\mathrm{d}t = [H_0, \rho_n(t)] + [H_{\mathrm{I}}(t), \rho_{n-1}(t)] \tag{3.56}$$
$$\cdots\cdots$$

Equation (3.53) is an identity, and its validity merely verifies that we have chosen the unperturbed density operator ρ_0 correctly. The subsequent three equations (3.54)–(3.56) represent the first-order, second-order and nth-order perturbations, respectively. Since ρ_0 is known, we may solve for the first-order perturbation (3.54) subject to the boundary condition $\rho_1(-\infty)=0$ and so obtain $\rho_1(t)$. Then, since $\rho_1(t)$ is thereby known, we may obtain $\rho_2(t)$ by solving (3.55) subject to the boundary condition $\rho_2(-\infty)=0$. By proceeding in this way, all the terms in the perturbation series (3.51) may be evaluated.

3.7.2 Time-development operators

In order to carry through this procedure it is convenient to imagine that we have succeeded in evaluating $\rho_{n-1}(t)$ for some particular value of n, and now wish to solve (3.56) for $\rho_n(t)$. The technique used to solve operator differential equations of this type is similar to that employed to solve ordinary differential equations of the form

$$\mathrm{d}y/\mathrm{d}t = f(t)y + g(t), \tag{3.57}$$

where $f(t)$ and $g(t)$ are known functions of t. To solve (3.57) we seek an 'integrating factor' $I(t)$ such that

$$I(t)\,\mathrm{d}y/\mathrm{d}t - I(t)f(t)y = \mathrm{d}[I(t)y]/\mathrm{d}t\,. \tag{3.58}$$

By carrying out the differentiation on the right-hand side of (3.58), we see that $I(t)$ must satisfy the differential equation

$$\mathrm{d}[I(t)]/\mathrm{d}t = -I(t)f(t). \tag{3.59}$$

Hence

$$I(t) = \exp\left[-\int_0^t f(\tau)\,\mathrm{d}\tau\right] \tag{3.60}$$

is an integrating factor of (3.57). Equation (3.57) may now be solved by multiplying through by $I(t)$ and making use of (3.58). Thus we obtain:

$$\mathrm{d}[I(t)y]/\mathrm{d}t = I(t)g(t), \tag{3.61}$$

and so y is given by

$$I(t)y = C\int_0^t I(\tau)g(\tau)\,\mathrm{d}\tau\,, \tag{3.62}$$

where C is determined by the boundary conditions.

The solution of (3.56) may be obtained in much the same way. However, we are now dealing with operators and it is not clear at the outset whether we should multiply on the left by an integrating factor $V_0(t)$, or multiply on the right by an integrating factor $U_0(t)$, or do both at the same time. The last possibility in fact embraces the other two, since we may always put either $V_0(t)=1$ or $U_0(t)=1$ if the need arises. Therefore, so as not to lose any generality, we seek a pair of integrating factors $V_0(t)$ and $U_0(t)$ such that

$$V_0(t)\left\{i\hbar\frac{\mathrm{d}}{\mathrm{d}t}\rho_n(t)-[H_0,\rho_n(t)]\right\}U_0(t) = i\hbar\frac{\mathrm{d}}{\mathrm{d}t}\left\{V_0(t)\,\rho_n(t)\,U_0(t)\right\}. \tag{3.63}$$

By carrying out the differentiations on the right-hand side, expanding the commutator on the left-hand side and rearranging the terms, we obtain

$$\left\{i\hbar\frac{\mathrm{d}}{\mathrm{d}t}V_0(t) + V_0(t)H_0\right\}\rho_n(t)\,U_0 + V_0(t)\,\rho_n(t)\left\{i\hbar\frac{\mathrm{d}}{\mathrm{d}t}U_0(t) - H_0\,U_0(t)\right\} = 0. \tag{3.64}$$

This equation will obviously be satisfied if $V_0(t)$ and $U_0(t)$ are chosen so that the quantities in braces both vanish. Thus we obtain a pair of uncoupled simultaneous differential equations for the integrating factors:

$$i\hbar\frac{\mathrm{d}}{\mathrm{d}t}V_0(t) = -V_0(t)H_0 \tag{3.65}$$

$$i\hbar\frac{\mathrm{d}}{\mathrm{d}t}U_0(t) = H_0\,U_0(t). \tag{3.66}$$

It is not difficult to show that (3.65) and (3.66) have the exponential solutions

$$U_0(t) = \exp(-iH_0t/\hbar) \tag{3.67}$$

$$V_0(t) = \exp(iH_0t/\hbar) = U_0(-t) \tag{3.68}$$

where the exponentials are regarded as being defined by their power series expansions (*cf.* (3.39)). The powers of H_0 which appear in these expansions obviously all commute with one another (two operators O and O' are said to commute if their commutator $[O, O']$, defined in (3.31), is zero). Consequently, $U_0(t)$ and $U_0(t')$, where t and t' are arbitrary times, commute with one another and indeed also commute with any operator which is defined by a power series in terms of H_0, *e.g.*, ρ_0. Moreover, $U_0(t)$ has many of the properties of an ordinary exponential. In particular we see that

$$U_0(t)\,U_0(t') = \exp(-iH_0 t/\hbar)\,\exp(-iH_0 t'/\hbar) = \exp[-iH_0(t+t')/\hbar] = U_0(t+t'). \tag{3.69}$$

The first and second equalities in (3.69) follow from the definition (3.67), the second equality may be verified by direct multiplication of the power series for $\exp(-iH_0 t/\hbar)$ and $\exp(-iH_0 t'/\hbar)$. We also see, from (3.67), that

$$U_0(0) = \exp(0) = 1. \tag{3.70}$$

Hence, by setting $t' = -t$ in (3.69), we obtain

$$U_0(t)\,U_0(-t) = 1; \tag{3.71}$$

i.e., $V_0(t) = U_0(-t) = \exp(itH_0/\hbar)$ is the inverse of $U_0(t) = \exp(-itH_0/\hbar)$. We shall make frequent use of (3.69), (3.70) and (3.71). Finally, by differentiating their power series expansions, we find that $U_0(t)$ and $V_0(t)$ do indeed satisfy the differential equations (3.65) and (3.66).

The operator $U_0(t)$ is usually called the 'unperturbed time-development operator', and the notation we have adopted for it is a common one. The name is derived from the observation that, by virtue of the differential equation (3.66) and the fact that $U_0(0) = 1$, the solution $\psi(t)$ of the unperturbed time-dependent Schrödinger equation

$$i\hbar\,\mathrm{d}\psi(t)/\mathrm{d}t = H_0\,\psi(t) \tag{3.72}$$

may be obtained by operating on the wave function $\psi(0)$ at time $t = 0$ with $U_0(t)$:

$$\psi(t) = U_0(t)\,\psi(0). \tag{3.73}$$

The latter may be verified readily by substituting (3.73) into (3.72), noting of course that $\mathrm{d}\psi(0)/\mathrm{d}t = 0$, to recover (3.66).

Apart from trivial multiplying constants, the integrating factors (3.67) and (3.68) are unique because (3.64) must hold good for arbitrary $\rho_n(t)$. We now use them to obtain the solution of (3.56). By operating on each side of that equation with $V_0(t) = U_0(-t)$ as a premultiplier, with $U_0(t)$ as a postmultiplier, and invoking (3.63), we obtain

$$i\hbar\,\frac{\mathrm{d}}{\mathrm{d}t}\left\{U_0(-t)\,\rho_n(t)\,U_0(t)\right\} = U_0(-t)\,[H_\mathrm{I}(t), \rho_{n-1}(t)]\,U_0(t). \tag{3.74}$$

Hence, by integrating (3.74),

$$i\hbar\,U_0(-t)\,\rho_n(t)\,U_0(t) = \int_{-\infty}^{t} \mathrm{d}\tau_1\,U_0(-\tau_1)\,[H_\mathrm{I}(\tau_1), \rho_{n-1}(\tau_1)]\,U_0(\tau_1), \tag{3.75}$$

where the lower limit of integration has been fixed in accordance with the boundary condition $\rho_n(-\infty)=0$. Equation (3.75) specifies $\rho_n(t)$ when $\rho_{n-1}(t)$ is known. The factors $U_0(-t)$ and $U_0(t)$ on the left-hand side may be removed by premultiplying by $U_0(t)$ and postmultiplying by $U_0(-t)$; however, for our purpose it is more convenient to leave them where they are for the moment. We may verify that $\rho_n(t)$, as given by (3.75), satisfies (3.74) simply by differentiating (3.75) with respect to t.

3.7.3 The interaction picture

The application of (3.75) to derive our final formulae is simplified by introducing the abbreviated notation:

$$H_{\rm I}'(t) = U_0(-t)H_{\rm I}(t)U_0(t). \tag{3.76}$$

The transformation of an operator by multiplying it on the right by an unperturbed time-development operator and on the left by its inverse is usually referred to as 'going into the interaction picture'. The Hamiltonian H_0 for the unperturbed system, being time-independent, is the same in both the Schrödinger and interaction pictures. We shall meet further examples of work in the interaction picture in Chapters 4 and 6.

By expanding the commutator on the right-hand side of (3.75) and inserting $U_0(\tau_1)U_0(-\tau_1)=1$ between $H_{\rm I}(\tau_1)$ and $\rho_{n-1}(\tau_1)$ in each of the two terms which arise, we see that

$$U_0(-t)\rho_n(t)U_0(t) = (i\hbar)^{-1}\int_{-\infty}^{t} {\rm d}\tau_1\,[H_{\rm I}'(\tau_1), U_0(-\tau_1)\rho_{n-1}(\tau_1)U_0(\tau_1)]. \tag{3.77}$$

By setting $n=1$ in this equation we obtain

$$U_0(-t)\rho_1(t)U_0(t) = (i\hbar)^{-1}\int_{-\infty}^{t} {\rm d}\tau_1\,[H_{\rm I}'(\tau_1), U_0(-\tau_1)\rho_0(\tau_1)U_0(\tau_1)]. \tag{3.78}$$

Now $\rho_0(\tau_1)$ is given by (3.40); it is, in fact, independent of τ_1 and, moreover, it is an exponential function of H_0 and therefore commutes with $U_0(\tau_1)$ (for the reasons stated following (3.68)). Consequently, $\rho_0(\tau_1) = \rho_0$ may be pulled through to the right of $U_0(\tau_1)$ in (3.78), and, using (3.71), we obtain the simpler result:

$$U_0(-t)\rho_1(t)U_0(t) = (i\hbar)^{-1}\int_{-\infty}^{t} {\rm d}\tau_1\,[H_{\rm I}'(\tau_1), \rho_0]. \tag{3.79}$$

This equation determines $\rho_1(t)$. To find $\rho_2(t)$ we set $n=2$ in (3.77) and use (3.79). Thus we obtain

$$U_0(-t)\,\rho_2(t)\,U_0(t)$$
$$= (i\hbar)^{-1}\int_{-\infty}^{t} d\tau_1\,[H_I'(\tau_1),\,U_0(-\tau_1)\,\rho_1(\tau_1)\,U_0(\tau_1)]$$
$$= (i\hbar)^{-1}\int_{-\infty}^{t} d\tau_1\,[H_I'(\tau_1),\,(i\hbar)^{-1}\int_{-\infty}^{t_1} d\tau_2\,[H_I'(\tau_2),\,\rho_0\,]]$$
$$= (i\hbar)^{-2}\int_{-\infty}^{t} d\tau_1\int_{-\infty}^{t_1} d\tau_2\,[H_I'(\tau_1),\,[H_I'(\tau_2),\,\rho_0\,]], \qquad (3.80)$$

where we have made use of the commutator identity (3.35) to pull $(i\hbar)^{-1}\int_{-\infty}^{t_1} d\tau_2$ out of the commutator brackets. By proceeding in this way we obtain the general result:

$$U_0(-t)\,\rho_n(t)\,U_0(t) = (i\hbar)^{-n}\int_{-\infty}^{t} d\tau_1\int_{-\infty}^{t_1} d\tau_2\cdots\int_{-\infty}^{t_{n-1}} d\tau_n$$
$$\times\,[H_I'(\tau_1),\,[H_I'(\tau_2),\,\cdots\,[H_I'(\tau_n),\,\rho_0\,]\cdots]], \qquad (3.81)$$

where the integrand is an nth-order multiple commutator and $n = 1, 2, ..., \infty$.

3.7.4 The end result

We have now obtained all the elements of a complete solution for the density operator $\rho(t)$ for a system initially in thermal equilibrium and subsequently perturbed by external forces, to infinite order in the perturbation $H_I(t)$. It is appropriate at this point briefly to draw together and summarise the various elements in the solution. The density operator for the system is expressed as a perturbation series:

$$\rho(t) = \rho_0 + \rho_1(t) + \rho_2(t) + \cdots + \rho_n(t) + \cdots, \qquad (3.51)$$

where the thermal-equilibrium density operator ρ_0 was given earlier:

$$\rho_0 = \eta\,\exp(-H_0/kT). \qquad (3.40)$$

The perturbation terms in (3.51) are derived by premultiplying (3.81) by $U_0(t)$ and postmultiplying by $U_0(-t)$, which, with the aid of (3.71) gives

$$\rho_n(t) = (i\hbar)^{-n}\,U_0(t)\int_{-\infty}^{t} d\tau_1\int_{-\infty}^{t_1} d\tau_2\cdots\int_{-\infty}^{t_{n-1}} d\tau_n$$
$$\times\,[H_I'(\tau_1),\,[H_I'(\tau_2),\,\cdots\,[H_I'(\tau_n),\,\rho_0\,]\cdots]]\,U_0(-t) \qquad (3.82)$$

for $n = 1, 2, ..., \infty$. In (3.82), which is the main result of the chapter, the unperturbed time-development operator is

$$U_0(t) = \exp(-iH_0t/\hbar)\,, \tag{3.67}$$

and

$$H_I'(t) = U_0(-t)H_I(t)U_0(t)\,, \tag{3.76}$$

where $H_I(t)$ is the perturbation of the Hamiltonian due to the external forces.

The response, both linear and nonlinear, of the system to the external forces is completely determined by the density operator. In the next chapter, the general formulae which have been derived here will be applied to the particular problem of calculating the nonlinear polarisation induced by optical fields. The power of the density-operator formalism for solving problems of this type was first emphasised by Kubo (1957).

4

The susceptibility tensors

In Chapter 2, the nonlinear susceptibility tensors were introduced and their general properties were found by considering some fundamental physical principles. We have now set up, in Chapter 3, all the formal apparatus required to derive explicit formulae for the susceptibility tensors of a medium. This is done in this chapter by considering the dynamical behaviour of the charged particles in the medium under the influence of an electric field. The formulae that we derive are fundamental and quite general; they provide the basis for the treatment of the nonlinear optical properties of any medium in the electric-dipole approximation. We apply the formulae to a simple case – an idealised molecular gas – which is conceptually straightforward and provides a quick route to an understanding of the formulae. We also consider the important, and often difficult, problem of passing from microscopic formulae (which apply to individual molecules or groups of molecules) to the macroscopic formulae which are required later when we consider the resulting wave propagation.

The approach taken in the early part of the chapter is to consider the energy associated with the electric-dipole moment in an electric field; this is probably the most readily understood picture, and leads to formulae which can be directly applied in a quantitative way to atoms and simple molecules. However, in a later section we cast the results into an alternative form in terms of the particle momenta. This latter approach is more appropriate for the discussion of condensed media in which the charged particles are free to move throughout the material instead of being tied to individual molecules; the results so obtained are used particularly in the later discussion of optical nonlinearities in semiconductors.

4.1 A general charged-particle distribution

4.1.1 The macroscopic polarisation

To obtain explicit formulae for the susceptibility tensors we consider a small volume V of the medium to which the macroscopic electric field $\mathbf{E}(t)$ is applied, and we suppose that the volume is sufficiently small that the spatial variation of the electric field can be ignored. Effects arising from the magnetic field associated with $\mathbf{E}(t)$ are also ignored. We suppose that V contains a number of charged particles (electrons and ion cores) and we denote the position vector of the jth electron (charge $-e$) by $\boldsymbol{\mu}_j$, and similarly the position vector of the kth nucleus of charge $Z_k e$ is denoted by $\boldsymbol{\mu}_k$. Then the dipole moment of the charged particle system within the small volume V is

$$\mathbf{Q} = -e\sum_j \boldsymbol{\mu}_j + e\sum_k Z_k \boldsymbol{\mu}_k . \tag{4.1}$$

We now wish to calculate the *macroscopic* polarisation, by which we mean the expectation value of the dipole moment $\mathbf{Q}$ per unit volume:

$$\mathbf{P}(t) = V^{-1}\langle\mathbf{Q}\rangle . \tag{4.2}$$

We suppose that there are enough particles in V to allow us to ignore the fluctuations of the dipole moment density. Using the formula (3.32), we can express (4.2) in terms of the density operator of the charged particle system, $\rho(t)$:

$$\mathbf{P}(t) = V^{-1}\mathrm{Tr}[\rho(t)\mathbf{Q}]. \tag{4.3}$$

To find $\rho(t)$ we use the results of §3.7. The energy of the dipole moment $\mathbf{Q}$ in the electric field $\mathbf{E}(t)$ is

$$\begin{aligned} H_{\mathrm{I}}(t) &= -\mathbf{Q}\cdot\mathbf{E}(t) \\ &= -Q_\alpha E_\alpha(t) , \end{aligned} \tag{4.4}$$

where the dummy suffix α is summed over x, y and z. This is precisely the perturbation of the Hamiltonian of the charged particle system due to the presence of the electric field. Consequently, $\rho(t)$ is given by (3.51) where, in evaluating $\rho_n(t)$, we replace $H_{\mathrm{I}}(t)$ by $-\mathbf{Q}\cdot\mathbf{E}(t)$. When (3.51) is substituted into (4.3) we see that the constitutive relation is reproduced in the form of the power series in the field (2.1), but we now have explicit formulae for the various terms in the series:

$$\mathbf{P}^{(0)} = V^{-1}\mathrm{Tr}[\rho_0 \mathbf{Q}] \tag{4.5}$$

$$\mathbf{P}^{(n)} = V^{-1}\mathrm{Tr}[\rho_n(t)\mathbf{Q}] , \quad n \neq 0. \tag{4.6}$$

It only remains to throw (4.6) into the canonical form (2.40):

$$\mathbf{P}^{(n)}(t) = \varepsilon_0 \int_{-\infty}^{+\infty} \mathrm{d}\omega_1 \cdots \int_{-\infty}^{+\infty} \mathrm{d}\omega_n\, \boldsymbol{\chi}^{(n)}(-\omega_\sigma ; \omega_1, ..., \omega_n) | \mathbf{E}(\omega_1) \cdots \mathbf{E}(\omega_n) \exp(-i\omega_\sigma t), \tag{4.7}$$

where

$$\omega_\sigma = \omega_1 + \omega_2 + \cdots + \omega_n . \tag{4.8}$$

The susceptibility tensors may then be determined by inspection, as we shall demonstrate.

The necessary manipulations are essentially the same for any value of n, but the notation is clumsy in the general case. For the sake of clarity we shall consider the case $n = 1$ in detail and quote the results for $n > 1$.

4.1.2 Linear susceptibility

By substituting for $\rho_1(t)$ in (4.6) from (3.79), we obtain

$$P_\mu^{(1)}(t) = V^{-1}\mathrm{Tr}\left\{(i\hbar)^{-1}U_0(t)\int_{-\infty}^{t} \mathrm{d}\tau\, [H_\mathrm{I}'(\tau), \rho_0]U_0(-t)Q_\mu\right\}. \quad (4.9)$$

Now $H_\mathrm{I}'(t)$ is defined by (3.76) and $H_\mathrm{I}(t)$ is given by (4.4). Hence, noting that $\mathbf{E}(t)$ is a classical vector which therefore commutes with $U_0(t)$, we have

$$H_\mathrm{I}'(t) = -Q_\alpha(t)E_\alpha(t), \quad (4.10)$$

where we have made use of the definition

$$\mathbf{Q}(t) = U_0(-t)\mathbf{Q}U_0(t). \quad (4.11)$$

(It will be recalled, from the comments following (3.76), that the transformation (4.11) takes the dipole-moment operator $\mathbf{Q}$ into the interaction picture.) It thus follows, with the aid of the commutator identity (3.35), that the commutator in (4.9) may be expressed in the form:

$$[H_\mathrm{I}'(\tau), \rho_0] = [-Q_\alpha(\tau)E_\alpha(\tau), \rho_0] = -E_\alpha(\tau)[Q_\alpha(\tau), \rho_0]. \quad (4.12)$$

Hence, pulling the factor $(i\hbar)^{-1}$ and the minus sign outside the trace symbol, (4.9) becomes

$$P_\mu^{(1)}(t) = -V^{-1}(i\hbar)^{-1}\mathrm{Tr}\left\{U_0(t)\int_{-\infty}^{t} \mathrm{d}\tau\, E_\alpha(\tau)[Q_\alpha(\tau), \rho_0]U_0(-t)Q_\mu\right\}. \quad (4.13)$$

We now observe that, since $\mathbf{E}(t)$ and $U_0(t)$ commute, the operator $U_0(t)$ may be pulled through to the right of $E_\alpha(\tau)$. The integration and the factor $E_\alpha(\tau)$ may be taken outside the trace symbol to yield

$$P_\mu^{(1)}(t) = -V^{-1}(i\hbar)^{-1}\int_{-\infty}^{t} \mathrm{d}\tau\, E_\alpha(\tau)\,\mathrm{Tr}\{U_0(t)[Q_\alpha(\tau), \rho_0]U_0(-t)Q_\mu\}. \quad (4.14)$$

The final result is obtained by expressing $E_\alpha(\tau)$ in terms of its Fourier transform with the aid of (2.25) and writing

$$\exp(-i\omega\tau) = \exp(-i\omega t)\exp[-i\omega(\tau - t)]. \tag{4.15}$$

After interchanging the order of the time and frequency integrations, we then find that $P_\mu^{(1)}(t)$ appears in the form (4.7), *i.e.*,

$$P_\mu^{(1)}(t) = \varepsilon_0 \int_{-\infty}^{+\infty} \mathrm{d}\omega\, \chi_{\mu\alpha}^{(1)}(-\omega_\sigma;\omega) E_\alpha(\omega) \exp(-i\omega_\sigma t) \tag{4.16}$$

where

$$\chi_{\mu\alpha}^{(1)}(-\omega_\sigma;\omega) = -(\varepsilon_0 V)^{-1}(i\hbar)^{-1} \times \int_{-\infty}^{t} \mathrm{d}\tau\, \mathrm{Tr}\{U_0(t)\,[Q_\alpha(\tau), \rho_0]\,U_0(-t) Q_\mu\} \exp[-i\omega(\tau - t)]. \tag{4.17}$$

Equation (4.17) provides us with an explicit formula for the first-order susceptibility tensor. We may simplify it a little by some further manipulations. At first sight the right-hand side of (4.17) appears to involve time t. However, we can now show in a straightforward way that in fact t may be eliminated. By expanding the commutator $[Q_\alpha(\tau), \rho_0]$ and inserting $U_0(-t)\,U_0(t) = 1$ between $Q_\alpha(\tau)$ and ρ_0 in each of the two terms, we find that

$$U_0(t)\,[Q_\alpha(\tau), \rho_0]\,U_0(-t) = [Q_\alpha(\tau - t), \rho_0]. \tag{4.18}$$

In deriving (4.18) we use the expressions (3.69) and (4.11) and the fact that $U_0(t)$ commutes with ρ_0. When (4.18) is substituted into (4.17), the change of integration variable $\tau \rightarrow \tau' + t$ removes t from the formula to yield

$$\chi_{\mu\alpha}^{(1)}(-\omega_\sigma;\omega) = -(\varepsilon_0 V)^{-1}(i\hbar)^{-1} \times \int_{-\infty}^{0} \mathrm{d}\tau'\, \mathrm{Tr}\{\,[Q_\alpha(\tau'), \rho_0]\,Q_\mu\} \exp(-i\omega\tau'). \tag{4.19}$$

Finally, it is convenient to pull ρ_0 out of the commutator brackets. To achieve this we observe that, since the trace of a product of operators is invariant under cyclic permutations of their order (as in (3.24)), we have

$$\begin{aligned} \mathrm{Tr}\{\,[Q_\alpha(\tau), \rho_0]\,Q_\mu\} &= \mathrm{Tr}\{Q_\alpha(\tau)\,\rho_0\,Q_\mu - \rho_0\,Q_\alpha(\tau)\,Q_\mu\} \\ &= \mathrm{Tr}\{\rho_0\,Q_\mu\,Q_\alpha(\tau) - \rho_0\,Q_\alpha(\tau)\,Q_\mu\} \\ &= \mathrm{Tr}\{\rho_0\,[Q_\mu, Q_\alpha(\tau)]\,\}. \end{aligned} \tag{4.20}$$

By using this identity in (4.19) we obtain the final expression for the first-order susceptibility tensor:

$$\chi_{\mu\alpha}^{(1)}(-\omega_\sigma;\omega) = -(\varepsilon_0 V)^{-1}(i\hbar)^{-1} \times \int_{-\infty}^{0} \mathrm{d}\tau\, \mathrm{Tr}\{\rho_0\,[Q_\mu, Q_\alpha(\tau)]\,\} \exp(-i\omega\tau). \tag{4.21}$$

4.1.3 The nonlinear susceptibilities

We now turn to the case when $n=2$. The manipulations follow essentially the same pattern as before with the exception that, in order to extract ρ_0 from the commutator brackets, we use a generalisation of (4.20), namely

$$\mathrm{Tr}\{[Q_\alpha(\tau_1),[Q_\beta(\tau_2),\rho_0]]Q_\mu\} = \mathrm{Tr}\{\rho_0[[Q_\mu,Q_\alpha(\tau_1)],Q_\beta(\tau_2)]\}. \tag{4.22}$$

This identity may be verified by repeated application of (4.20). (We note that, as in (3.24), the relations (4.20) and (4.22) are identities which do not depend on the nature of the operators involved.) Thus we find that $P_\mu^{(2)}(t)$ appears in the canonical form (4.7), *i.e.*,

$$P_\mu^{(2)}(t) = \varepsilon_0 \int_{-\infty}^{+\infty} d\omega_1 \int_{-\infty}^{+\infty} d\omega_2\, \chi^{(2)}_{\mu\alpha\beta}(-\omega_\sigma;\omega_1,\omega_2)E_\alpha(\omega_1)E_\beta(\omega_2)\exp(-i\omega_\sigma t) \tag{4.23}$$

where

$$\chi^{(2)}_{\mu\alpha\beta}(-\omega_\sigma;\omega_1,\omega_2) = (\varepsilon_0 V)^{-1}(i\hbar)^{-2} \times \int_{-\infty}^{0} d\tau_1 \int_{-\infty}^{\tau_1} d\tau_2\, \mathrm{Tr}\{\rho_0[[Q_\mu,Q_\alpha(\tau_1)],Q_\beta(\tau_2)]\}\exp[-i(\omega_1\tau_1+\omega_2\tau_2)]. \tag{4.24}$$

This is the second-order susceptibility tensor in the sense that it gives correctly the second-order polarisation when substituted into (4.23). However, it evidently does not possess intrinsic permutation symmetry, *i.e.*, it is not invariant under the interchange of the pairs (α,ω_1) and (β,ω_2). It will be recalled that the use of the susceptibility tensors with intrinsic permutation symmetry was a matter of choice. The second-order susceptibility tensor which appears in (4.23) need not have intrinsic permutation symmetry (and (4.24) does not), but it may always be modified so as to have this property. To achieve this end we interchange the dummy suffices α and β in (4.23) and also interchange the integration variables ω_1 and ω_2. Then we have

$$P_\mu^{(2)}(t) = \varepsilon_0 \int_{-\infty}^{+\infty} d\omega_1 \int_{-\infty}^{+\infty} d\omega_2\, \chi^{(2)}_{\mu\beta\alpha}(-\omega_\sigma;\omega_2,\omega_1)E_\beta(\omega_2)E_\alpha(\omega_1)\exp(-i\omega_\sigma t). \tag{4.25}$$

By adding together (4.23) and (4.25) and dividing by 2 we find that (4.23) is reproduced but with $\chi^{(2)}_{\mu\alpha\beta}(-\omega_\sigma;\omega_1,\omega_2)$ replaced by

$$\tfrac{1}{2}[\chi^{(2)}_{\mu\alpha\beta}(-\omega_\sigma;\omega_1,\omega_2) + \chi^{(2)}_{\mu\beta\alpha}(-\omega_\sigma;\omega_2,\omega_1)], \tag{4.26}$$

which obviously has the required intrinsic permutation symmetry. Thus we see that the second-order susceptibility tensor in its symmetrised form is given by

$$\chi^{(2)}_{\mu\alpha\beta}(-\omega_\sigma;\omega_1,\omega_2) = \tfrac{1}{2}\mathbf{S}\,(\varepsilon_0 V)^{-1}(i\hbar)^{-2}$$
$$\times\int_{-\infty}^{0}\mathrm{d}\tau_1\int_{-\infty}^{\tau_1}\mathrm{d}\tau_2\,\mathrm{Tr}\{\rho_0\,[[Q_\mu, Q_\alpha(\tau_1)], Q_\beta(\tau_2)]\}\exp[-i(\omega_1\tau_1+\omega_2\tau_2)], \tag{4.27}$$

where the symmetrising operation $\mathbf{S}$ indicates that the expression which follows it is to be summed over the two possible permutations of the pairs (α,ω_1) and (β,ω_2).

This same procedure can be followed to obtain the symmetrised form of the nth-order susceptibility $\boldsymbol{\chi}^{(n)}$ for arbitrary n, and we quote the result:

$$\chi^{(n)}_{\mu\alpha_1\cdots\alpha_n}(-\omega_\sigma;\omega_1,\ldots,\omega_n) = \frac{1}{n!}\mathbf{S}\,(\varepsilon_0 V)^{-1}(-i\hbar)^{-n}$$
$$\times\int_{-\infty}^{0}\mathrm{d}\tau_1\int_{-\infty}^{\tau_1}\mathrm{d}\tau_2\cdots\int_{-\infty}^{\tau_{n-1}}\mathrm{d}\tau_n\,\mathrm{Tr}\Big\{\rho_0\,[\cdots[[Q_\mu, Q_{\alpha_1}(\tau_1)], Q_{\alpha_2}(\tau_2)],\cdots$$
$$\cdots Q_{\alpha_n}(\tau_n)]\Big\}\exp\left[-i\sum_{r=1}^{n}\omega_r\tau_r\right], \tag{4.28}$$

where the symmetrising operation $\mathbf{S}$ now indicates that the expression which follows it is to be summed over all $n!$ permutations of the n pairs $(\alpha_1,\omega_1),(\alpha_2,\omega_2),\ldots,(\alpha_n,\omega_n)$. It will be noticed that the nature of the symmetrising operation depends on the order of the susceptibility n and one should perhaps write $\mathbf{S}_n$ rather than $\mathbf{S}$; however the value of n is always clear from the context, and there is no need to complicate the notation in this way.

The extraction of ρ_o from the commutator brackets to obtain the form (4.28) was performed with the aid of a generalisation of the identities (4.20) and (4.22) to arbitrary order, namely:

$$\mathrm{Tr}\{[Q_{\alpha_1}(\tau_1), [Q_{\alpha_2}(\tau_2),\ldots, [Q_{\alpha_n}(\tau_n), \rho_0]]\cdots]Q_\mu\} =$$
$$\mathrm{Tr}\{\rho_0\,[\cdots[[Q_\mu, Q_{\alpha_1}(\tau_1)], Q_{\alpha_2}(\tau_2)],\cdots Q_{\alpha_n}(\tau_n)]\}. \tag{4.29}$$

This identity may be readily proved by induction for arbitrary n.

4.2 Explicit formulae for an assembly of independent molecules

The formulae for the susceptibility tensors derived in the previous section are fundamental; they apply to *any* medium in the electric-dipole approximation. They provide a general basis for the treatment of the optical properties of material media which do not involve spatial dispersion. However, the physical content of the formulae is concealed by the compactness of the notation we have introduced in order to handle the general case. In this section we rewrite the formulae in a more transparent form for a medium which is an assembly of distinguishable, independent and similarly-oriented molecules. This is not a very good model of any real medium, but it is conceptually simple and provides a quick route to an understanding of the physical content of the formulae.

The operators which occur in the expressions given in the previous section for the susceptibility tensors relate to the entire system of particles in a small volume V of the medium. However, when the particle system consists of an assembly of independent molecules, as we are assuming, the many-particle operators may readily be expressed in terms of monomolecular operators which relate to the individual molecules. In doing so we shall make frequent use of the fact that operators associated with different molecules commute with one another.

We also assume that the molecular system is sufficiently dilute that each of the molecules in the small volume V experiences the same electric field, which is equal to the macroscopic field $\mathbf{E}(t)$ (*i.e.* the interaction between the molecular dipoles is neglected). We return to the question of local fields, and also molecular orientation averaging, in §4.4.

We suppose that there are M molecules in the small volume V, and we denote the unperturbed Hamiltonian and dipole moment of the mth molecule by H_m and $e\mathbf{r}_m$, respectively. Then the unperturbed Hamiltonian and dipole moment of the whole assembly are given by

$$H_0 = \sum_m H_m \tag{4.30}$$

and

$$\mathbf{Q} = \sum_m e\mathbf{r}_m , \tag{4.31}$$

respectively. Moreover, since the individual molecular Hamiltonians commute with one another, the thermal-equilibrium density operator of the whole assembly is given by (3.40) as

$$\begin{aligned} \rho_0 &= \eta \exp(-H_0/kT) \\ &= \rho_1 \rho_2 \cdots \rho_m \cdots \rho_M , \end{aligned} \tag{4.32}$$

where

$$\rho_m = \eta^{1/M} \exp(-H_m/kT) \tag{4.33}$$

is the thermal-equilibrium density operator for the mth molecule, and η is the normalisation constant introduced in (3.37). Similarly, the unperturbed time-development operator of the whole assembly is given, from (3.67), by

$$\begin{aligned} U_0(t) &= \exp(-iH_0 t/\hbar) \\ &= U_1(t)\, U_2(t) \cdots U_m(t) \cdots U_M(t) , \end{aligned} \tag{4.34}$$

where

$$U_m(t) = \exp(-iH_m t/\hbar) \tag{4.35}$$

is the unperturbed time-development operator of the mth molecule. Furthermore, by making use of the facts that $U_l(t)$ commutes with $e\mathbf{r}_m$ when $l \neq m$, and $U_l(-t)U_l(t) = 1$, we see that $\mathbf{Q}(t)$, the dipole moment of the whole assembly in the interaction picture (4.11), is given by

$$\begin{aligned}\mathbf{Q}(t) &= U_0(-t)\sum_m e\mathbf{r}_m U_0(t)\\ &= \sum_m e\mathbf{r}_m(t).\end{aligned} \tag{4.36}$$

Here $e\mathbf{r}_m(t)$ is the dipole-moment operator for the mth molecule in the interaction picture, given by

$$e\mathbf{r}_m(t) = U_m(-t)e\mathbf{r}_m U_m(t). \tag{4.37}$$

It only remains to consider the multiple commutators which occur in the expressions for the susceptibility tensors. We consider first of all $[Q_\mu, Q_\alpha(\tau)]$ which occurs in the formula (4.21) for the first-order susceptibility tensor. By making use of the commutator identity (3.35) and the fact that operators associated with different molecules commute with one another, we see that

$$\begin{aligned}[Q_\mu, Q_\alpha(\tau)] &= \left[\sum_m er_{m\mu}, \sum_m er_{m\alpha}(\tau)\right]\\ &= \sum_m [er_{m\mu}, er_{m\alpha}(\tau)].\end{aligned} \tag{4.38}$$

Next we consider the second-order commutator $[[Q_\mu, Q_\alpha(\tau_1)], Q_\beta(\tau_2)]$ which occurs in the formula (4.27) for the second-order susceptibility tensor. By making use of (4.38) we see that

$$\begin{aligned}[[Q_\mu, Q_\alpha(\tau_1)], Q_\beta(\tau_2)] &= \left[\sum_m [er_{m\mu}, er_{m\alpha}(\tau_1)], \sum_m er_{m\beta}(\tau_2)\right]\\ &= \sum_m \left[[er_{m\mu}, er_{m\alpha}(\tau_1)], er_{m\beta}(\tau_2)\right].\end{aligned} \tag{4.39}$$

It is clear from these two cases that the nth order commutator, which occurs in the general formula (4.28) for the nth-order susceptibility tensor, can be expressed as the sum over all molecules of the corresponding nth-order commutator of monomolecular dipole-moment operators.

The traces involved in the formulae may be evaluated by using any representation for the whole assembly of M molecules in the small volume V. Since we are assuming that the molecules are distinguishable and independent, it is convenient to use a representation in which the many-molecular wave functions are obtained by multiplying together monomolecular wave functions. Let $\boldsymbol{\Theta}_m$ denote the totality of internal coordinates of the mth molecule (we assume that the centre of mass is at rest) and let the set of monomolecular wave functions $\{u_{i_m}(\boldsymbol{\Theta}_m)\}$ be a representation for the mth molecule. Here, as in (3.5), the label i is used to distinguish the various members of the set. Then a representation for the whole assembly of molecules is provided by the set of product functions $\{\Phi_I\}$ given by

$$\Phi_I = u_{i_1}(\boldsymbol{\Theta}_1)u_{i_2}(\boldsymbol{\Theta}_2)\cdots u_{i_m}(\boldsymbol{\Theta}_m)\cdots u_{i_M}(\boldsymbol{\Theta}_M), \tag{4.40}$$

where I denotes the set of monomolecular labels $i_1, i_2, ..., i_M$, each of which runs over all possible values.

Now, all the traces with which we are concerned (as, for example, in (4.28)), take the form

$$F = \mathrm{Tr}(\rho_0 \sum_m C_m) \tag{4.41}$$

where C_m denotes a multiple commutator of the dipole-moment operators associated with the mth molecule. Hence the commutator C_m is itself an operator associated with the mth molecule. By substituting for ρ_0 from (4.32) and evaluating the trace in the representation (4.40), using (3.17), we find that

$$\begin{aligned} F &= \sum_I \int \mathrm{d}\varsigma_1 \int \mathrm{d}\varsigma_2 \cdots \int \mathrm{d}\varsigma_M\, \Phi_I^* \rho_1 \rho_2 \cdots \rho_M \sum_m C_m \Phi_I \\ &= \sum_m \sum_I \int \mathrm{d}\varsigma_1 \int \mathrm{d}\varsigma_2 \cdots \int \mathrm{d}\varsigma_M\, \Phi_I^* \rho_1 \rho_2 \cdots \rho_M C_m \Phi_I \end{aligned} \tag{4.42}$$

where $\mathrm{d}\varsigma_m$ denotes the volume element in the configuration space of the mth molecule. Since the monomolecular density operators commute with one another, we may bring ρ_m through to the left of C_m in (4.42):

$$F = \sum_m \sum_I \int \mathrm{d}\varsigma_1 \int \mathrm{d}\varsigma_2 \cdots \int \mathrm{d}\varsigma_M\, \Phi_I^* \rho_1 \cdots \rho_{m-1} \rho_{m+1} \cdots \rho_M \rho_m C_m \Phi_I. \tag{4.43}$$

Now, since the wave functions $\{\Phi_I\}$ are a product of monomolecular wave functions including the set $\{u_{i_m}(\boldsymbol{\Theta}_m)\}$, and recalling that the trace of an operator is independent of the representation used for its evaluation, we can use (3.17) to write:

$$\mathrm{Tr}(\rho_m) = \sum_I \int \mathrm{d}\varsigma_m\, \Phi_I^* \rho_m \Phi_I. \tag{4.44}$$

Hence the many-molecular integral in (4.43) factors into a product of monomolecular integrals, and since the sum over I involves a summation over all $i_1, i_2, ..., i_M$, we see that (4.42) reduces to

$$F = \sum_m \mathrm{Tr}(\rho_1)\mathrm{Tr}(\rho_2)\cdots \mathrm{Tr}(\rho_{m-1})\mathrm{Tr}(\rho_m C_m)\mathrm{Tr}(\rho_{m+1})\cdots \mathrm{Tr}(\rho_M). \tag{4.45}$$

The traces appearing in (4.45) are of course monomolecular traces, *i.e.*, for any operator O_m associated with the mth molecule:

$$\mathrm{Tr}(O_m) = \sum_i \int d\varsigma_m\, u_{i_m}{}^*(\boldsymbol{\Theta}_m)\, O_m\, u_{i_m}(\boldsymbol{\Theta}_m). \qquad (4.46)$$

Now, in the previous chapter (§3.4) we noted that for any density operator ρ, we must have $\mathrm{Tr}(\rho)=1$. This property followed directly from setting the arbitrary operator O equal to 1 in the expectation-value formula (3.32). Thus (4.45) immediately simplifies to

$$F = \sum_m \mathrm{Tr}(\rho_m C_m), \qquad (4.47)$$

without making any assumptions about the nature of the elements in the summation. A further simplification can be made, however, by observing that $\mathrm{Tr}(\rho_m C_m)$ is in fact independent of m in the simple case which we are considering because all the M molecules have been assumed to be identical and similarly oriented. Consequently all the terms in the sum (4.47) are identical, and we may rewrite (4.47) in the simple form:

$$F = M\,\mathrm{Tr}(\rho_1 C_1), \qquad (4.48)$$

where we have arbitrarily chosen to express the trace in terms of the monomolecular operators for the molecule labelled 1.

Thus, by comparing (4.41) and (4.48), we see that we may replace the many-molecular density operator and multiple commutators in the formulae for the susceptibility tensors by corresponding monomolecular operators, provided that we multiply by the number of molecules. We may simply use the formulae derived in the previous section with the factor V^{-1} replaced by $M/V=N$, the number density of molecules, and the understanding that all the operators and traces are now monomolecular. Moreover, while (4.48) is expressed in terms of molecule number 1, we could just as well have chosen any other molecule since they are all identical. Thus the molecular suffix really conveys no information and henceforth we shall omit it. In what follows we denote the unperturbed Hamiltonian, thermal-equilibrium density operator, dipole moment, configuration coordinates and monomolecular wave functions of a molecule (any molecule) by H_0, ρ_0, $e\mathbf{r}$, $\boldsymbol{\Theta}$ and $u_i(\boldsymbol{\Theta})$, respectively. Moreover, we denote the unperturbed time-development operator and the dipole moment in the interaction picture, for a molecule, by $U_0(t)$ and $e\mathbf{r}(t)$, respectively. There is thus some risk of confusion since some of these symbols (H_0, $U_0(t)$ and ρ_0) have previously referred to the whole molecular assembly. However, the risk is small since the formulae in their monomolecular form always contain the factor N instead of V^{-1} which appears in the many-molecular formulae. It seems worthwhile to accept this small possibility of confusion in order to avoid further complication of the notation. (We note, however, that by a choice of notation we have deliberately distinguished clearly between the macroscopic dipole-moment operators, $\mathbf{Q}$ and $\mathbf{Q}(t)$, and the corresponding molecular operators $e\mathbf{r}$ and $e\mathbf{r}(t)$, since this will assist our discussion in later sections of this chapter and in Chapter 6.)

The monomolecular representation $\{u_i(\boldsymbol{\Theta})\}$ remains at our disposal. To complete the evaluation of the traces, it is convenient to use the energy representation which was discussed in §3.5. Thus we define $u_i(\boldsymbol{\Theta})$, and the energy E_i to which it belongs, by the eigenvalue equation

$$H_0 u_i(\boldsymbol{\Theta}) = \mathrm{E}_i u_i(\boldsymbol{\Theta}). \tag{4.49}$$

We saw in §3.5 that the chief merit of the energy representation is that it gives a simple diagonal form to the matrices of operators which are functions of H_0. Thus, by applying the relation (3.48) to the thermal-equilibrium density operator (4.32), we obtain

$$(\rho_0)_{ai} = \rho_0(a)\,\delta_{ai}, \tag{4.50}$$

where we have introduced the notation $\rho_0(a)$ for the diagonal thermal-equilibrium density-matrix element for the state labelled a. It is given by (4.32) and (4.49) as

$$\rho_0(a) = \eta \exp(-\mathrm{E}_a/kT). \tag{4.51}$$

Thus $\rho_0(a)$ is equal to the statistical fraction of the total molecular population which, in thermal equilibrium, occupies the energy state E_a. We shall see below that the expressions for the susceptibilities contain the term $\sum_a \rho_0(a)$, where the label a ranges over all the molecular energy eigenstates; this summation term picks out the molecular ground states for which the density-matrix element $\rho_0(a)$ is nonzero.

We can now use (4.50) to derive some important and useful relations which apply in the energy representation. In calculating the nth-order susceptibility, we are required to evaluate the trace of the product of $n+2$ operators, the first of which is the equilibrium density operator ρ_0. To do this, we need expressions for the diagonal elements of such a product. For the case $n=1$, we can directly apply the matrix product rule in the form stated in (3.12) to obtain the relation

$$\begin{aligned}(\rho_0 O O')_{aa} &= \sum_i \sum_b (\rho_0)_{ai} O_{ib} O'_{ba} \\ &= \rho_0(a) \sum_b O_{ab} O'_{ba}\end{aligned} \tag{4.52}$$

where, in the second line, we have used (4.50). By repeated applications of the matrix product rule, we obtain the general form:

$$(\rho_0 O O' O'' \cdots O^{n-1\prime} O^{n\prime})_{aa} = \rho_0(a) \sum_{b_1, b_2, \ldots, b_n} O_{ab_1} O'_{b_1 b_2} O''_{b_2 b_3} \cdots O^{n-1\prime}_{b_{n-1} b_n} O^{n\prime}_{b_n a} \quad (4.53)$$

where the operator distinguished by n primes is written thus: $O^{n\prime}$.

The final manipulation required in the energy representation is to express the the unperturbed time-development operator (4.35) in diagonal form, using (3.48):

$$[U_0(t)]_{ab} = \exp(-i\mathrm{E}_a t/\hbar)\,\delta_{ab}. \quad (4.54)$$

It follows from the definition (4.37) for $e\mathbf{r}_m(t)$, together with (4.54) and the matrix product rule, that

$$\begin{aligned}[er_\alpha(t)]_{ab} &= [U_0(-t)\, er_\alpha\, U_0(t)]_{ab} \\ &= \exp(i\Omega_{ab} t)\, er^\alpha_{ab}, \end{aligned} \quad (4.55)$$

where er^α_{ab} denotes the (ab)th matrix element of er_α, and Ω_{ab} is a molecular transition frequency defined in terms of molecular energy levels E_a and E_b by $\Omega_{ab} = (\mathrm{E}_a - \mathrm{E}_b)/\hbar$.

We have now set up all the apparatus required to throw the formulae for the susceptibility tensors, derived in the previous section, into a more transparent form. The remaining calculations are now quite straightforward. Beginning with the first-order susceptibility tensor, the formula (4.21), when expressed in terms of monomolecular operators, becomes

$$\chi^{(1)}_{\mu\alpha}(-\omega_\sigma;\omega) = -\frac{N}{\varepsilon_0}\frac{e^2}{i\hbar}\int_{-\infty}^{0} d\tau\, \mathrm{Tr}\{\rho_0\,[r_\mu, r_\alpha(\tau)]\}\exp(-i\omega\tau) \quad (4.56)$$

where N is the number density of molecules. By first expanding the commutator, we can use (4.52) together with (4.55) to evaluate the trace in (4.56):

$$\mathrm{Tr}\{\rho_0\,[r_\mu, r_\alpha(\tau)]\} = \sum_a \rho_0(a) \sum_b \{r^\mu_{ab} r^\alpha_{ba} \exp(i\Omega_{ba}\tau) - r^\alpha_{ab} r^\mu_{ba} \exp(-i\Omega_{ba}\tau)\}. \quad (4.57)$$

In (4.57) we made use of the fact that $\Omega_{ab} = -\Omega_{ba}$. Now, the transition frequency Ω_{ba} is real, consequently the integrand in (4.56) does not converge when ω is real. However, we took care in Chapter 2 to put ω in the upper half of the complex plane (*i.e.*, the imaginary part of ω is greater than zero), in which case the integral converges. The main reason why complex frequencies were introduced at the beginning of the discussion of the susceptibility tensors (§2.2) was to secure the convergence of the integrals which arise here. Thus, after a straightforward integration, we obtain the familiar expression for the linear susceptibility:

$$\chi^{(1)}_{\mu\alpha}(-\omega_\sigma;\omega) = \frac{N}{\varepsilon_0}\frac{e^2}{\hbar}\sum_{ab}\rho_0(a)\left[\frac{r^\mu_{ab}\,r^\alpha_{ba}}{\Omega_{ba}-\omega}+\frac{r^\alpha_{ab}\,r^\mu_{ba}}{\Omega_{ba}+\omega}\right]. \tag{4.58}$$

The final evaluation of the expression (4.27) for the second-order susceptibility tensor proceeds on similar lines. When expressed in terms of monomolecular operators (4.27) becomes

$$\chi^{(2)}_{\mu\alpha\beta}(-\omega_\sigma;\omega_1,\omega_2) = \frac{N}{\varepsilon_0}\frac{e^3}{2(i\hbar)^2}\,\mathbf{S}\int_{-\infty}^{0}\mathrm{d}\tau_1\int_{-\infty}^{\tau_1}\mathrm{d}\tau_2$$
$$\times\,\mathrm{Tr}\{\rho_0\,[[r_\mu,r_\alpha(\tau_1)],r_\beta(\tau_2)]\}\exp[-i(\omega_1\tau_1+\omega_2\tau_2)]. \tag{4.59}$$

Again, by first expanding the commutator, the relation (4.53) for $n=2$, together with (4.55), is used to evaluate the trace which appears in (4.59):

$$\mathrm{Tr}\{\rho_0\,[[r_\mu,r_\alpha(\tau_1)],r_\beta(\tau_2)]\} =$$
$$\sum_a\rho_0(a)\sum_{bc}\Big\{r^\mu_{ab}\,r^\alpha_{bc}\,r^\beta_{ca}\exp[i(\Omega_{bc}\tau_1+\Omega_{ca}\tau_2)]$$
$$-\,r^\alpha_{ab}\,r^\mu_{bc}\,r^\beta_{ca}\exp[i(\Omega_{ab}\tau_1+\Omega_{ca}\tau_2)] - r^\beta_{ab}r^\mu_{bc}r^\alpha_{ca}\exp[i(\Omega_{ca}\tau_1+\Omega_{ab}\tau_2)]$$
$$+\,r^\beta_{ab}r^\alpha_{bc}r^\mu_{ca}\exp[i(\Omega_{bc}\tau_1+\Omega_{ab}\tau_2)]\Big\}. \tag{4.60}$$

The integrals in (4.59) converge because ω_1 and ω_2 both lie in the upper half-plane. Thus, after a simple integration we obtain:

$$\chi^{(2)}_{\mu\alpha\beta}(-\omega_\sigma;\omega_1,\omega_2) = \frac{N}{\varepsilon_0}\frac{e^3}{2\hbar^2}\,\mathbf{S}\sum_{abc}\rho_0(a)$$
$$\times\left[\frac{r^\mu_{ab}\,r^\alpha_{bc}\,r^\beta_{ca}}{(\Omega_{ab}+\omega_1+\omega_2)(\Omega_{ac}+\omega_2)}-\frac{r^\alpha_{ab}\,r^\mu_{bc}\,r^\beta_{ca}}{(\Omega_{bc}+\omega_1+\omega_2)(\Omega_{ac}+\omega_2)}\right.$$
$$\left.-\frac{r^\beta_{ab}\,r^\mu_{bc}\,r^\alpha_{ca}}{(\Omega_{bc}+\omega_1+\omega_2)(\Omega_{ba}+\omega_2)}+\frac{r^\beta_{ab}\,r^\alpha_{bc}\,r^\mu_{ca}}{(\Omega_{ca}+\omega_1+\omega_2)(\Omega_{ba}+\omega_2)}\right]. \tag{4.61}$$

In (4.61) we have made repeated use of relations of the type $\Omega_{bc}+\Omega_{ca}=\Omega_{ba}=-\Omega_{ab}$. The presence of the symmetrising operation $\mathbf{S}$ in (4.61) allows us to combine together the middle two terms. Since $\mathbf{S}$ indicates that the expression which follows it is to be summed over the two possible permutations of the pairs (α,ω_1) and (β,ω_2), it follows that we may permute these pairs in any of the terms in (4.61) without altering the final result. By interchanging (α,ω_1) and (β,ω_2) in the third term and combining it with the second term, we obtain:

$$- \frac{r^{\alpha}_{ab}\, r^{\mu}_{bc}\, r^{\beta}_{ca}}{(\Omega_{bc} + \omega_1 + \omega_2)(\Omega_{ac} + \omega_2)} - \frac{r^{\alpha}_{ab}\, r^{\mu}_{bc}\, r^{\beta}_{ca}}{(\Omega_{bc} + \omega_1 + \omega_2)(\Omega_{ba} + \omega_1)}$$

$$= - \frac{r^{\alpha}_{ab}\, r^{\mu}_{bc}\, r^{\beta}_{ca}}{(\Omega_{ba} + \omega_1)(\Omega_{ac} + \omega_2)} \tag{4.62}$$

since $(\Omega_{ac} + \omega_2) + (\Omega_{ba} + \omega_1) = \Omega_{bc} + \omega_1 + \omega_2$. The presence of the symmetrising operation also allows us to interchange the pairs (α,ω_1) and (β,ω_2). It is useful to do so because the two coordinate labels α and β then appear in dictionary order in every term in the final formula. It is also useful to manipulate the signs in the denominators so that the transition energies $\hbar\Omega_{ij} = \mathrm{E}_i - \mathrm{E}_j$ are always referred to the energy E_a (*i.e.*, make $j = a$). By carrying out these rearrangements we obtain the desired formula for the second-order susceptibility tensor:

$$\chi^{(2)}_{\mu\alpha\beta}(-\omega_\sigma ; \omega_1 , \omega_2) = \frac{N}{\varepsilon_0} \frac{e^3}{2\hbar^2} \mathbf{S} \sum_{abc} \rho_0(a)$$

$$\times \left[\frac{r^{\mu}_{ab}\, r^{\alpha}_{bc}\, r^{\beta}_{ca}}{(\Omega_{ba} - \omega_1 - \omega_2)(\Omega_{ca} - \omega_2)} + \frac{r^{\alpha}_{ab}\, r^{\mu}_{bc}\, r^{\beta}_{ca}}{(\Omega_{ba} + \omega_1)(\Omega_{ca} - \omega_2)} + \frac{r^{\alpha}_{ab}\, r^{\beta}_{bc}\, r^{\mu}_{ca}}{(\Omega_{ba} + \omega_1)(\Omega_{ca} + \omega_1 + \omega_2)} \right]. \tag{4.63}$$

Finally, we quote below the formula for the third-order susceptibility tensor. This is obtained from (4.28) for $n = 3$, by evaluating the trace in the energy representation, with the use of (4.53) for $n = 3$ and (4.55). The expansion of the third-order commutator yields eight terms, but the presence of the symmetrising operation allows us to combine together the three terms in which r^{μ}_{bc} appears as the second factor in the numerator and also the three terms in which r^{μ}_{cd} appears as the third factor in the numerator. Thus we find four terms in the final formula, and after rearranging so that the three labels α β and γ appear in dictionary order, and referring all transition energies to E_a, we obtain:

$$\chi^{(3)}_{\mu\alpha\beta\gamma}(-\omega_\sigma ; \omega_1 , \omega_2 , \omega_3) = \frac{N}{\varepsilon_0} \frac{e^4}{3!\,\hbar^3} \mathbf{S}$$

$$\times \sum_{abcd} \rho_0(a) \left[\frac{r^{\mu}_{ab}\, r^{\alpha}_{bc}\, r^{\beta}_{cd}\, r^{\gamma}_{da}}{(\Omega_{ba} - \omega_1 - \omega_2 - \omega_3)(\Omega_{ca} - \omega_2 - \omega_3)(\Omega_{da} - \omega_3)} + \frac{r^{\alpha}_{ab}\, r^{\mu}_{bc}\, r^{\beta}_{cd}\, r^{\gamma}_{da}}{(\Omega_{ba} + \omega_1)(\Omega_{ca} - \omega_2 - \omega_3)(\Omega_{da} - \omega_3)} \right.$$

$$+ \frac{r^{\alpha}_{ab}\, r^{\beta}_{bc}\, r^{\mu}_{cd}\, r^{\gamma}_{da}}{(\Omega_{ba} + \omega_1)(\Omega_{ca} + \omega_1 + \omega_2)(\Omega_{da} - \omega_3)}$$

$$\left. + \frac{r^{\alpha}_{ab}\, r^{\beta}_{bc}\, r^{\gamma}_{cd}\, r^{\mu}_{da}}{(\Omega_{ba} + \omega_1)(\Omega_{ca} + \omega_1 + \omega_2)(\Omega_{da} + \omega_1 + \omega_2 + \omega_3)} \right]. \tag{4.64}$$

An example of the full expansion of (4.64) in the case of a four-level system is given by Bloembergen *et al* (1978).

The generalisation of the formulae (4.58), (4.63) and (4.64) to arbitrary order will become clear after we have introduced the overall permutation-symmetry property in the next section.

To derive these explicit formulae for the susceptibility tensors we have used the density-matrix approach developed in Chapter 3. We should point out, however, that this is not the only approach that can be taken. Similar results are obtained using a time-dependent perturbation analysis (Ducuing, 1969).

4.3 Overall permutation symmetry

The formulae for the susceptibility tensors derived in the previous section may be thrown into a form which reveals an important permutation-symmetry property of the tensor components. The *intrinsic* permutation-symmetry property, first introduced in §2.1 and used extensively in the previous sections of this chapter (as implied by the operator $\mathbf{S}$), indicates that $\chi^{(n)}_{\mu\alpha_1\cdots\alpha_n}(-\omega_\sigma;\omega_1,\ldots,\omega_n)$ is invariant under all $n!$ permutations of the n pairs $(\alpha_1,\omega_1),(\alpha_2,\omega_2),\ldots,(\alpha_n,\omega_n)$. *Overall* permutation symmetry, which we introduce here, is a more general property, at least in the sense that it encompasses intrinsic permutation symmetry. However, it is important to appreciate that intrinsic permutation symmetry is a completely rigorous property which arises from the principles of causality and reality, and which applies universally. The overall permutation-symmetry property, on the other hand, is an approximation which applies only under particular conditions; namely, when all of the optical frequencies involved in the susceptibility formulae (*i.e.,* the primary frequencies $\omega_\sigma,\omega_1,\ldots,\omega_n$ and the combinations of these which occur in the denominators) are far removed from molecular transition frequencies – in other words, when the medium can be assumed to be loss-free at all the relevant optical frequencies. The reasons for this restriction become clearer when, in §4.5, we introduce transition damping factors into the susceptibility formulae. The physical consequences of overall permutation symmetry are discussed in §5.1.

Here we shall simply demonstrate how overall permutation symmetry arises, and introduce a compact notation for its description. The usefulness of this notation is that it allows us to write down readily a general formula for the nth-order susceptibility tensor.

4.3.1 Compact formulae for nonresonant susceptibilities

We begin by considering the previous formula for the first-order susceptibility tensor, namely:

$$\chi^{(1)}_{\mu\alpha}(-\omega_\sigma;\omega) = \frac{N}{\varepsilon_0}\frac{e^2}{\hbar}\sum_{ab}\rho_0(a)\left[\frac{r^\mu_{ab}r^\alpha_{ba}}{\Omega_{ba}-\omega}+\frac{r^\alpha_{ab}r^\mu_{ba}}{\Omega_{ba}+\omega}\right]. \tag{4.65}$$

It will be seen that the second term in square brackets may be derived from the first by interchanging the coordinate labels μ and α and replacing $-\omega$ by ω. Consequently, if we interchange μ and α and interchange $-\omega_\sigma$ and ω in $\chi^{(1)}_{\mu\alpha}(-\omega_\sigma;\omega)$, and remembering that for the first-order susceptibility ω_σ is defined as $\omega_\sigma = \omega$, it follows that the only result is to interchange the order in which the two terms appear in the square brackets in (4.65) and their sum is unaltered. Thus we have

$$\begin{aligned}\chi^{(1)}_{\alpha\mu}(\omega;-\omega_\sigma) &= \frac{N}{\varepsilon_0}\frac{e^2}{\hbar}\sum_{ab}\rho_0(a)\left[\frac{r^\alpha_{ab}r^\mu_{ba}}{\Omega_{ba}+\omega}+\frac{r^\mu_{ab}r^\alpha_{ba}}{\Omega_{ba}-\omega}\right]\\ &= \chi^{(1)}_{\mu\alpha}(-\omega_\sigma;\omega).\end{aligned} \tag{4.66}$$

This is the overall permutation-symmetry property of the first-order susceptibility tensor. Thus (4.65) may be rewritten in a more compact form:

$$\chi^{(1)}_{\mu\alpha}(-\omega_\sigma;\omega) = \frac{N}{\varepsilon_0}\frac{e^2}{\hbar}\mathbf{S}_\mathrm{T}\sum_{ab}\rho_0(a)\left[\frac{r^\mu_{ab}r^\alpha_{ba}}{\Omega_{ba}-\omega}\right], \tag{4.67}$$

where the 'total symmetrisation operator' $\mathbf{S}_\mathrm{T}$ indicates that the expression which follows it is to be summed over all possible permutations of the pairs $(\mu,-\omega_\sigma)$ and (α,ω). There are in fact two possible permutations, which are: (i) the identity permutation $(\mu,-\omega_\sigma)\to(\mu,-\omega_\sigma)$, which merely reproduces the term stated explicitly in (4.67), *i.e.*, the first term in square brackets in (4.65); (ii) the interchange $(\mu,-\omega_\sigma)\leftrightarrow(\alpha,\omega)$, which gives the second term in square brackets in (4.65). Since $\mathbf{S}_\mathrm{T}$ indicates that a summation is to be made over all permutations of the pairs $(\mu,-\omega_\sigma)$ and (α,ω), it is clear that $\chi^{(1)}_{\mu\alpha}(-\omega_\sigma;\omega)$ is invariant under such permutations. This is an expression of the overall permutation-symmetry property in first order, and in this form the property generalises to all orders.

We now turn to a consideration of the formula (4.63) for the second-order susceptibility tensor. We again associate the frequency $-\omega_\sigma$ with the coordinate suffix μ. It can be seen by inspection of (4.63) that the second term in square brackets may be derived from the first by interchanging μ and α and replacing $-\omega_\sigma(=-\omega_1-\omega_2)$ by ω_1, *i.e.*, by permuting the pairs $(\mu,-\omega_\sigma)$, (α,ω_1), (β,ω_2) as follows:

$$(\mu,-\omega_\sigma),(\alpha,\omega_1),(\beta,\omega_2)\to(\alpha,\omega_1),(\mu,-\omega_\sigma),(\beta,\omega_2). \tag{4.68}$$

Similarly, the last term in square brackets may be derived from the first by permuting the pairs as follows:

$$(\mu,-\omega_\sigma),(\alpha,\omega_1),(\beta,\omega_2)\rightarrow(\alpha,\omega_1),(\beta,\omega_2),(\mu,-\omega_\sigma). \tag{4.69}$$

Finally, the first term in square brackets may of course be derived from itself by applying the identity permutation

$$(\mu,-\omega_\sigma),(\alpha,\omega_1),(\beta,\omega_2)\rightarrow(\mu,-\omega_\sigma),(\alpha,\omega_1),(\beta,\omega_2). \tag{4.70}$$

Altogether there are six possible permutations of the three pairs $(\mu,-\omega_\sigma)$, (α,ω_1) and (β,ω_2). Equations (4.68), (4.69) and (4.70) specify the three permutations in which $(\mu,-\omega_\sigma)$ occupies the second, third and first place in the sequence while the order of the pairs (α,ω_1) and (β,ω_2) is left undisturbed. The other three permutations are derived from these by interchanging the pairs (α,ω_1) and (β,ω_2). Now the formula (4.63) already involves the symmetrising operation $\mathbf{S}$ which implies that the expression following it is to be summed over the two possible permutations of the pairs (α,ω_1) and (β,ω_2). Hence, when this symmetrising operation is carried out, there are six terms in the square brackets and they may all be derived from the first term by the application of the six permutations of the pairs $(\mu,-\omega_\sigma)$, (α,ω_1), (β,ω_2). Thus (4.63) may be written in the more compact form:

$$\chi^{(2)}_{\mu\alpha\beta}(-\omega_\sigma;\omega_1,\omega_2)=\frac{Ne^3}{\varepsilon_0\,2\hbar^2}\,\mathbf{S}_\mathrm{T}\sum_{abc}\rho_0(a)\left[\frac{r^\mu_{ab}\,r^\alpha_{bc}\,r^\beta_{ca}}{(\Omega_{ba}-\omega_1-\omega_2)(\Omega_{ca}-\omega_2)}\right], \tag{4.71}$$

where the total symmetrisation operation $\mathbf{S}_\mathrm{T}$ indicates that the expression which follows it is to be summed over all six permutations of the pairs $(\mu,-\omega_\sigma)$, (α,ω_1), (β,ω_2). Since $\mathbf{S}_\mathrm{T}$ involves a summation over all possible permutations, it is clear that $\chi^{(2)}_{\mu\alpha\beta}(-\omega_\sigma;\omega_1,\omega_2)$ is invariant under any of them. This is the overall permutation-symmetry property in second order.

The presence of the symmetrising operation $\mathbf{S}$ in (4.63) already implies that $\chi^{(2)}_{\mu\alpha\beta}(-\omega_\sigma;\omega_1,\omega_2)$ has intrinsic permutation symmetry. The new feature introduced by the presence of the total symmetrisation operation $\mathbf{S}_\mathrm{T}$ in (4.71) is that the invariance is preserved under permutations involving the extra pair $(\mu,-\omega_\sigma)$. As a particular example of overall permutation symmetry in second order, we consider the permutation (4.68); thus we have

$$\chi^{(2)}_{\mu\alpha\beta}(-\omega_\sigma;\omega_1,\omega_2)=\chi^{(2)}_{\alpha\mu\beta}(\omega_1;-\omega_\sigma,\omega_2). \tag{4.72}$$

We may write down six relations of this type, one of which is an identity.

The above considerations are easily extended to the formula (4.64) for the third-order susceptibility tensor $\chi^{(3)}_{\mu\alpha\beta\gamma}(-\omega_\sigma;\omega_1,\omega_2,\omega_3)$, for which ω_σ is defined to be $\omega_\sigma = \omega_1+\omega_2+\omega_2$. The four terms in the square brackets in (4.64) may all be derived from the first by application of the four permutations of the pairs $(\mu,-\omega_\sigma)$, (α,ω_1), (β,ω_2) and (γ,ω_3) in which $(\mu,-\omega_\sigma)$ appears in the first, second, third and fourth place in the sequence while the pairs (α,ω_1), (β,ω_2), and (γ,ω_3) appear in natural order. The symmetrisation operation $\mathbf{S}$ permutes the pairs (α,ω_1), (β,ω_2), and (γ,ω_3) in all six possible ways and sums the result. Hence (4.64) may be written in the compact form:

$$\chi^{(3)}_{\mu\alpha\beta\gamma}(-\omega_\sigma;\omega_1,\omega_2,\omega_3) = \frac{N}{\varepsilon_0}\frac{e^4}{3!\hbar^3}\mathbf{S}_\mathrm{T} \times \sum_{abcd}\rho_0(a)\left[\frac{r^\mu_{ab}\,r^\alpha_{bc}\,r^\beta_{cd}\,r^\gamma_{da}}{(\Omega_{ba}-\omega_1-\omega_2-\omega_3)(\Omega_{ca}-\omega_2-\omega_3)(\Omega_{da}-\omega_3)}\right], \tag{4.73}$$

where the total symmetrisation operation $\mathbf{S}_\mathrm{T}$ indicates that the expression which follows it is to be summed over all 24 permutations of the pairs $(\mu,-\omega_\sigma)$, (α,ω_1), (β,ω_2) and (γ,ω_3). Clearly, $\chi^{(3)}_{\mu\alpha\beta\gamma}(-\omega_\sigma;\omega_1,\omega_2,\omega_3)$ is invariant under all such permutations.

The generalisation of the formulae (4.67), (4.71) and (4.73) to arbitrary order is now obvious by inspection. In making this generalisation, the frequency $-\omega_\sigma$ is defined as previously, *i.e.*,

$$-\omega_\sigma = -(\omega_1+\omega_2+\cdots+\omega_n), \tag{4.74}$$

and is always associated with the cartesian coordinate μ. (We note that the pair $(\mu,-\omega_\sigma)$, and not (μ,ω_σ), is involved in overall permutation symmetry; hence the convention of writing $\chi^{(n)}_{\mu\alpha_1\cdots\alpha_n}(-\omega_\sigma;\omega_1,\ldots,\omega_n)$ rather than $\chi^{(n)}_{\mu\alpha_1\cdots\alpha_n}(\omega_\sigma;\omega_1,\ldots,\omega_n)$.) We thus obtain the general (but rather dense) formula:

$$\chi^{(n)}_{\mu\alpha_1\cdots\alpha_n}(-\omega_\sigma;\omega_1,\ldots,\omega_n) = \frac{N}{\varepsilon_0}\frac{e^{n+1}}{n!\hbar^n}\mathbf{S}_\mathrm{T}\sum_{ab_1\cdots b_n}\rho_0(a) \times\left[\frac{r^\mu_{ab_1}\,r^{\alpha_1}_{b_1b_2}\cdots r^{\alpha_{n-1}}_{b_{n-1}b_n}\,r^{\alpha_n}_{b_na}}{(\Omega_{b_1a}-\omega_1-\cdots-\omega_n)(\Omega_{b_2a}-\omega_2-\cdots-\omega_n)\cdots(\Omega_{b_na}-\omega_n)}\right]. \tag{4.75}$$

In (4.75) $\mathbf{S}_\mathrm{T}$ is the total symmetrisation operation in nth order; it implies that the expression which follows it is to be summed over all $(n+1)!$ permutations of the pairs $(\mu,\omega_\sigma),(\alpha_1,\omega_1),\ldots,(\alpha_n,\omega_n)$. Clearly $\boldsymbol{\chi}^{(n)}$ is invariant under all such permutations. This is the most general expression of the overall permutation-symmetry property.

It will be noticed that the definition of $\mathbf{S}_\mathrm{T}$, like that of the intrinsic-permutation-symmetry operator $\mathbf{S}$ in (4.28), is different in different orders n and strictly, therefore, should carry a superscript n. However, the order will always be clear from the context and, as for $\mathbf{S}$, there is no need to complicate the notation in this way.

Lastly, it is often useful to simplify the appearance of the final formulae (4.67), (4.71), (4.73) and (4.75) by writing them in the scalar form defined by (2.58); we state below the resulting expression in nth order:

$$\chi^{(n)}(-\omega_\sigma\,;\omega_1,\ldots,\omega_n) = \frac{N\ e^{n+1}}{\varepsilon_0\ n!\hbar^n}\,\mathbf{S}_\mathrm{T}\sum_{ab_1\cdots b_n}\rho_0(a)$$

$$\times\left[\frac{\mathbf{e}_\sigma^*\cdot\mathbf{r}_{ab_1}\ \mathbf{e}_1\cdot\mathbf{r}_{b_1b_2}\ \cdots\ \mathbf{e}_n\cdot\mathbf{r}_{b_na}}{(\Omega_{b_1a}-\omega_1-\cdots-\omega_n)(\Omega_{b_2a}-\omega_2-\cdots-\omega_n)\cdots(\Omega_{b_na}-\omega_n)}\right], \quad (4.76)$$

noting, once again, that the polarisation vector $\mathbf{e}_i^*$ is associated with the negative frequency $-\omega_i$. When expressed in this form, the total symmetrisation operation $\mathbf{S}_\mathrm{T}$ requires that the expression following it is to be summed over all permutations of the pairs $(\mathbf{e}_\sigma^*, -\omega_\sigma)$, $(\mathbf{e}_1, \omega_1), \ldots, (\mathbf{e}_n, \omega_n)$.

The full expansion of (4.75) or (4.76), when the terms generated by $\mathbf{S}_\mathrm{T}$ are written out explicitly, contains $(n+1)!$ terms within the square brackets, although if some of the pairs $(\mathbf{e}_i, \omega_i)$ are in fact equal, then there will be fewer distinct terms. This is of no consequence, however, since the operation $\mathbf{S}_\mathrm{T}$ specifies that *all* the permutations are to be summed, regardless of whether or not they are distinct.

4.3.2 Third-harmonic generation

These points are best clarified by reference to an example, which will also demonstrate the application of the general formulae to a particular nonlinear process. The example considered here is third-harmonic generation (others occur in later sections). Third-harmonic generation can occur when an incident field at the frequency ω induces a polarisation at the frequency 3ω; the process is governed by the third-order susceptibility $\boldsymbol{\chi}^{(3)}(-3\omega;\omega,\omega,\omega)$. (The nonlinear polarisation so induced may generate a travelling wave at the third-harmonic frequency, as described in Chapter 7.) Because three of the frequency arguments of this susceptibility are identical, the 24 terms generated by $\mathbf{S}_\mathrm{T}$ in the formula (4.73) reduce to four distinct terms, each of which occurs six times. Thus, using the scalar form (4.76), we have

$$\chi^{(3)}(-3\omega;\omega,\omega,\omega) = \frac{N}{\varepsilon_0}\frac{e^4}{3!\hbar^3}\sum_{abcd}\rho_0(a)$$
$$\times 6\left[\frac{\mathbf{e}_{3\omega}{}^{*}\cdot\mathbf{r}_{ab}\ \mathbf{e}_{\omega}\cdot\mathbf{r}_{bc}\ \mathbf{e}_{\omega}\cdot\mathbf{r}_{cd}\ \mathbf{e}_{\omega}\cdot\mathbf{r}_{da}}{(\Omega_{ba}-3\omega)(\Omega_{ca}-2\omega)(\Omega_{da}-\omega)} + \frac{\mathbf{e}_{\omega}\cdot\mathbf{r}_{ab}\ \mathbf{e}_{3\omega}{}^{*}\cdot\mathbf{r}_{bc}\ \mathbf{e}_{\omega}\cdot\mathbf{r}_{cd}\ \mathbf{e}_{\omega}\cdot\mathbf{r}_{da}}{(\Omega_{ba}+\omega)(\Omega_{ca}-2\omega)(\Omega_{da}-\omega)}\right.$$
$$\left.+\frac{\mathbf{e}_{\omega}\cdot\mathbf{r}_{ab}\ \mathbf{e}_{\omega}\cdot\mathbf{r}_{bc}\ \mathbf{e}_{3\omega}{}^{*}\cdot\mathbf{r}_{cd}\ \mathbf{e}_{\omega}\cdot\mathbf{r}_{da}}{(\Omega_{ba}+\omega)(\Omega_{ca}+2\omega)(\Omega_{da}-\omega)} + \frac{\mathbf{e}_{\omega}\cdot\mathbf{r}_{ab}\ \mathbf{e}_{\omega}\cdot\mathbf{r}_{bc}\ \mathbf{e}_{\omega}\cdot\mathbf{r}_{cd}\ \mathbf{e}_{3\omega}{}^{*}\cdot\mathbf{r}_{da}}{(\Omega_{ba}+\omega)(\Omega_{ca}+2\omega)(\Omega_{da}+3\omega)}\right]. \tag{4.77}$$

The singularities that occur when any of the frequency denominators in (4.77) approach zero are considered in detail in §4.5.

4.3.3 Virtual transitions and Feynman diagrams

Diagrammatic techniques can be used to keep track of the various terms in (4.75) and (4.76) which are generated by $\mathbf{S}_{\mathrm{T}}$ (Ward, 1965), and these also provide a pictorial interpretation of the matter-field interaction in terms of photon creation and annihilation (Loudon, 1983). Since we use the classical description of fields throughout this book, we should regard the polarisation vectors $\mathbf{e}_j$ and $\mathbf{e}_j{}^*$ which appear in the susceptibility formulae as commuting annihilation and creation operators, respectively. (It will be recalled that in the definition (2.58), the vector $\mathbf{e}_i{}^*$ is to be associated with a negative frequency $-\omega_i$ which appears as an argument of $\chi^{(n)}(-\omega_\sigma;\omega_1,\ldots,\omega_n)$.) To take an example, the four terms in the formula (4.77) for the third-harmonic susceptibility correspond, in order, to the four Feynman diagrams drawn in Fig. 4.1. The first diagram (a) represents the process in which a molecule, initially in a state labelled a, emits a third-harmonic photon 3ω whilst making a virtual transition to a state labelled b. By absorbing a fundamental photon ω the molecule then makes a virtual transition from b to c. The absorption of two further ω photons takes the molecule into d and finally back to the initial state a. The summation in (4.77) signifies that each of the labels a, b, c, d ranges over all of the molecular energy eigenstates, and that all combinations of the pathways $a\rightarrow b\rightarrow c\rightarrow d\rightarrow a$ contribute to the term depicted in Fig. 4.1(a). The process involves virtual transitions; the molecule undergoes no real net change of energy and is restored to its initial state when the process is complete. Energy conservation need not (and in general will not) hold at each virtual transition in the sequence depicted in the Feynman diagram. However, energy is conserved overall since $\hbar(\omega+\omega+\omega)=\hbar 3\omega$. These remarks hold for all four diagrams drawn in Fig. 4.1 in which the virtual transitions occur in various sequences corresponding to the four terms in (4.77). The factor of 6 arises from the indistinguishable permutations of the ω photons in each diagram.

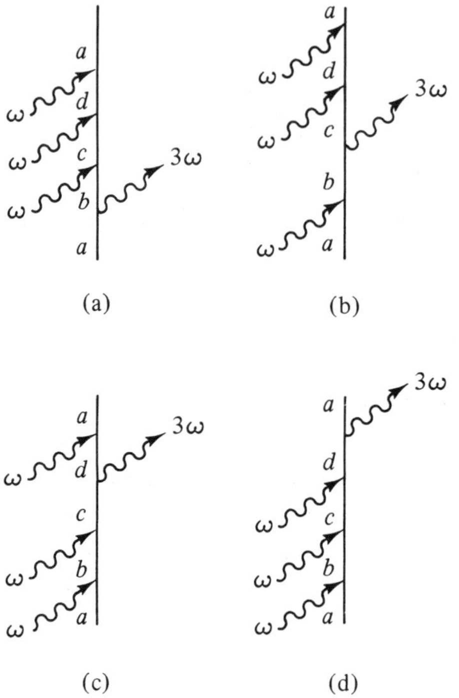

Fig. 4.1 Feynman diagrams illustrating the process of third-harmonic generation (the optical frequencies involved are assumed to be far removed from resonances). Photons (indicated by wavy lines) incident from the left represent absorption, and emission processes are represented by photons leaving to the right. The sequence of virtual molecular transitions involved in the whole process is indicated by the upward progression through states labelled a, b,... on the vertical line.

Since energy is generally not conserved during a virtual transition (such as $a \to b$ in Fig. 4.1(a), for example), the uncertainty principle dictates that the transition is possible only if the molecule remains in the state b for a time shorter than the inverse of the detuning (which is $(\Omega_{ba} - 3\omega)^{-1}$ in this case). Virtual transitions therefore occur on an ultrafast time scale (see §6.6) and may be assumed to be instantaneous for many practical purposes.

4.4 The passage from microscopic to macroscopic formulae

In §4.1 we derived fundamental formulae for the susceptibility tensors in terms of electric-dipole operators and their matrix elements (equivalent formulae in terms of current operators and their matrix elements are derived in §4.6). The general expressions (4.28) and (4.131) for $\boldsymbol{\chi}^{(n)}$ in the many-particle energy representation apply to *any* medium in the electric-dipole approximation. However, to calculate the properties of a medium consisting of an assembly of molecules (or microscopic units), it is usually more practical to calculate the expectation value of the dipole moment $e\mathbf{r}$ of an individual molecule in the single-particle energy representation. A simple example of this, given in §4.2, is the derivation of formulae for the susceptibility tensors for an assembly of identical molecules. In making this derivation we made sweeping assumptions about the nature of the assembly: in particular, the molecules were taken to be independent and similarly oriented. These assumptions were made in the interest of clarity, but the transition between monomolecular and many-molecular formulae was thus oversimplified. In this section we reexamine that transition. In so doing, we move a step closer to removing some of the restrictive assumptions and also more clearly identify the situations in which some of the assumptions can be justified.

4.4.1 Induced molecular dipole moment

We begin this discussion by considering the expectation value of the dipole moment of the mth molecule, denoted by $e\mathbf{r}_m$, in terms of a power series in the electric field, similar to the earlier series (2.1) for the macroscopic polarisation $\mathbf{P}(t)$:

$$e\mathbf{r}_m = e\mathbf{r}_m^{(0)} + e\mathbf{r}_m^{(1)} + \cdots + e\mathbf{r}_m^{(n)} + \cdots , \tag{4.78}$$

where $e\mathbf{r}^{(0)}$ is the permanent moment, $e\mathbf{r}^{(1)}$ is linear in the field, $e\mathbf{r}^{(2)}$ is quadratic, and so on. Here we have reverted to including explicitly the label for the mth molecule, since this will allow us to consider the situation of an assembly of dissimilar molecules, but also, more importantly, we can thus distinguish between molecules and consider the effect of interactions between them. The term $e\mathbf{r}^{(0)}$ in (4.78) represents the permanent electric-dipole moment possessed by polar molecules. The various higher-order terms ($n = 1,2,\ldots$) in the series can be related to the electric field $\mathbf{E}_m(t)$ acting locally at the site of the mth molecule by

$$e\mathbf{r}_m^{(n)} = (n!)^{-1} \int_{-\infty}^{+\infty} d\omega_1 \cdots \int_{-\infty}^{+\infty} d\omega_n \, \boldsymbol{\gamma}_m^{(n)}(-\omega_\sigma;\omega_1,\ldots,\omega_n) \,|\, \mathbf{E}_m(\omega_1) \cdots \mathbf{E}_m(\omega_n) \exp(-i\omega_\sigma t) , \tag{4.79}$$

where the Fourier components $\mathbf{E}_m(\omega)$ are related to the field $\mathbf{E}_m(t)$ by the transform (2.26), and $\omega_\sigma = \omega_1 + \omega_2 + \cdots + \omega_n$ as before. The linear ($n = 1$) term in the series (4.78) involves the molecular polarisability $\boldsymbol{\gamma}^{(1)}$, which is a second-rank tensor. The higher-order molecular polarisabilities $\boldsymbol{\gamma}^{(n)}$, for $n=2, 3,...$, are tensors of rank $n+1$ and are termed hyperpolarisabilities. (An ambiguity exists in the literature where different authors refer to the hyperpolarisability $\boldsymbol{\gamma}^{(n)}$ variously as the nth-order polarisability or the $(n-1)$th-order hyperpolarisability; we prefer to avoid this confusing terminology.) The units of $\boldsymbol{\gamma}^{(n)}$ are $\mathrm{C\,m(V\,m^{-1})^{-n}}$, which for $\boldsymbol{\gamma}^{(1)}$ can be expressed as $\mathrm{F\,m^2}$.

We note that the expression (4.79) is very similar to the earlier expression (2.40) for the nth-order macroscopic polarisation $\mathbf{P}^{(n)}(t)$, with the molecular hyperpolarisability taking the place of the macroscopic nonlinear susceptibility, except that (4.79) is in terms of the Fourier components of the *local* field $\mathbf{E}_m(t)$ rather than the macroscopic field. The principles of time-invariance, causality and reality, which were discussed in §§2.1 and 2.2 in relation to the macroscopic polarisation, apply also to the expectation values of the induced dipole moments at the molecular level, and thus it follows that the permutation-symmetry properties of the susceptibilities apply equally well to the molecular hyperpolarisabilities. Therefore, when considering the situation in which the applied field is a superposition of monochromatic waves (as in §2.3), we can write the induced dipole moment as

$$e\mathbf{r}_{\omega_\sigma}^{(n)m} = (n!)^{-1}K(-\omega_\sigma\,;\omega_1,...,\omega_n) \times \boldsymbol{\gamma}_m^{(n)}(-\omega_\sigma\,;\omega_1,...,\omega_n) \,|\, \mathbf{E}_{\omega_1}^m \cdots \mathbf{E}_{\omega_n}^m, \quad (4.80)$$

where the monochromatic-wave amplitudes for the local field $\mathbf{E}_\omega^m$ are defined in the same way as for the macroscopic fields in (2.49), and similarly the expectation value of the induced dipole moment has been expressed in terms of its Fourier frequency components as

$$e\mathbf{r}_m^{(n)} = \tfrac{1}{2}\sum_{\omega\geq 0}\left[e\mathbf{r}_\omega^{(n)m}\exp(-i\omega t) + e\mathbf{r}_{-\omega}^{(n)m}\exp(i\omega t)\right]; \quad (4.81)$$

$$e\mathbf{r}_{-\omega}^{(n)m} = (e\mathbf{r}_\omega^{(n)m})^*. \quad (4.82)$$

The latter expression results from the fact that the expectation value $e\mathbf{r}_m^{(n)}$ is real. The factor K in (4.80), which arises from the intrinsic permutation symmetry of the hyperpolarisabilities, is identical to that described and listed in §2.3.

For a d.c. (zero-frequency) field, the factors $K(-\omega_\sigma\,;\omega_1,...,\omega_n)$ are equal to unity in all orders, and in this case the series (4.78) reduces to the familiar expression for the static dipole moment:

$$e\mathbf{r}_m = e\mathbf{r}_m^{(0)} + \boldsymbol{\alpha}_m\cdot\mathbf{E}_m + \tfrac{1}{2}\boldsymbol{\beta}_m\!:\!\mathbf{E}_m\mathbf{E}_m + \tfrac{1}{6}\boldsymbol{\gamma}_m\,\vdots\,\mathbf{E}_m\mathbf{E}_m\mathbf{E}_m + \cdots \quad (4.83)$$

where we have adopted the commonly-used notation for the lowest-order tensors $\boldsymbol{\gamma}^{(n)}$:

$$\begin{aligned}&\boldsymbol{\alpha}(-\omega;\omega)\equiv\boldsymbol{\gamma}^{(1)}(-\omega;\omega)\,,\\&\boldsymbol{\beta}(-\omega_\sigma;\omega_1,\omega_2)\equiv\boldsymbol{\gamma}^{(2)}(-\omega_\sigma;\omega_1,\omega_2)\,,\\&\boldsymbol{\gamma}(-\omega_\sigma;\omega_1,\omega_2,\omega_3)\equiv\boldsymbol{\gamma}^{(3)}(-\omega_\sigma;\omega_1,\omega_2,\omega_3).\end{aligned}\tag{4.84}$$

It is a common practice to omit the frequency arguments when describing the d.c. response, as in (4.83).

Unfortunately there is no universal convention for the numerical factors in (4.79) and (4.80). Several alternative forms of simple expressions such as (4.83) occur in the literature. The factor $(n!)^{-1}$ is usual in the nonlinear-optics literature, but is sometimes omitted by chemists. The practice of including the numerical factor K explicitly in (4.80), as we prefer, rather than absorbing it into the hyperpolarisability, has certain advantages as described previously in §2.3 for the susceptibilities.

Before proceeding further, we should briefly mention other (potentially confusing) terminology which occurs in the literature. One sometimes sees expressions for the molecular dipole moment $e\mathbf{r}_m^{(n)}$ which are similar to (4.79), but with the local fields $\mathbf{E}_m(\omega)$ replaced by macroscopic fields $\mathbf{E}(\omega)$, and the corresponding response tensors (similar to $\boldsymbol{\gamma}_m^{(n)}$) are referred to as 'macroscopic hyperpolarisabilities'. Another terminology is to refer to the 'microscopic susceptibilities' $\boldsymbol{\chi}_{\text{mic}}^{(n)}$ of individual molecules, commonly defined as

$$\boldsymbol{\chi}_{\text{mic}}^{(n)}=\boldsymbol{\gamma}^{(n)}/\varepsilon_0 n!.\tag{4.85}$$

Our preference, however, is to work with the (microscopic) molecular hyperpolarisabilities $\boldsymbol{\gamma}^{(n)}$ and (macroscopic) susceptibilities $\boldsymbol{\chi}^{(n)}$, as defined by (4.79) and (2.40), respectively; the relations between these two tensors are given in §4.4.3.

4.4.2 The local field

We now consider the relation between the microscopic and macroscopic formulae. The general formulae for the macroscopic susceptibilities $\boldsymbol{\chi}^{(n)}$ ((4.28), and the alternative form (4.126)) are written in the many-particle energy representation for a volume element V of the medium. The expression (2.40) for the polarisation $\mathbf{P}^{(n)}(t)$ in V, in terms of $\boldsymbol{\chi}^{(n)}$, involves the components of the macroscopic electric field $\mathbf{E}(\omega)$ in V. It is assumed that V is big enough to contain a very large number of molecules, but also small enough that the macroscopic field $\mathbf{E}(\omega)$ can be considered uniform within it. Despite the generality of that approach, it is often more practical to use the single-particle energy representation to calculate the expectation value of the dipole moment of the mth molecule, $e\mathbf{r}_m$, and then find the macroscopic polarisation $\mathbf{P}^{(n)}(t)$ by summing over the dipole moments per unit volume in V. However, we have already noted that the dipole moment $e\mathbf{r}_m$ is dependent upon the components $\mathbf{E}_m(\omega)$ of the *local* field acting on the molecule; the appropriate expression (4.79) is in terms of the molecular hyperpolarisability. Therefore, in order to extract the expressions for the (macroscopic) susceptibilities from those of the (microscopic) hyperpolarisabilities, the relation between the local and macroscopic fields must be determined.

This problem arises because the field acting on a molecule, and effective in polarising it, is a field specified with microscopic precision, whereas the macroscopic electric field is intentionally defined in such a way that it makes no explicit reference to microscopic coordinates, and indeed is an average field over the loosely-defined region V which includes not only many molecular sites, but also the regions of space between them. In determining the local field we must, in general, take into account the depolarising field acting on any particular molecule due to the influence of surrounding molecules. (This problem is side-stepped in §4.2, for simplicity there, by assuming that the molecules are non-interacting; the depolarising fields are thus neglected and the local field acting on each molecule within the volume element V is therefore taken to be equal to the macroscopic field in V.)

The local-field problem, even for a dielectric with an idealised linear response, is a difficult one which has received the attention of theoreticians over many decades, although this has yielded practical results in only a few simple (but important) cases; these will be summarised shortly. First, however, we describe an empirical approach to the problem of the local field in nonlinear optics (Armstrong *et al*, 1962). As we shall see, the very useful aspect of this approach is that it allows the available expressions for the linear case to be extended readily to the nonlinear situation.

We assume that the molecules interact mainly through the dipolar field of the induced dipoles; this is certainly a reasonable assumption when the distance between molecules is much larger than their intrinsic dimensions. In that case, the dipolar field effective at the site of a particular molecule can also be reasonably assumed to be uniform over the molecule. In condensed phases where the molecular distances are of the same order as their dimensions, there may be some contribution from higher multipole fields. However, due to symmetry, these multipole fields are unlikely to induce an appreciable dipole moment, and in any case, the dipole term is already sufficiently complicated and uncertain to make the inclusion of higher-order terms of little value.

In the dipole approximation the field acting locally at the site of the mth molecule can be written:

$$\mathbf{E}_\omega^m = \mathbf{E}_\omega + (\varepsilon_0 V)^{-1}\sum_l \mathbf{L}_\omega^{ml}\cdot e\mathbf{r}_\omega^l, \tag{4.86}$$

where the macroscopic field is taken to comprise a superposition of monochromatic waves, and we use the notation of §2.3; thus $\mathbf{E}_\omega$ is a monochromatic-wave amplitude of the macroscopic field. The Lorentz tensor $\mathbf{L}_\omega^{ml}$ is of second rank, and its elements are dimensionless – the factor $(\varepsilon_0 V)^{-1}$ in (4.86) ensures that this is so. The summation is over all molecules in a small volume V surrounding the mth molecule which, as noted earlier, is assumed to be large enough to contain a very large number of other molecules, but is also small enough that the macroscopic field $\mathbf{E}_\omega$ can be considered uniform within it. To simplify the notation, we shall drop the explicit frequency dependence in the following steps; this will focus attention on the subscripts l, m and so on, which label individual molecules. We now define a set of 3×3 matrices $\{\mathbf{M}^{lm}\}$ such that

$$\mathbf{M}^{lm} = \delta_{lm}\mathbf{I} - (\varepsilon_0 V)^{-1}\boldsymbol{\alpha}_l\cdot\mathbf{L}^{lm}, \tag{4.87}$$

where $\mathbf{I}$ is the identity matrix, and hence we can write an expression:

$$\sum_l (\mathbf{M}^{lm})^{\mathrm{T}}\cdot\mathbf{E}^l = \sum_l \left(\delta_{lm}\mathbf{I} - (\varepsilon_0 V)^{-1}\mathbf{L}^{ml}\cdot\boldsymbol{\alpha}_l\right)\cdot\mathbf{E}^l. \tag{4.88}$$

Here we have made use of the fact that energy considerations require the symmetry of the polarisability and Lorentz tensors, such that $(\boldsymbol{\alpha}_l)^{\mathrm{T}} = \boldsymbol{\alpha}_l$ and $\mathbf{L}^{ml} = \mathbf{L}^{lm} = (\mathbf{L}^{lm})^{\mathrm{T}}$, where the superscript ^{T} denotes the matrix transpose. Also we have used the matrix identity $(\mathbf{AB})^{\mathrm{T}} = \mathbf{B}^{\mathrm{T}}\mathbf{A}^{\mathrm{T}}$. We now make the approximation that the dominant contribution to the local field is that due to the linear polarisability of surrounding molecules, and thus (4.86) becomes:

$$\mathbf{E}^m \simeq \mathbf{E} + (\varepsilon_0 V)^{-1}\sum_l \mathbf{L}^{ml}\cdot\boldsymbol{\alpha}_l\cdot\mathbf{E}^l. \tag{4.89}$$

Substituting (4.89) into (4.88) leads to the expression:

$$\sum_l (\mathbf{M}^{lm})^{\mathrm{T}}\cdot\mathbf{E}^l = \mathbf{E}. \tag{4.90}$$

The 3×3 matrices $\mathbf{M}^{lj}$, when taken altogether, can be considered to be a 'supermatrix' of size $3N$×$3N$, where N is the number of molecules in V. It follows that this supermatrix has an inverse, which allows us to define a new set of 3×3 matrices, $\mathbf{R}^{jk}$, such that

$$\sum_j \mathbf{M}^{lj}\cdot\mathbf{R}^{jk} = \sum_j \mathbf{R}^{lj}\cdot\mathbf{M}^{jk} = \delta_{lk}\mathbf{I}. \tag{4.91}$$

(We note that the Roman labels j, k, l and m are used to distinguish between molecules, and should not be confused as being Greek labels for the cartesian coordinate axes.) If we now premultiply both sides of (4.90) by $\sum_j (\mathbf{R}^{jk})^{\mathrm{T}}$, and use the transpose of (4.91) with $l = k$, we obtain a useful relation between the local field for the kth molecule and the macroscopic field, namely:

$$\mathbf{E}^k = \mathbf{f}_k(\omega)\cdot\mathbf{E}, \tag{4.92}$$

where $\mathbf{f}_k(\omega) = \sum_j (\mathbf{R}^{jk})^{\mathrm{T}}$ is a 3x3 tensor whose components are known as 'local-field factors'. Also, by manipulating the various definitions (4.86), (4.87) and (4.91), we can express the series (4.78) in the form:

$$\begin{aligned} e\mathbf{r}^m_{\omega_\sigma} &= \delta_{ml}\, e\mathbf{r}^l_{\omega_\sigma} \\ &= \sum_k \mathbf{R}^{mk}\cdot\mathbf{M}^{kl}\cdot e\mathbf{r}^l_{\omega_\sigma} \\ &= [\mathbf{f}_m(\omega_\sigma)]^{\mathrm{T}}\cdot\Big[\boldsymbol{\alpha}_k(-\omega_\sigma;\omega_\sigma)\cdot\mathbf{E}_{\omega_\sigma} + \boldsymbol{\beta}_k(-\omega_\sigma;\omega_1,\omega_2):\mathbf{E}^k_{\omega_1}\mathbf{E}^k_{\omega_2} \\ &\qquad + \boldsymbol{\gamma}_k(-\omega_\sigma;\omega_1',\omega_2',\omega_3'):\mathbf{E}^k_{\omega_1'}\mathbf{E}^k_{\omega_2'}\mathbf{E}^k_{\omega_3'} + \cdots\Big], \end{aligned} \tag{4.93}$$

with $\omega_\sigma = \omega_1 + \omega_2 = \omega_1' + \omega_2' + \omega_3'$. Finally, by comparing (4.80) with the various terms in the final line of (4.93), we obtain

$$e\mathbf{r}^{(1)}_\omega = \mathbf{f}(\omega)\cdot\boldsymbol{\alpha}(-\omega;\omega)\cdot\mathbf{E}_\omega, \tag{4.94}$$

and

$$\begin{aligned} e\mathbf{r}^{(n)}_{\omega_\sigma} &= (n!)^{-1}K(-\omega_\sigma;\omega_1,\dots,\omega_n) \\ &\quad\times[\mathbf{f}(\omega_\sigma)]^{\mathrm{T}}\cdot\boldsymbol{\gamma}^{(n)}(-\omega_\sigma;\omega_1,\dots,\omega_n)\,|\,\mathbf{f}(\omega_1)\cdot\mathbf{E}_{\omega_1}\cdots\mathbf{f}(\omega_n)\cdot\mathbf{E}_{\omega_n} \end{aligned} \tag{4.95}$$

for $n = 2,3,\dots$. Here we have assumed that the molecules have identical hyperpolarisabilities, so that the labels m and k can be omitted.

This final expression (4.95) may appear rather forbidding. It does, in fact, provide a simple and useful recipe: first, the relationship (4.92) between the local and macroscopic fields is determined from the linear induced dipole moment (4.94), and then the appropriate relation is introduced into the nonlinear terms by using (4.95). The expression (4.95) for the nth-order dipole moment (with $n = 2,3,\dots$) contains $n+1$ linear local-field factors; one factor is associated with each of the frequencies $\omega_\sigma, \omega_1, \dots, \omega_n$. The local-field description in nonlinear optics is thus reduced to the fundamental problem for linear dielectrics.

The calculation of the local field, even for linear dielectric media, is, however, a difficult problem which, as mentioned earlier, is made amenable in only a few special (yet important) cases. The classical approach to the problem was established over 70 years ago by Lorentz (1916). Previously we noted that it can be shown, from general energy considerations, that the Lorentz tensor is symmetric in the electric-dipole approximation. A property of any second-rank symmetric tensor is that it may be diagonalised by a suitable choice of principal axes. Lorentz showed that for atoms or nonpolar molecules with well-localised bound electrons and distributed either on a cubic lattice or randomly in space, the diagonal elements of the Lorentz tensor **L** are equal and sum to unity, *i.e.*, **L** can be represented as a scalar, $L = \frac{1}{3}$. Then (4.86) leads to the well-known Lorentz formula:

$$\mathbf{E}_\omega^m = \mathbf{E}_\omega + \mathbf{P}_\omega / 3\varepsilon_0 . \tag{4.96}$$

By making the same approximation which previously lead to (4.89), namely that the dominant contribution to the local field is that due to the linear polarisability of the surrounding molecules, (4.96) can be expressed as $\mathbf{E}_\omega^m = [2+\varepsilon(\omega)]\mathbf{E}_\omega/3$, where the dielectric constant $\varepsilon(\omega) = 1+\chi^{(1)}(-\omega;\omega)$ is a scalar in an isotropic medium or one with cubic symmetry (see §5.3). In this case, the local-field tensor $\mathbf{f}(\omega)$ defined by (4.92) may also therefore be replaced by a scalar, which we denote by $f(\omega)$; *i.e.*,

$$\mathbf{f}(\omega)\cdot\mathbf{E}_\omega = f(\omega)\,\mathbf{E}_\omega = [2+\varepsilon(\omega)]\,\mathbf{E}_\omega/3 . \tag{4.97}$$

Therefore (4.94) and (4.95) become

$$e\mathbf{r}_\omega^{(1)} = [2+\varepsilon(\omega)]\,\boldsymbol{\alpha}(-\omega;\omega)\cdot\mathbf{E}_\omega/3 . \tag{4.98}$$

and

$$e\mathbf{r}_{\omega_\sigma}^{(n)} = (3^{n+1}n!)^{-1}K(-\omega_\sigma;\omega_1,\ldots,\omega_n)\,[2+\varepsilon(\omega_\sigma)] \\ \times \boldsymbol{\gamma}^{(n)}(-\omega_\sigma;\omega_1,\ldots,\omega_n)\,|\,[2+\varepsilon(\omega_1)]\,\mathbf{E}_{\omega_1}\cdots[2+\varepsilon(\omega_n)]\,\mathbf{E}_{\omega_n}, \tag{4.99}$$

respectively.

The Lorentz relation (4.97) applies well to isotropic nonpolar liquids and to cubic ionic crystals, such as NaCl, which have well-separated atoms with strongly localised electrons. For other crystals, as the electronic distribution associated with each atom or ion becomes more extended in space, the value of L is expected to decrease and the local field should approach more and more closely the macroscopic field. In some cubic crystals, such as AgCl, there is known to be appreciable overlap of the electronic wave functions extending over several atoms, so that $L < \frac{1}{3}$.

The method used by Lorentz to obtain the relation (4.96) was to transform the summation over V in (4.86) into a surface integral; the validity of this approach rests on the hypothesis that the polarisation within the dense medium is uniform. Provided the dimensions of the volume V surrounding a molecule are small compared to an optical wavelength, the surface integral can be evaluated using the standard methods of electrostatics (Bottcher, 1952). It is found that the value obtained depends, not on the size of the cavity, but on its shape. Ideally, the shape should be chosen to represent the environment of the molecule. In liquids and cubic crystals – the case considered by Lorentz – the environment is spherically symmetric and the result (4.96) assumes a spherical volume *V*. Calculations more sophisticated than the Lorentz model have been tried, such as those using ellipsoidal volumes adapted to the shape of the molecules, to pay some tribute to actual symmetries (Chemla, 1975; Cojan *et al*, 1977). However, these approaches are rough estimates which can only account for the general trends of local field determination (Chemla, 1980).

Although a general theory for all crystal symmetries is not yet available, it is likely that for noncubic crystal symmetries and – very importantly – provided that the electrons remain well-localised, **L** approximates to a symmetric tensor whose diagonal elements sum to unity. For the most commonly-occurring crystal lattices it is expected that the off-diagonal elements of **L** would be small and that the diagonal elements would be not far from $\frac{1}{3}$. The extreme case of the strongly delocalised nearly-free electrons of a metal was considered by Darwin (1934). He showed that, in this case, the rapid spatial fluctuations of the microscopic field should be completely averaged out over the extended electronic wave functions, and that the local and macroscopic fields should be equal, *i.e.*, $\mathbf{L}\rightarrow 0$. This would also apply to the Bloch (band-state) electrons in metallic and semiconducting crystals, whose wave functions extend over very many unit cells.

So far in this discussion of local fields we have considered only induced dipoles; however external fields, especially static fields, can also orient molecules possessing permanent electric-dipole moments. For gases and liquids which contain freely rotating molecules, there are refinements to the Lorentz model which can be made – for example, the treatment by Onsanger (1936) – to take into account the orientational reactive forces exerted between molecules. These forces can be represented as reaction fields. The reaction field acting on a molecule depends on the dipole of the molecule itself and therefore cannot contribute to its own orientation. Onsanger showed that the local-field tensor at frequency ω which appears in (4.94) and (4.95) can be approximated, for a polar liquid, by the scalar:

$$f(\omega) = \frac{(\varepsilon^{\infty}+2)\,\varepsilon(\omega)}{\varepsilon^{\infty}+2\varepsilon(\omega)}, \tag{4.100}$$

where ε^{∞} denotes a value of the dielectric constant obtained by extra-polating the results of electrical measurements to high frequencies, and $\varepsilon(\omega)$ is the dielectric constant at frequency ω. At optical frequencies, $f(\omega)$

given by (4.100) is almost identical to (4.97), since $\varepsilon^{\infty} \simeq \varepsilon(\omega)$ provided ω is far-removed from resonances. However, the local-field factor $f(0)$ given by (4.100) is the one that should be associated with a d.c. (or very low frequency) field (Levine and Bethea, 1975).

In summary, in media displaying induced electronic or ionic polarisation (but not orientation polarisation), the first approximation to the local-field correction is of the form (4.89). In rarified media such as gases, where the factors $[2+\varepsilon(\omega)]/3$ which appear in (4.98) and (4.99) are almost equal to one, the local-field corrections are so small as to be negligible. The corrections are much greater in the case of bound charges in condensed matter, however, and can have a significant effect on the comparison of experimentally-determined nonlinear optical coefficients with those calculated from microscopic theories. (For further details and methods of dealing with liquids and solids, the reader is referred to Chemla (1975) and the books by Fröhlich (1958) and Anderson (1964).) Nevertheless, the form and essential properties of the hyperpolarisabilities and susceptibilities are, in fact, unaffected by local field considerations, as is apparent from (4.95) for example.

4.4.3 Macroscopic polarisation and orientation average

For all media, the macroscopic nonlinear polarisation $\mathbf{P}_{\omega}^{(n)}$ is expressed in terms of the expectation values of the induced microscopic dipole moments as:

$$\mathbf{P}_{\omega}^{(n)} = \sum_{t} N_t < e\mathbf{r}_{\omega}^{(n)t} > , \qquad (4.101)$$

where N_t is the number density of polarisable units of the species t and the summation is over all species (molecules in the case of non-associating liquids or gases; electrons, atoms, ions or groups, as appropriate, in the case of crystals). For example, most organic crystals are 'molecular solids' held together by relatively weak van der Waals and dipole-dipole intermolecular forces; the molecules largely retain their individual physical properties and therefore the appropriate species in (4.101) are the constituent molecules. In forming the sum in (4.101) for the case of crystalline media, we have to remember that the vector axes are fixed axes in space and not, as might be supposed, the principal axes for each group. The macroscopic susceptibilities are determined by equating (4.95) and (4.101) with the corresponding expression from Chapter 2, namely:

$$\mathbf{P}_{\omega_\sigma}^{(n)} = \varepsilon_0 K(-\omega_\sigma;\omega_1,\ldots,\omega_n)\boldsymbol{\chi}^{(n)}(-\omega_\sigma;\omega_1,\ldots,\omega_n) \,|\, \mathbf{E}_{\omega_1}\cdots\mathbf{E}_{\omega_n} , \qquad (4.102)$$

and the result for $n = 2, 3, \ldots$ is:

$$\boldsymbol{\chi}^{(n)}(-\omega_\sigma;\omega_1,\ldots,\omega_n)\,|\,\mathbf{e}_1\cdots\mathbf{e}_n$$
$$= (\varepsilon_0 n!)^{-1}\sum_t N_t < [\mathbf{f}_t(\omega_\sigma)]^{\mathrm{T}}\cdot\boldsymbol{\gamma}_t^{(n)}(-\omega_\sigma;\omega_1,\ldots,\omega_n)\,|$$
$$\mathbf{f}_t(\omega_1)\cdot\mathbf{e}_1\cdots\mathbf{f}_t(\omega_n\cdot\mathbf{e}_n > , \quad (4.103)$$

where we have used the notation of (2.57): $\mathbf{E}_{\omega_j} = \mathbf{e}_j E_j$. In the case when the local-field tensors $\mathbf{f}(\omega)$ can be represented by a scalar $f(\omega)$, (4.103) assumes the simpler form:

$$\boldsymbol{\chi}^{(n)}(-\omega_\sigma;\omega_1,\ldots,\omega_n) = (\varepsilon_0 n!)^{-1}\sum_t N_t <\boldsymbol{\gamma}_t^{(n)}(-\omega_\sigma;\omega_1,\ldots,\omega_n)>$$
$$\times f_t(\omega_\sigma) f_t(\omega_1)\cdots f_t(\omega_n). \quad (4.104)$$

The last items to be defined in (4.101), (4.103) and (4.104) are the angle brackets, which denote an orientation average. According to the transformation law for polar tensors, which is derived in §5.3.1, we can write the orientationally-averaged hyperpolarisability for a particular species as:

$$<\gamma^{(n)}_{\mu\alpha_1\cdots\alpha_n}(-\omega_\sigma;\omega_1,\ldots,\omega_n)> =$$
$$<R_{\mu u} R_{\alpha_1 a_1}\cdots R_{\alpha_n a_n}> \gamma^{(n)}_{u a_1\cdots a_n}(-\omega_\sigma;\omega_1,\ldots,\omega_n). \quad (4.105)$$

Here $u a_1 \cdots a_n$ take the values of the axes of the cartesian coordinate system fixed with respect to the molecule, and $R_{\alpha a} = \cos\theta_{\alpha a}$ where $\theta_{\alpha a}$ is the angle between the laboratory-frame axis $\boldsymbol{\alpha}$ and the axis $\mathbf{a}$ in the molecular frame. The average value of the product of direction cosines in (4.105) is defined as:

$$<R_{\mu u} R_{\alpha_1 a_1}\cdots R_{\alpha_n a_n}> =$$
$$\frac{1}{8\pi^2}\int_0^{2\pi}\int_0^{\pi}\int_0^{2\pi} R_{\mu u}(\phi,\theta,\psi)\prod_{p=1}^{n} R_{\alpha_p a_p}(\phi,\theta,\psi)\sin\theta\,\mathrm{d}\phi\,\mathrm{d}\theta\,\mathrm{d}\psi\,, \quad (4.106)$$

where ϕ,θ,ψ are the Euler angles for the coordinate-frame transformation.

To calculate the orientation average, knowledge of the spatial-symmetry properties of the medium is required (see §5.3). The evaluation of the orientation average for cartesian tensors is discussed in detail by Cyvin *et al* (1965) and Andrews and Thirunamachandran (1977). The expression (4.105) involves an integral over 3^{2n+2} direction cosines, and with increasing n quickly becomes intractable. In some cases the use of irreducible-tensor techniques leads to a more convenient formalism with analytical solutions (Chemla and Bonneville, 1978).

It will be recalled that in the simple artificial situation considered in §§4.2 and 4.3, the molecules in the assembly were assumed to share an identical orientation. The explicit formulae for the nonlinear susceptibilities in that case did, however, contain the factor $\sum_a \rho_0(a)$, which represents a weighting over the thermal-equilibrium distribution of initial states labelled a, including degenerate states. It turns out that for an assembly of independent freely-rotating molecules, the average over degenerate initial states itself implies the macroscopic isotropy of the medium (see §5.3). Therefore, the angle brackets which appear in (4.101), (4.103) and (4.104) can be omitted justifiably from the formulae for the nonlinear susceptibilities of the idealised molecular gas, given in §§4.2 and 4.3, even when we drop the artificial assumption that the molecules have an identical orientation.

It follows from these various considerations that *general* formulae for the molecular hyperpolarisabilities $\boldsymbol{\gamma}^{(n)}$ are obtained from the special-case formulae for $\boldsymbol{\chi}^{(n)}$ given in §§4.2, 4.3 and 4.5 by using the simple relation:

$$\boldsymbol{\gamma}^{(n)}(-\omega_\sigma;\omega_1,\ldots,\omega_n) = \varepsilon_0\, n!\, \boldsymbol{\chi}^{(n)}(-\omega_\sigma;\omega_1,\ldots,\omega_n)/N\,, \tag{4.107}$$

and similarly, with the aid of (4.85), we obtain:

$$\boldsymbol{\chi}^{(n)}_{\text{mic}}(-\omega_\sigma;\omega_1,\ldots,\omega_n) = \boldsymbol{\chi}^{(n)}(-\omega_\sigma;\omega_1,\ldots,\omega_n)/N\,. \tag{4.107a}$$

Finally, for a detailed and careful account of the transition between microscopic and macroscopic descriptions in electromagnetism, the reader is referred to Robinson (1973).

4.5 Resonant susceptibilities

We now return to consider in further detail the form of the nonlinear susceptibilities derived in §§4.2 and 4.3. We shall examine, in particular, the very important situation when certain of the optical frequencies fall in the vicinity of transition frequencies of the medium.

For simplicity, we retain the assumption that the medium consists of an assembly of microscopic polarisable units which are independent and noninteracting. We are thus able to neglect the local-field factors. This is done merely to simplify the appearance of our formulae. Local-field corrections can be incorporated as described in the previous section, if necessary, without altering the following discussion.

4.5.1 Resonance enhancement

We take as an example the general third-order susceptibility. By expressing the previous result (4.64) in the scalar form of (2.58), we have:

$$\chi^{(3)}(-\omega_\sigma;\omega_1,\omega_2,\omega_3) = \frac{N}{\varepsilon_0}\frac{e^4}{3!\hbar^3}\,\mathbf{S}\sum_{abcd}\rho_0(a)$$
$$\times\left[\frac{\mathbf{e}_\sigma^{*}\cdot\mathbf{r}_{ab}\;\mathbf{e}_1\cdot\mathbf{r}_{bc}\;\mathbf{e}_2\cdot\mathbf{r}_{cd}\;\mathbf{e}_3\cdot\mathbf{r}_{da}}{(\Omega_{ba}-\omega_\sigma)(\Omega_{ca}-\omega_2-\omega_3)(\Omega_{da}-\omega_3)}\right.$$
$$+\frac{\mathbf{e}_1\cdot\mathbf{r}_{ab}\;\mathbf{e}_\sigma^{*}\cdot\mathbf{r}_{bc}\;\mathbf{e}_2\cdot\mathbf{r}_{cd}\;\mathbf{e}_3\cdot\mathbf{r}_{da}}{(\Omega_{ba}+\omega_1)(\Omega_{ca}-\omega_2-\omega_3)(\Omega_{da}-\omega_3)}$$
$$+\frac{\mathbf{e}_1\cdot\mathbf{r}_{ab}\;\mathbf{e}_2\cdot\mathbf{r}_{bc}\;\mathbf{e}_\sigma^{*}\cdot\mathbf{r}_{cd}\;\mathbf{e}_3\cdot\mathbf{r}_{da}}{(\Omega_{ba}+\omega_1)(\Omega_{ca}+\omega_1+\omega_2)(\Omega_{da}-\omega_3)}$$
$$\left.+\frac{\mathbf{e}_1\cdot\mathbf{r}_{ab}\;\mathbf{e}_2\cdot\mathbf{r}_{bc}\;\mathbf{e}_3\cdot\mathbf{r}_{cd}\;\mathbf{e}_\sigma^{*}\cdot\mathbf{r}_{da}}{(\Omega_{ba}+\omega_1)(\Omega_{ca}+\omega_1+\omega_2)(\Omega_{da}+\omega_\sigma)}\right]. \qquad (4.108)$$

Written in this form, the intrinsic symmetrisation operation $\mathbf{S}$ requires that the expression following it is to be summed over the 3! possible permutations of the pairs $(\mathbf{e}_1,\omega_1)$, $(\mathbf{e}_2,\omega_2)$ and $(\mathbf{e}_3,\omega_3)$. This expression for the susceptibility has singularities in the frequency domain which occur when any one of the frequency denominators of the type $\Omega-\omega'$ approaches zero. Thus the susceptibility may be 'resonantly enhanced' when one or more of the input optical frequencies, or particular combinations of them, are in close coincidence with *certain* molecular transitions of the medium; the transitions in question are those for which the product of matrix elements in the numerator is nonzero. For an nth-order susceptibility, one-photon, two-photon and higher-order resonances can occur, up to n-photon. As described in Chapter 1, resonance enhancement is one of the most important features of the susceptibilities, and is the basis for applications of nonlinear optics in spectroscopy. For dilute media, such as atomic vapours, some nonlinear effects are prominent *only* for input fields occurring in particular ranges of frequencies which exploit resonance enhancement. It is clear that for the largest nonlinear susceptibilities we require large values for the matrix elements of the electric-dipole operator $e\mathbf{r}$ and small frequency denominators.

It is useful to take a specific example: once again we consider third-harmonic generation. The third-order susceptibility for this process is obtained from the general expression (4.108) by making the substitutions $\sigma\rightarrow 3\omega$ and $1,2,3\rightarrow\omega$. The result is (4.77) which we repeat here for convenience:

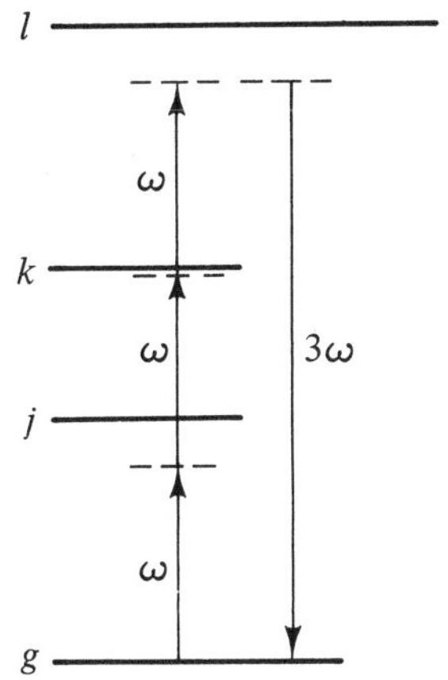

Fig. 4.2 Molecular energy-level diagram depicting the process of third-harmonic generation. Resonances occur when any of the frequencies ω, 2ω and 3ω are in close coincidence with molecular transitions.

$$\chi^{(3)}(-3\omega;\omega,\omega,\omega) = \frac{N}{\varepsilon_0}\frac{e^4}{3!\hbar^3}\sum_{abcd}\rho_0(a)$$
$$\times 6\left[\frac{\mathbf{e}_{3\omega}{}^{*}\cdot\mathbf{r}_{ab}\ \mathbf{e}_{\omega}\cdot\mathbf{r}_{bc}\ \mathbf{e}_{\omega}\cdot\mathbf{r}_{cd}\ \mathbf{e}_{\omega}\cdot\mathbf{r}_{da}}{(\Omega_{ba}-3\omega)(\Omega_{ca}-2\omega)(\Omega_{da}-\omega)} + \frac{\mathbf{e}_{\omega}\cdot\mathbf{r}_{ab}\ \mathbf{e}_{3\omega}{}^{*}\cdot\mathbf{r}_{bc}\ \mathbf{e}_{\omega}\cdot\mathbf{r}_{cd}\ \mathbf{e}_{\omega}\cdot\mathbf{r}_{da}}{(\Omega_{ba}+\omega)(\Omega_{ca}-2\omega)(\Omega_{da}-\omega)}\right.$$
$$\left. + \frac{\mathbf{e}_{\omega}\cdot\mathbf{r}_{ab}\ \mathbf{e}_{\omega}\cdot\mathbf{r}_{bc}\ \mathbf{e}_{3\omega}{}^{*}\cdot\mathbf{r}_{cd}\ \mathbf{e}_{\omega}\cdot\mathbf{r}_{da}}{(\Omega_{ba}+\omega)(\Omega_{ca}+2\omega)(\Omega_{da}-\omega)} + \frac{\mathbf{e}_{\omega}\cdot\mathbf{r}_{ab}\ \mathbf{e}_{\omega}\cdot\mathbf{r}_{bc}\ \mathbf{e}_{\omega}\cdot\mathbf{r}_{cd}\ \mathbf{e}_{3\omega}{}^{*}\cdot\mathbf{r}_{da}}{(\Omega_{ba}+\omega)(\Omega_{ca}+2\omega)(\Omega_{da}+3\omega)}\right]. \tag{4.109}$$

Figure 4.2 shows a typical energy-level scheme, in which the unperturbed molecular ground state is labelled g and three other energy levels of the molecule are depicted, having the labels j, k and l. For simplicity we assume here that the ground energy level E_g is nondegenerate and that the medium is at a temperature of 0 K, so that $\rho_0(a)$ in (4.109) is equal to unity when $a = g$ (the ground state) and is zero otherwise. Then the sum over a is simple; we have only to set $a = g$ everywhere. Resonances occur when the various optical frequencies and combinations of them are nearly coincident with appropriate molecular transitions. With the energy levels and frequencies as depicted in Fig. 4.2, the first term in (4.109) taken with the substitutions $a\rightarrow g$, $b\rightarrow l$, $c\rightarrow k$, $d\rightarrow j$ will be large, and indeed is usually dominant. (This also corresponds to the first Feynman diagram in Fig. 4.1.)

To get some feeling for the benefit from resonance enhancement, we now take some roughly typical values for the parameters in (4.109). Let us suppose, to begin with, that all of the optical frequencies are well removed from molecular resonances so that all the denominator terms of the type $\Omega \pm \omega'$ are of very roughly the same magnitude: $\sim 10^{15}$ s^{-1}. We also take all the matrix elements $|\mathbf{e}\cdot\mathbf{r}|$ to be ~0.1 nm. Then from (4.109) we find $\chi^{(3)}/N \simeq 6\times10^{-48}$ m^5 V^{-2} $(4\times10^{-34}$ esu). Suppose we now allow the third-harmonic frequency 3ω to come closer to resonance with a molecular transition Ω_{lg}, so that $|\Omega_{lg}-3\omega| = 10^{13}$ s^{-1} (this implies, for example, a detuning from resonance of ~1 nm for a fundamental wavelength of 1 μm). In this case, the resonantly-enhanced susceptibility reaches a value of $\sim 6\times10^{-45}$ m^5 V^{-2} $(4\times10^{-32}$ esu). These rough estimates are in good agreement with detailed calculations and measurements of the susceptibility $\chi^{(3)}$ for some atomic systems, an example of which is shown in Fig. 4.3. Third-harmonic generation in atomic vapours is a well-established method for generating ultraviolet and vacuum-ultraviolet radiation (Hanna *et al,* 1979; Reintjes, 1984).

The use of resonances to enhance a specific susceptibility often has the effect of similarly enhancing other competing or undesirable processes. For example, by allowing an input frequency to approach resonance with a single-photon molecular transition, that input field may suffer a corresponding increase in absorption; this occurs because the first-order susceptibility $\chi^{(1)}$ is also enhanced. Similarly, the field at the frequency 3ω which is generated by the induced nonlinear polarisation may also suffer significant loss if 3ω approaches resonance with a single-photon molecular transition. By contrast, the two-photon absorption cross-section is often small enough that a resonance $\Omega - 2\omega \to 0$ may be used to enhance $\chi^{(3)}$ without the process becoming significantly lossy. (Two-photon resonant processes are considered further in §4.5.2.) For some applications in nonlinear optics (*e.g.,* switching devices which use the nonlinear refractive index), a greater degree of absorption can be tolerated. In that case it may be possible to bring the optical frequency into closer resonance than considered in the example above. Ultimately, 'pumping' of the molecular population becomes important, and it is then more appropriate to treat the matter-field interaction in terms of an effective two-level system (as discussed in Chapter 6).

For very close resonance, the mathematical divergences in formulae such as (4.109) are clearly unphysical, and strictly they occur only because higher-order nonlinearities have been neglected. When excited very close to resonant transitions, the molecules undergo large perturbations – thus invalidating the small-perturbation approximations made previously – and the transition frequencies which occur in the denominator terms become field-dependent themselves. When these strong field-dependent perturbations, or level shifts, are taken into account, the resulting induced polarisation remains finite. We return to a

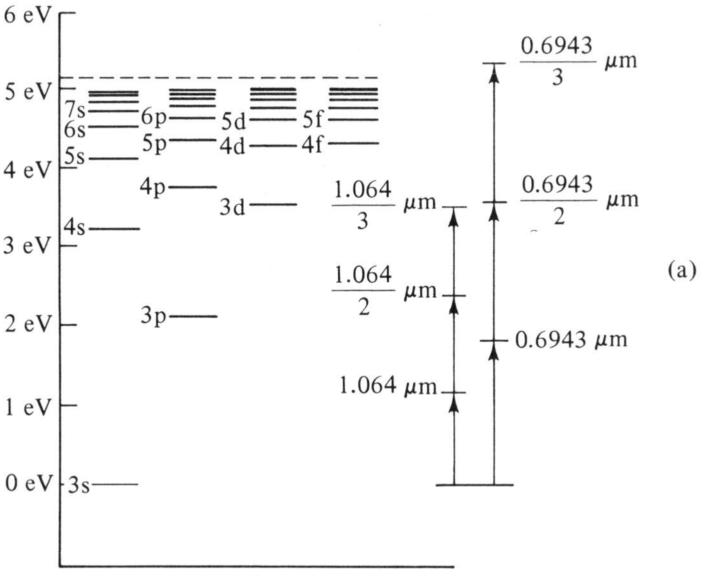

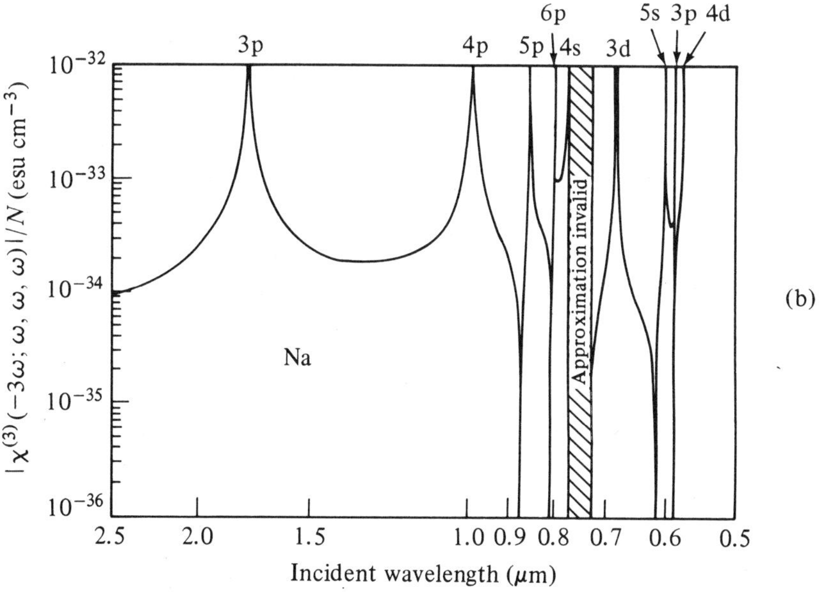

Fig. 4.3 (a) Energy levels of sodium. (b) Third-harmonic susceptibility per atom $\chi^{(3)}/N$ calculated from (4.109) using values of transition frequencies and matrix elements derived from spectroscopic data. Notice, for example, the resonance enhancement of $\chi^{(3)}$ when the third-harmonic frequency 3ω approaches resonance with the 3s–4p electronic transition; this occurs for an input wavelength of about 1 μm. To convert to SI units ($m^5\ V^{-2}$), the values shown must be multipled by 1.40×10^{-14} (see Appendix 2). Experimentally-determined values agree within 15%. (After Miles and Harris, 1973.) © IEEE

discussion of strongly-resonant nonlinearities in Chapter 6. In many cases, however, the resonant nonlinearities are dominated by various transition-line broadening processes, perhaps due to interactions between the molecules, which also ensure that the resonant susceptibilities do not diverge. A particular resonant process can then be represented by a single order of nonlinearity derived using the small-perturbation analysis, as previously, but with the addition of appropriate damping terms $\pm i\Gamma$ in the frequency denominators of the susceptibilities, as for example: $(\Omega_{ba} \pm i\Gamma_{ba} - \omega_\sigma)$. The phenomenological damping factor Γ_{ba} is thus identified as a dephasing parameter appropriate to weak collisions, and as such represents a spectral linewidth for the transition of frequency Ω_{ba}. The smallest values which the damping factors Γ can take are determined by spontaneous emission, and in this case Γ_{ba}^{-1} is a natural lifetime in the absence of collisions and other perturbations. In practice there may be several significant and complex damping mechanisms contributing simultaneously, such as various forms of collisional dephasing; the usual practice is to combine these in an empirically-determined value of Γ for each of the relevant transitions. The *signs* associated with the terms $i\Gamma$ are, however, predetermined in each case by the principle of causality, as we shall now explain.

Early in our treatment of the material response functions in Chapter 2 we recognised that in the real world there always exist relaxation processes which ensure that the response will eventually tend to zero at long times after the medium is subjected to an impulse excitation. It was convenient in Chapter 2 and in earlier sections of the present chapter to neglect these relaxation processes (because the calculations and resulting formulae were thus simplified), but in order to ensure the convergence of the various time integrals involved we had recourse to an artificial device: namely, the frequencies ω were taken to lie in the upper half of the complex plane (see § 2.2). In the real world, of course, frequencies are real quantities, and so the guiding rule for the correct introduction of damping terms is that the poles of the susceptibility should lie in the *lower* half-plane. This rule applies to the *general* form of the susceptibility of a given order n: $\boldsymbol{\chi}^{(n)}(-\omega_\sigma; \omega_1, \ldots, \omega_n)$ (as opposed to the susceptibility describing a particular nonlinear process, obtained by assigning specific values to the frequency arguments $\omega_1, \ldots, \omega_n$). When determining the poles of the general susceptibility, it must be remembered that the frequency arguments $\omega_1, \ldots, \omega_n$ each range over all positive and negative values.

The guiding principle behind this rule can be appreciated by considering the linear susceptibility: when the damping factor $\pm i\Gamma_{ba}$ is introduced into the expressions (4.56) and (4.57), we find integrals such as

$$\int_{-\infty}^{0} d\tau \, \exp[i(\Omega_{ba} \pm i\Gamma_{ba} - \omega)\tau] \rightarrow [i(\Omega_{ba} - i\Gamma_{ba} - \omega)]^{-1} ; \qquad (4.110)$$

the integral converges to the result given only if the ambiguous sign associated with $i\Gamma_{ba}$ is taken to be negative (assuming ω is real), otherwise the integral diverges and is unphysical. Thus, in place of (4.58), the expression for the linear susceptibility becomes

$$\chi^{(1)}_{\mu\alpha}(-\omega_\sigma;\omega) = \frac{N}{\varepsilon_0}\frac{e^2}{\hbar}\sum_{ab}\rho_0(a)\left[\frac{r^\mu_{ab}\,r^\alpha_{ba}}{\Omega_{ba} - i\Gamma_{ba} - \omega} + \frac{r^\alpha_{ab}\,r^\mu_{ba}}{\Omega_{ba} + i\Gamma_{ba} + \omega}\right]. \qquad (4.111)$$

In the vicinity of a resonance $\omega = \Omega_{ba}$, the susceptibility given by (4.111) exhibits a 'Lorentzian' line shape, as depicted in Fig. 1.2. The susceptibility is thus a complex quantity, and its real and imaginary parts can be identified with physically distinct phenomena; the real part of $\chi^{(1)}(-\omega_\sigma;\omega)$ is proportional to the refractive index, whilst the imaginary part is proportional to the coefficient of absorption (or emission, depending on its sign). This is considered further in §6.3.

The above rule for the correct introduction of damping terms applies to all orders of the nonlinear susceptibility in the same way. Thus for example, when damping terms are inserted into the expression (4.108) for the general third-order susceptibility, we obtain:

$$\chi^{(3)}(-\omega_\sigma;\omega_1,\omega_2,\omega_3) = \frac{N}{\varepsilon_0}\frac{e^4}{3!\hbar^3}\,\mathbf{S}\sum_{abcd}\rho_0(a)$$
$$\times\left[\frac{\mathbf{e}_\sigma^{*}\cdot\mathbf{r}_{ab}\;\mathbf{e}_1\cdot\mathbf{r}_{bc}\;\mathbf{e}_2\cdot\mathbf{r}_{cd}\;\mathbf{e}_3\cdot\mathbf{r}_{da}}{(\Omega_{ba} - i\Gamma_{ba} - \omega_\sigma)(\Omega_{ca} - i\Gamma_{ca} - \omega_2 - \omega_3)(\Omega_{da} - i\Gamma_{da} - \omega_3)}\right.$$
$$+\frac{\mathbf{e}_1\cdot\mathbf{r}_{ab}\;\mathbf{e}_\sigma^{*}\cdot\mathbf{r}_{bc}\;\mathbf{e}_2\cdot\mathbf{r}_{cd}\;\mathbf{e}_3\cdot\mathbf{r}_{da}}{(\Omega_{ba} + i\Gamma_{ba} + \omega_1)(\Omega_{ca} - i\Gamma_{ca} - \omega_2 - \omega_3)(\Omega_{da} - i\Gamma_{da} - \omega_3)}$$
$$+\frac{\mathbf{e}_1\cdot\mathbf{r}_{ab}\;\mathbf{e}_2\cdot\mathbf{r}_{bc}\;\mathbf{e}_\sigma^{*}\cdot\mathbf{r}_{cd}\;\mathbf{e}_3\cdot\mathbf{r}_{da}}{(\Omega_{ba} + i\Gamma_{ba} + \omega_1)(\Omega_{ca} + i\Gamma_{ca} + \omega_1 + \omega_2)(\Omega_{da} - i\Gamma_{da} - \omega_3)}$$
$$\left.+\frac{\mathbf{e}_1\cdot\mathbf{r}_{ab}\;\mathbf{e}_2\cdot\mathbf{r}_{bc}\;\mathbf{e}_3\cdot\mathbf{r}_{cd}\;\mathbf{e}_\sigma^{*}\cdot\mathbf{r}_{da}}{(\Omega_{ba} + i\Gamma_{ba} + \omega_1)(\Omega_{ca} + i\Gamma_{ca} + \omega_1 + \omega_2)(\Omega_{da} + i\Gamma_{da} + \omega_\sigma)}\right]. \qquad (4.112)$$

Clearly, if the frequency (or combination of frequencies) which appears in a particular denominator term is far removed from molecular resonances, then the corresponding damping term can be safely neglected.

By inspection of the above equations it is now apparent why, as stated in §4.3, the overall permutation-symmetry property applies only when none of the frequencies $\omega_\sigma, \omega_1, ..., \omega_n$ or their combinations are resonant with transition frequencies in the medium: only under these conditions can all of the damping factors be neglected and thus the frequencies can be freely permuted according to the overall symmetry rule. However, as we have commented earlier, the intrinsic permutation-symmetry property applies rigorously even when damping is included.

As explained in Chapter 6, a price paid for the resonant enhancement of optical nonlinearity is the slowing-down of the speed of response, which may be a serious limitation for some applications such as optical switching and signal processing.

4.5.2 Two-photon resonances

We now consider in more detail two-photon-resonant third-order nonlinearities. Such processes have widespread practical applications; this is because two-photon resonance can significantly enhance the desired nonlinearity, whilst at the same time competing absorption processes can be minimised by avoiding coincidences between the optical frequencies and *single*-photon resonances. As an example, we examine the case of stimulated Raman scattering. As described in §7.3, this process leads to amplification of a signal at a frequency ω_S in the presence of a strong pump field at ω_P, provided that the two-photon resonance condition $\omega_P - \omega_S \simeq \Omega_{fg}$ is satisfied (illustrated in Fig. 4.4(a)); this condition is known as a 'Raman resonance' and ω_S is called the Stokes frequency. Here Ω_{fg} denotes the transition frequency between a molecular ground state (g) and a higher-lying final state (f). Thus Ω_{fg} is positive and the Stokes frequency is lower than the pump frequency. The susceptibility tensor which describes stimulated Raman scattering is the third-order tensor $\boldsymbol{\chi}^{(3)}(-\omega_S; \omega_P, -\omega_P, \omega_S)$. It may be derived from (4.112) by selecting $\omega_\sigma, \omega_1, \omega_2, \omega_3$ from the required frequency arguments $-\omega_S, \omega_P, -\omega_P, \omega_S$, and a good approximation is made by keeping only the resonant terms. For simplicity we begin by assuming, as in the previous example, that the ground energy level E_g is nondegenerate and that the medium is at 0 K, so that $\rho_0(a)$ is equal to unity when $a = g$ and is zero otherwise. Therefore we can set $a = g$ everywhere in (4.112). Since $\omega_P - \omega_S \simeq \Omega_{fg}$, the Raman-resonant terms are those which contain the denominator $\Omega_{fg} \pm i\Gamma_{fg} - \omega_P + \omega_S$. From (4.112) we see that this denominator must arise from either $(\Omega_{ca} - i\Gamma_{ca} - \omega_2 - \omega_3)$ or $(\Omega_{ca} + i\Gamma_{ca} + \omega_\sigma - \omega_3)$,

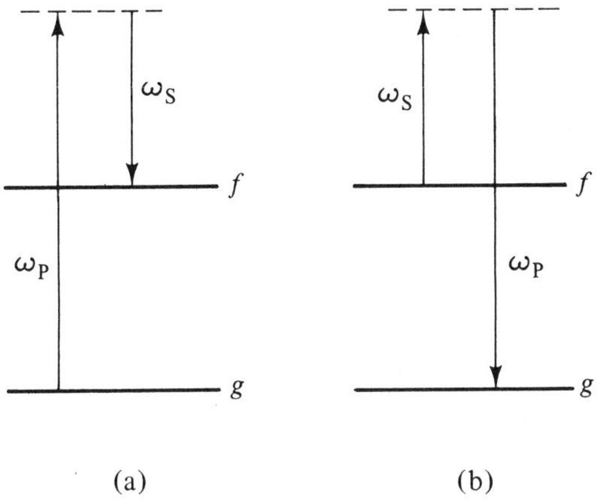

Fig. 4.4 Molecular energy-level diagram depicting (a) stimulated Raman scattering on the two-photon transition $\mathbb{E}_g \rightarrow \mathbb{E}_f$, and (b) the reverse process $\mathbb{E}_f \rightarrow \mathbb{E}_g$

where $a = g$ and $c = f$, and also where $\omega_\sigma, \omega_1, \omega_2, \omega_3$ are to be selected from the frequency arguments of $\chi^{(3)}(-\omega_S; \omega_P, -\omega_P, \omega_S)$. It is easy to show that the required resonant terms come from the last two terms in square brackets in (4.112), in which the pairs $(\mathbf{e}_1, \omega_1)$, $(\mathbf{e}_2, \omega_2)$ and $(\mathbf{e}_3, \omega_3)$ are permuted (by means of the symmetrising operation $\mathbf{S}$) according to each of the schemes:

$$\begin{aligned} &(\mathbf{e}_1, \omega_1), (\mathbf{e}_2, \omega_2), (\mathbf{e}_3, \omega_3) \rightarrow (\mathbf{e}_2, \omega_2), (\mathbf{e}_3, \omega_3), (\mathbf{e}_1, \omega_1) \\ &(\mathbf{e}_1, \omega_1), (\mathbf{e}_2, \omega_2), (\mathbf{e}_3, \omega_3) \rightarrow (\mathbf{e}_3, \omega_3), (\mathbf{e}_2, \omega_2), (\mathbf{e}_1, \omega_1). \end{aligned} \tag{4.113}$$

Thus we arrive at

$$\begin{aligned} &\chi^{(3)}(-\omega_S; \omega_P, -\omega_P, \omega_S) = \\ &\frac{N e^4 / 6\hbar^3 \varepsilon_0}{\Omega_{fg} + i\Gamma_{fg} - \omega_P + \omega_S} \sum_{bd} \left[\frac{\mathbf{e}_P{}^* \cdot \mathbf{r}_{gb}\, \mathbf{e}_S \cdot \mathbf{r}_{bf}\, \mathbf{e}_S{}^* \cdot \mathbf{r}_{fd}\, \mathbf{e}_P \cdot \mathbf{r}_{dg}}{(\Omega_{bg} - \omega_P)(\Omega_{dg} - \omega_P)} \right. \\ &+ \frac{\mathbf{e}_S \cdot \mathbf{r}_{gb}\, \mathbf{e}_P{}^* \cdot \mathbf{r}_{bf}\, \mathbf{e}_S{}^* \cdot \mathbf{r}_{fd}\, \mathbf{e}_P \cdot \mathbf{r}_{dg}}{(\Omega_{bg} + \omega_S)(\Omega_{dg} - \omega_P)} + \frac{\mathbf{e}_P{}^* \cdot \mathbf{r}_{gb}\, \mathbf{e}_S \cdot \mathbf{r}_{bf}\, \mathbf{e}_P \cdot \mathbf{r}_{fd}\, \mathbf{e}_S{}^* \cdot \mathbf{r}_{dg}}{(\Omega_{bg} - \omega_P)(\Omega_{dg} + \omega_S)} \\ &\left. + \frac{\mathbf{e}_S \cdot \mathbf{r}_{gb}\, \mathbf{e}_P{}^* \cdot \mathbf{r}_{bf}\, \mathbf{e}_P \cdot \mathbf{r}_{fd}\, \mathbf{e}_S{}^* \cdot \mathbf{r}_{dg}}{(\Omega_{bg} + \omega_S)(\Omega_{dg} + \omega_S)} \right]. \end{aligned} \tag{4.114}$$

In this equation we have assumed that neither ω_P nor ω_S is close to a transition frequency, and we have neglected the linewidths in the nonresonant denominators. We notice also that, by the process of selecting the required resonant terms, the sign of the damping factor $i\Gamma$ in the two-photon-resonant denominator has been fixed (positive in this particular example process, stimulated Raman scattering). In some cases, however, the sign of the damping factor can be determined by direct physical interpretation; thus, for stimulated Raman scattering involving ground-state molecules, the requirement that the gain at the Stokes frequency be positive immediately stipulates that the imaginary part of the susceptibility be negative (see §7.3).

We can simplify the appearance of (4.114) and other two-photon-resonant susceptibilities by introducing the first-order transition hyper-polarisability $\alpha_{fg}(\omega_1;\omega_2)$, which we define as

$$\alpha_{fg}(\omega_1;\omega_2) = \frac{e^2}{\hbar}\sum_i \left[\frac{\mathbf{e}_1\cdot\mathbf{r}_{fi}\ \mathbf{e}_2\cdot\mathbf{r}_{ig}}{(\Omega_{ig}-\omega_2)} + \frac{\mathbf{e}_2\cdot\mathbf{r}_{fi}\ \mathbf{e}_1\cdot\mathbf{r}_{ig}}{(\Omega_{ig}-\omega_1)}\right], \qquad (4.115)$$

and where, for simplicity, the damping terms have been omitted from the frequency denominators. By interchanging the pairs $(\mathbf{e}_1,\omega_1)$ and $(\mathbf{e}_2,\omega_2)$ in (4.115), we see that

$$\alpha_{fg}(\omega_1;\omega_2) = \alpha_{fg}(\omega_2;\omega_1). \qquad (4.116)$$

(In §6.5, the transition hyperpolarisabilities will be related to the classical theory of electronic polarisability.) It is not difficult to show that the Raman susceptibility (4.114) can be expressed in the simpler form:

$$\chi^{(3)}(-\omega_S;\omega_P,-\omega_P,\omega_S) = \frac{N/6\hbar\varepsilon_0}{\Omega_{fg}+i\Gamma_{fg}-\omega_P+\omega_S}\,\alpha_{fg}(-\omega_S;\omega_P)\,\alpha_{fg}(-\omega_S;\omega_P)^*. \qquad (4.117)$$

To verify (4.117) we have only to write out the last two terms on the right-hand side as the product of two sums, each given by (4.115), and compare the result with (4.114) with the aid of the property $(\mathbf{e}\cdot\mathbf{r}_{ij})^* = \mathbf{e}^*\cdot\mathbf{r}_{ji}$ from (3.14).

Having thus simplified the appearance of the Raman susceptibility, we can now backtrack a little way by removing the previous restrictive assumptions that the initial energy level E_g is non-degenerate and that $\rho_0(g)=1$. We are concerned with the two-photon resonant transition $E_g \leftrightarrow E_f$; therefore the general sum over $\rho_0(a)$ in (4.112) is restricted to $\rho_0(g)$ and $\rho_0(f)$. Previously, when we assumed that $\rho_0(g)=1$ (and therefore $\rho_0(f)=0$), it was sufficient to put $a=g$ everywhere in (4.112). If, however, we now consider the situation when $\rho_0(f)>0$, we should take into account the probability that the fraction of molecules which are initially in the states at energy E_f may make the reverse transition $E_f \rightarrow E_g$, as illustrated in Fig. 4.4(b). We find in this case that, in (4.112), there are terms containing the Raman-resonant denominator $\Omega_{gf}-i\Gamma_{gf}+\omega_P-\omega_S$. When these additional terms are evaluated as described above and added to (4.114), we obtain the final result:

$$\chi^{(3)}(-\omega_S;\omega_P,-\omega_P,\omega_S) = \frac{N/6\hbar\varepsilon_0}{\Omega_{fg}+i\Gamma_{fg}-\omega_P+\omega_S}\sum_{\{g,f\}}\left[\rho_0(g)-\rho_0(f)\right]<|\alpha_{fg}(-\omega_S;\omega_P)|^2>. \tag{4.118}$$

The angle brackets denote an orientation average, and the symbol $\Sigma_{\{g,f\}}$ denotes sums taken over the degenerate states of energy E_g and E_f, and $\rho_0(g)$ and $\rho_0(f)$ are the fractional populations of the degenerate states at these energies. We are thus able to take into account all transitions between the lower and upper levels which constitute the two-photon resonance. From symmetry considerations (see §5.3) we can assume that degenerate states of a given energy are equally populated. The second term in the square brackets in (4.118) corresponds to the reverse Raman process, shown in Fig. 4.4(b), in which a Stokes photon is absorbed and a pump photon is emitted. The Raman susceptibility may therefore be reduced in magnitude in the case when $\rho_0(g)-\rho_0(f)$ is significantly smaller than $\rho_0(g)$. If, however, the Stokes energy shift, corresponding to the transition energy Ω_{fg}, is much larger than kT then the equilibrium thermal population in the upper level E_f will be negligible. In that case a significant occupancy of the upper level would occur only if there were substantial resonant pumping of the molecular population, in which case a large-perturbation treatment of resonant nonlinearities (as described in Chapter 6) may be more appropriate. However, if $\Omega_{fg} \lesssim kT$ (as may be the case for Raman scattering on a rotational molecular transition, for example), then the term $\rho_0(f)$ in (4.118) may be significant in determining the magnitude of the Raman susceptibility.

The general expression (4.112) for the third-order susceptibility can be shown to factorise into a product of two first-order transition hyperpolarisabilities for *all* two-photon-resonant third-order processes:

$$\chi^{(3)}(-\omega_\sigma;\omega_1,\omega_2,\omega_3) = \frac{fN/\hbar\varepsilon_0}{\Delta \pm i\Gamma}\sum_{\{g,f\}}\left[\rho_0(g)-\rho_0(f)\right]<\alpha_{fg}(\text{i})^*\,\alpha_{fg}(\text{ii})>, \tag{4.119}$$

where $\Delta \pm i\Gamma$ is the two-photon-resonant denominator, the first-order transition hyperpolarisability is as defined in (4.115), (i) and (ii) refer to pairs of frequencies chosen from $-\omega_\sigma, \omega_1, \omega_2, \omega_3$, and f is a simple numerical factor. Table 4.1 (which is adapted from Hanna *et al* (1979), with corrections) enumerates the various parameters of (4.119) for the most important cases. Although the order of the frequency arguments in $\chi^{(3)}$ is usually not significant, for resonant susceptibilities it is convenient to order the arguments so as to match those of the transition

Table 4.1: Two-photon-resonant third-order nonlinear susceptibilities

Process	Reference to Fig. 4.5	Susceptibility	Resonance $\Delta \pm i\Gamma \simeq 0$	f	Pair(i)	Pair(ii)
Raman-resonant difference mixing	(a)	$\chi^{(3)}(-\omega_4;-\omega_3,-\omega_2,\omega_1)$ $\chi^{(3)}(-\omega_3;-\omega_4,-\omega_2,\omega_1)$	$\Omega-i\Gamma+\omega_2-\omega_1$	$\frac{1}{6}$	$\omega_4;\omega_3$	$-\omega_2;\omega_1$
Coherent anti-Stokes Raman scattering	(b)	$\chi^{(3)}(-\omega_4;\omega_3,-\omega_2,\omega_1)$ $\chi^{(3)}(-\omega_{AS};\omega_P,-\omega_S,\omega_P)$	$\Omega-i\Gamma+\omega_2-\omega_1$ $\Omega-i\Gamma+\omega_S-\omega_P$	$\frac{1}{6}$ $\frac{1}{3}$	$\omega_4;-\omega_3$ $\omega_{AS};-\omega_P$	$-\omega_2;\omega_1$ $-\omega_S;\omega_P$
Biharmonic pumping	(b)	$\chi^{(3)}(-\omega_3;\omega_4,\omega_2,-\omega_1)$	$\Omega+i\Gamma+\omega_2-\omega_1$	$\frac{1}{6}$	$\omega_3;-\omega_4$	$-\omega_2;\omega_1$
Stimulated Raman scattering	(b)	$\chi^{(3)}(-\omega_S;\omega_P,-\omega_P,\omega_S)$	$\Omega+i\Gamma+\omega_S-\omega_P$	$\frac{1}{6}$	$-\omega_S;\omega_P$	$-\omega_S;\omega_P$
TPA-resonant difference mixing	(c)	$\chi^{(3)}(-\omega_4;-\omega_3,\omega_2,\omega_1)$ $\chi^{(3)}(-\omega_3;-\omega_4,\omega_2,\omega_1)$	$\Omega-i\Gamma-\omega_2-\omega_1$	$\frac{1}{6}$	$\omega_4;\omega_3$	$\omega_2;\omega_1$
Two-photon absorption	(c)	$\chi^{(3)}(-\omega_1;-\omega_2,\omega_2,\omega_1)$ $\chi^{(3)}(-\omega;-\omega,\omega,\omega)$	$\Omega-i\Gamma-\omega_2-\omega_1$ $\Omega-i\Gamma-2\omega$	$\frac{1}{6}$	$\omega_1;\omega_2$ $\omega;\omega$	$\omega_2;\omega_1$ $\omega;\omega$
Up-conversion	(d)	$\chi^{(3)}(-\omega_4;\omega_3,\omega_2,\omega_1)$ $\chi^{(3)}(-\omega_4;\omega_3,\omega_1,\omega_1)$	$\Omega-i\Gamma-\omega_2-\omega_1$ $\Omega-i\Gamma-2\omega_1$	$\frac{1}{6}$	$\omega_4;-\omega_3$	$\omega_2;\omega_1$ $\omega_1;\omega_1$
Third-harmonic generation	(d)	$\chi^{(3)}(-3\omega;\omega,\omega,\omega)$	$\Omega-i\Gamma-2\omega$	$\frac{1}{2}$	$3\omega;-\omega$	$\omega;\omega$
TPA-resonant difference mixing	(d)	$\chi^{(3)}(-\omega_3;\omega_4,-\omega_2,-\omega_1)$	$\Omega+i\Gamma-\omega_2-\omega_1$	$\frac{1}{6}$	$\omega_3;-\omega_4$	$\omega_2;\omega_1$

The resonant denominator $\Delta \pm i\Gamma$, the numerical factor f and the pairs of frequencies (i) and (ii) are to be inserted in (4.119).

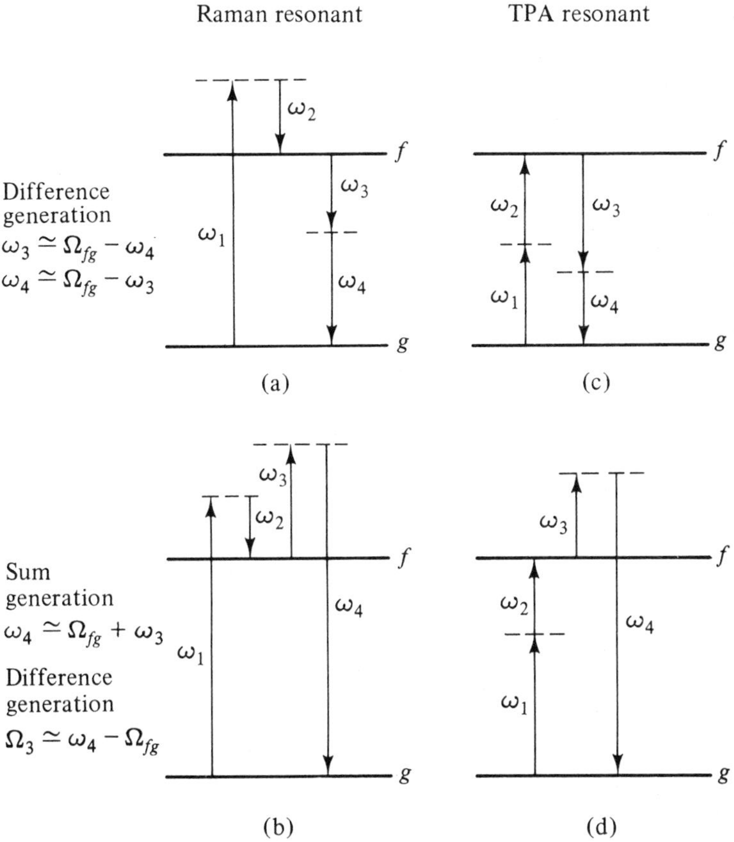

Fig. 4.5 Classification of two-photon-resonant third-order nonlinear susceptibilities according to the use of Raman and two-photon-absorption (TPA) types of resonance. The input fields at ω_1, ω_2, and ω_3 (or ω_4) generate a field at ω_4 (or ω_3).

(a) Raman-resonant difference-frequency generation of ω_3 or ω_4.

(b) Raman-resonant sum-frequency generation of ω_4. With $\omega_1 = \omega_3$, the process is known as coherent anti-Stokes Raman scattering (CARS) and ω_4 is the anti-Stokes frequency.
Raman-resonant difference-frequency generation of ω_3; known as coherent Raman mixing or biharmonic pumping.
Stimulated Raman scattering in the degenerate case $\omega_1 = \omega_4 \equiv \omega_P$ and $\omega_2 = \omega_3 \equiv \omega_S$.

(c) TPA-resonant difference-frequency generation of ω_3 or ω_4.
Two-photon absorption itself is the degenerate case $\omega_1 = \omega_4$ and $\omega_2 = \omega_3$.

(d) TPA-resonant sum-frequency generation of ω_4; known as up-conversion.
The degenerate case $\omega_1 = \omega_2 = \omega_3 \equiv \omega$ is third-harmonic generation.
TPA-resonant difference-frequency generation of ω_3.

hyperpolarisabilities appearing in (4.119) (*i.e.*, the arguments of $\chi^{(3)}$ are conventionally written in the same order as the pairs (i) and (ii) in the corresponding entry of Table 4.1).

Any two-photon-resonant third-order susceptibility may be classified into one of the four general types illustrated in Fig. 4.5, as indicated for each of the examples listed in Table 4.1. To gain some physical appreciation of these different types, it is useful to recall the classical picture of spontaneous vibrational Raman scattering (see §6.5.1). This is based on the idea that the nonlinear medium possesses a natural polarisability which oscillates at a vibrational frequency Ω_{fg}; an incident wave at the frequency ω_P is thus modulated such that new sideband frequencies are generated at $\omega_P - \Omega_{fg}$ and $\omega_P + \Omega_{fg}$, known as the Stokes and anti-Stokes frequencies respectively. With this classical picture in mind, we now turn to Fig. 4.5 and examine in particular the Raman resonant processes, labelled (a) and (b). Consider the generation in the medium of a polarisation at the frequency ω_4, induced by the incident fields at ω_1, ω_2 and ω_3. The effect of the two fields ω_1 and ω_2, which are in two-photon Raman resonance with the medium, is to set up an excitation in the medium which oscillates at the resonance frequency Ω_{fg}. The state of the medium so excited is a superposition of the unperturbed states labelled g and f, and this superposition of states constitutes a 'macroscopic *polarisability*'. (A clearer picture of the meaning of the 'excitation' induced in the medium emerges in Chapter 6 where resonant nonlinear processes are considered in greater detail.) The third wave ω_3 is now analogous to the pump wave in the classical Raman scattering picture; it interacts with the macroscopic polarisability to generate a *polarisation* at the sideband frequencies $\Omega_{fg} - \omega_3$ and $\Omega_{fg} + \omega_3$, which correspond to the Raman-resonant difference- and sum-frequency generation processes depicted in Figs. 4.5(a) and (b) respectively. (We use italics to emphasise the distinction between the macroscopic *polarisability* of the medium which is a two-photon-resonant excitation induced by the fields at ω_1 and ω_2, and the nonlinear *polarisation* which may then result when the third field at ω_3 is also present.)

This interpretation of the processes (a) and (b) is not unique, however. We could equally well have considered the case where the rôles of ω_3 and ω_4 were reversed: a polarisation at ω_3 would be generated if an input field at ω_4 were to interact with the macroscopic polarisability. For the case depicted in Fig. 4.5(a), the generated polarisation would be at the frequency $\Omega_{fg} - \omega_4$, whereas for the case of Fig. 4.5(b), the generated polarisation would be at the difference frequency $\omega_4 - \Omega_{fg}$. In fact, it is customary to distinguish between these various processes by giving them names, such as 'CARS' or 'coherent Raman mixing', as indicated in the caption to Fig. 4.5.

Figures 4.5(c) and (d) show the other type of two-photon resonance, sometimes called a 'two-photon-absorption' (TPA) resonance, in which the sum frequency $\omega_1+\omega_2$ is in resonance with Ω_{fg}. A macroscopic polarisability is thus created, analogous to the Raman-resonant case, and the third field at ω_3 (or ω_4) then induces a nonlinear polarisation at the various sum and difference frequencies, as indicated.

It is conventional for the arrows in energy-level diagrams, such as Fig. 4.5, to be drawn so as to form a closed loop, in the manner of a vector diagram, thus indicating the validity of the relation $\omega_\sigma-\omega_1-\cdots-\omega_n=0$. However the directions of the arrows themselves assume no physical significance; a common error is to try to interpret the direction of an arrow in an energy-level diagram as meaning that the field at a particular frequency experiences loss or gain (this is a confusion with the diagrammatic techniques of §4.3.3). Indeed, in the description above we have neglected the possibility that a field generated by the induced polarisation (at ω_4, say) may in some circumstances grow to such an extent that it is capable of driving the reverse process in which the complementary field (ω_3 in this case) begins to experience gain; depending on the circumstances, a stable balance or an oscillatory exchange of energy between the various interacting fields may occur.

It can be shown that any resonant nonlinear susceptibility will factorise into products of transition hyperpolarisabilities of the appropriate order. For example, three-photon-resonant fifth-order susceptibilities will factorise into a product of two second-order transition hyperpolarisabilities; details are given by Hanna *et al* (1979).

4.5.3 Secular resonances

Resonance enhancement of the susceptibilities can occur when a frequency denominator approaches zero (or its limiting value as determined by damping); this is a resonant singularity. Another type of singularity – termed a secular singularity – appears to arise when a denominator approaches zero because *both* the transition frequency *and* the combination of field frequencies that occurs in it are zero. An example of such a secular term would be the frequency denominator $(\Omega_{aa}+\omega-\omega)^{-1}$ which occurs in the fully-expanded expression for the optical-Kerr susceptibility $\boldsymbol{\chi}^{(3)}(-\omega;\omega,-\omega,\omega)$, as derived from (4.73). These secular terms are potentially confusing, since it may be thought that the susceptibility could be enhanced by exploiting these apparent resonances. In fact, secular singularities are nonexistent. The expression

(4.76) for the nth-order susceptibility (and similar expressions for susceptibilities of specific order $n = 2,3,...$) is useful for its compactness and simplicity, yet it contains a number of self-cancelling terms. When this redundancy is eliminated, the secular divergences vanish. As an illustration of this, Orr and Ward (1971) obtained (by a derivation different from the one which we have used) an expression for the third-order susceptibility in the form:

$$\chi^{(3)}(-\omega_\sigma;\omega_1,\omega_2,\omega_3) = \frac{N}{\varepsilon_0}\frac{e^4}{3!\hbar^3}\mathbf{S}_T\sum_a \rho_0(a)$$

$$\times\left[\sum_{\not{a}bcd}\frac{\mathbf{e}_\sigma^*\cdot\mathbf{r}_{ab}\;\mathbf{e}_1\cdot(\mathbf{r}_{bc}-\mathbf{r}_{aa})\;\mathbf{e}_2\cdot(\mathbf{r}_{cd}-\mathbf{r}_{aa})\;\mathbf{e}_3\cdot\mathbf{r}_{da}}{(\Omega_{ba}-\omega_\sigma)(\Omega_{ca}-\omega_2-\omega_3)(\Omega_{da}-\omega_3)}\right.$$

$$\left.-\sum_{\not{a}cd}\frac{\mathbf{e}_\sigma^*\cdot\mathbf{r}_{ac}\;\mathbf{e}_1\cdot\mathbf{r}_{ca}\;\mathbf{e}_2\cdot\mathbf{r}_{ad}\;\mathbf{e}_3\cdot\mathbf{r}_{da}}{(\Omega_{ca}-\omega_\sigma)(\Omega_{da}-\omega_3)(\Omega_{da}+\omega_2)}\right], \tag{4.120}$$

where the significant initial states (*i.e.*, those for which $\rho_0(a) > 0$) are assumed to be nondegenerate, the symbol $\not{a}$ indicates that a is to be omitted in the summations over intermediate states, and damping has been neglected. Yuratich (Hanna *et al*, 1979), showed that the two forms (4.73) and (4.120) for $\chi^{(3)}$ are in fact identical, being merely algebraic rearrangements of one another. Thus, whereas (4.73) appears to contain secular divergences, the same susceptibility cast in the form (4.120) is manifestly nonsingular. Despite these potential pitfalls, the use of (4.73) and (4.76) is usually preferred because of the simplicity of form in all orders.

4.5.4 The Born-Oppenheimer approximation: electronic, vibrational and hybrid resonances

The formulae for the molecular susceptibilities given in earlier sections of this chapter involve matrix elements of the electric-dipole operator $e\mathbf{r}$; for example, the matrix element $e\mathbf{r}_{ab}$ is defined by (3.7) in terms of the wave functions u_a and u_b as $e\mathbf{r}_{ab} = \int d\varsigma\, u_a^*\, e\mathbf{r}\, u_b$. The formulae for the susceptibilities are correct provided the exact wave functions are used, which fully specify the quantum states of the electrons *and* nuclei in the medium. However, the mechanisms involved in the interaction of light with electrons and nuclei are different and, in the Born-Oppenheimer approximation, the electronic and nuclear contributions to the molecular polarisability can be separated. Flytzanis (1975) showed that by making this separation, the resonant terms in the susceptibilities fall into three categories: those that are purely electronic, purely vibrational and 'hybrid' resonances. The purely electronic resonances involve the interaction of high-frequency components of the field (visible and ultraviolet waves and their high-frequency combinations – harmonics and sum frequencies) with the electronic clouds. The purely vibrational resonances involve interactions between low-frequency components of the field (infrared waves and difference-frequency components of high-frequency fields) with the nuclear charges (or ionic lattice in the case of a solid). As described more fully below, the hybrid resonances arise when some low-frequency components of the field interact with the nuclear charges so as to modulate the electronic interaction experienced by some higher-frequency components. Whereas electronic and vibrational resonances are well known in linear spectroscopy, Flytzanis pointed out that hybrid resonances occur only in nonlinear optics.

We now show how these different resonances arise. As described in many quantum-mechanics texts (*e.g.*, Schiff (1968) p.448), the Born-Oppenheimer approximation allows the molecular wave functions, such as u_a, to be factorised into their separate electronic and nuclear parts; thus,

$$u_a = \phi_a(\boldsymbol{\mu}_j, \mathbf{q}_l)\, V_a(\mathbf{q}_l)\,, \tag{4.121}$$

where ϕ_a is the electronic part, which depends on both the position coordinates of the electrons $\boldsymbol{\mu}_j$ and the nuclear coordinates. The latter are expressed conveniently in terms of the coordinates of the normal modes of vibration $\mathbf{q}_l$ (Flytzanis, 1975). The nuclear part of the wave function, V_a, depends on $\mathbf{q}_l$, but is independent of the electronic coordinates. The energy of the electrons, which we denote by $\hbar\Omega_a^{\mathrm{e}}$, can be expanded in the form:

$$\hbar\Omega_a^{\mathrm{e}}(q_l) = \hbar\Omega_a^{\mathrm{e}}(0) + \sum_l \tfrac{1}{2}\Omega_l^2 q_l^2 + \cdots\,, \tag{4.122}$$

where $\hbar\Omega_a^{\mathrm{e}}(0)$ is the energy of the electrons when the nuclei are fixed at their equilibrium positions, and Ω_l are the normal mode frequencies. When only the first two terms in (4.122) are retained, the motion of the nuclei is equivalent to a set of harmonic oscillators with the energy

$$\hbar\Omega_a^{\mathrm{v}} = \sum_l (n_l + \tfrac{1}{2})\hbar\Omega_l\,, \tag{4.123}$$

where n_l is a quantum number associated with the lth normal mode of vibration. The denominators in the various formulae for the susceptibilities contain transition energies, such as $\hbar\Omega_{ba}$, which may now be written as

$$\hbar\Omega_{ba} = \hbar\,[\Omega_b^{\mathrm{e}}(0) - \Omega_a^{\mathrm{e}}(0) + \Omega_b^{\mathrm{v}} - \Omega_a^{\mathrm{v}}]. \tag{4.124}$$

In addition, the electric-dipole operator may be separated into its electronic and nuclear parts (*cf.* (4.1)):

$$e\mathbf{r} = e\mathbf{r}_e + e\mathbf{r}_v = -e\sum_j \boldsymbol{\mu}_j + e\sum_k Z_k \boldsymbol{\mu}_k , \tag{4.125}$$

where $\boldsymbol{\mu}_j$ denotes the position vector of the jth electron with charge $-e$, and $\boldsymbol{\mu}_k$ denotes the position of the kth nucleus with charge $Z_k e$, and the summation is over all electrons and nuclei in the microscopic unit. It is convenient also to expand the dipole matrix element in the electronic ground state in terms of the normal modes:

$$e\mathbf{r}_{aa} = e\mathbf{r}_{aa}^{(0)} + \sum_l (e\mathbf{r}_{aa}^{(1)})_l q_l + \cdots . \tag{4.126}$$

The various expressions for the hyperpolarisabilities and microscopic susceptibilities contain a summation over molecular states; in performing this summation, there are two distinct situations that occur. The first is when a pair of electronic states differ, so that $\hbar\Omega_b^e(0) \neq \hbar\Omega_a^e(0)$. Usually, in that case, the energy separation between the electronic levels is much greater than the separation between vibrational levels, *i.e.*, $|\Omega_b^e(0) - \Omega_a^e(0)| >> |\Omega_b^v - \Omega_a^v|$, and $\Omega_{ba} \simeq \Omega_b^e(0) - \Omega_a^e(0)$. This gives rise to terms in the formulae for the hyperpolarisabilities which may be designated as being purely electronic. These electronic terms are found by replacing the electric-dipole operator $e\mathbf{r}$ in the formulae with the electronic operator $e\mathbf{r}_e$, and in addition, to a good approximation, the matrix elements of $e\mathbf{r}_e$ may be defined with respect to the electronic wave functions only, *i.e.*, $e\mathbf{r}_{ab} \rightarrow \int d\varsigma\, \phi_a{}^* e\mathbf{r}_e \phi_b$.

The second situation occurs when the electronic states involved in a transition Ω_{ba} are identical, so that $\Omega_{ba} = \Omega_b^v - \Omega_a^v$, thus giving rise to terms which are purely vibrational. Consider, for example, the polarisability $\boldsymbol{\alpha}(-\omega;\omega)$ obtained from (4.58) together with (4.107); as shown by Flytzanis (1975), this contains purely-vibrational terms of the form:

$$\boldsymbol{\alpha}_{aa}(-\omega;\omega) = \cdots + \sum_l \left[\frac{\mathbf{e}^* \cdot (e\mathbf{r}_{aa}^{(1)})_l \; \mathbf{e} \cdot (e\mathbf{r}_{aa}^{(1)})_l}{\Omega_l^2 - \omega^2} \right] + \cdots . \tag{4.127}$$

Thus, in the Born-Oppenheimer approximation, the linear polarisability is a simple sum of purely electronic and purely vibrational terms. However, in the case of the hyperpolarisabilities, both electronic and vibrational contributions may appear in the same term, giving rise to 'hybrid resonances'. To illustrate this, we consider the hyperpolarisability $\boldsymbol{\gamma}(-\omega_\sigma;\omega_1, -\omega_2, \omega_3)$ obtained from (4.112) together with (4.107). The formula (4.112) contains a sum over all molecular states, which are labelled *a,b,c,d*. For states u_a, u_b, u_c and u_d all having different electronic wave functions, one obtains the purely electronic terms discussed above. Similarly, when all the states share the same electronic wave function, one obtains the purely vibrational terms. The hybrid terms in (4.112) arise from the contributions such as

$$\frac{\mathbf{e}_\sigma^{*}\cdot\mathbf{r}_{ab}\ \mathbf{e}_1\cdot\mathbf{r}_{bc}\ \mathbf{e}_2\cdot\mathbf{r}_{cd}\ \mathbf{e}_3\cdot\mathbf{r}_{da}}{(\Omega_{ba}-i\Gamma_{ba}-\omega_\sigma)(\Omega_{ca}-i\Gamma_{ca}+\omega_2-\omega_3)(\Omega_{da}-i\Gamma_{da}-\omega_3)},$$

in which, for example, the states u_a and u_c share the same electronic wave function, but the electronic parts of u_b and u_d differ from that of u_a and u_c. The frequency denominator in this case is given approximately by

$$(\Omega_b^{\mathrm{e}}(0)-\Omega_a^{\mathrm{e}}(0)-i\Gamma_{ba}-\omega_\sigma)\,(\Omega_c^{\mathrm{v}}-\Omega_a^{\mathrm{v}}-i\Gamma_l+\omega_2-\omega_3)\,(\Omega_d^{\mathrm{e}}(0)-\Omega_a^{\mathrm{e}}(0)-i\Gamma_{da}-\omega_3)$$

where, for simplicity, we have assumed that there is only one vibrational mode Ω_l. Thus resonance enhancement occurs when $\omega_\sigma\to\Omega_{ba}$, or $\omega_3\to\Omega_{da}$, or when $\omega_3-\omega_2\to\Omega_c^{\mathrm{v}}-\Omega_a^{\mathrm{v}}$. The latter is an example of a hybrid resonance. Its physical meaning is that the optical difference frequency $\omega_3-\omega_2$ drives a nuclear vibrational mode, which in turn modulates the electronic interactions involving the frequencies ω_3 and ω_σ. Thus, in the case of the hyperpolarisabilities, the interaction of a *combination* of optical frequencies with the mode of molecular vibration can give rise to new resonant terms.

We should briefly note two final points: first, that the hyperpolarisabilities are temperature-dependent through the mode populations $\rho_0(a)$ which appear in the formulae such as (4.75), and second, that the vibration modes may reveal forbidden electronic transitions through the breaking of symmetry selection rules (§5.3) by the motion of the nuclei.

4.6 Mobile charged particles in solids

The formulae for the susceptibility tensors developed in the previous sections of this chapter are expressed in terms of electric-dipole operators and their matrix elements. We chose this formalism because it is conceptually simple. There are, however, alternative formulae expressed in terms of electric-current operators and their matrix elements. These are particularly convenient when the charged particles are free to move through the entire volume V of the system instead of being tied to individual molecules (an important example being the motion of electrons and holes in semiconductors). In that case the energy eigenfunctions extend throughout V. This means that the matrix elements of the electric-dipole operators depend on the size and shape of V and, in particular, are ill defined when $V\to\infty$. On the other hand, the matrix elements of the current operators are always well defined.

We use the alternative formulae in our discussion of optical nonlinearity in semiconductors (Chapters 8 and 9). In the next sub-section we write down, without proof, the analogue of the fundamental equation (4.28) in terms of current operators and use it to derive the analogues of (4.75) and (4.76). Two methods of deriving the alternative fundamental equation are indicated in §4.6.2 and one of them is worked out in detail. We also refer to the nth-order nonlinear conductivity tensor $\boldsymbol{\sigma}^{(n)}$, and relate this to $\boldsymbol{\chi}^{(n)}$.

4.6.1 Susceptibility formulae in terms of current operators

In the absence of applied fields the current operator in V is

$$\boldsymbol{\Pi} = -\frac{e}{m}\sum_i \mathbf{p}_j + e\sum_k \frac{Z_k}{M_k}\mathbf{p}_k \tag{4.128}$$

where, as in the expression (4.1) for the dipole moment $\mathbf{Q}$, j labels electrons (with charge $-e$, mass m, position vector $\boldsymbol{\mu}_j$ and momentum $\mathbf{p}_j$) and k labels ion cores (with charge $Z_k e$, mass M_k, position vector $\boldsymbol{\mu}_k$ and momentum $\mathbf{p}_k$). We see that $\boldsymbol{\Pi}$ is simply charge times velocity summed over all the charged particles in V. For want of a better name we call it the current operator although, strictly speaking, it is the average current density ($\boldsymbol{\Pi}/V$) times V. The alternative formulae for the susceptibilities are expressed in terms of $\boldsymbol{\Pi}$ and the corresponding operator in the interaction picture:

$$\boldsymbol{\Pi}(t) = U_0(-t)\,\boldsymbol{\Pi}\,U_0(t)\,, \tag{4.129}$$

where $U_0(-t)=\exp(-iH_0t/\hbar)$ with H_0 denoting the field-free, many-particle Hamiltonian.

We show in §4.6.2 that (4.28) remains valid for $n>2$ when Q_μ, $Q_{\alpha_1}(\tau_1)$, $Q_{\alpha_2}(\tau_2), \ldots, Q_{\alpha_n}(\tau_n)$ are replaced respectively by $\Pi_\mu/(-i\omega_\sigma)$, $\Pi_{\alpha_1}(\tau_1)/i\omega_1$, $\Pi_{\alpha_2}(\tau_2)/i\omega_2, \ldots, \Pi_{\alpha_n}(\tau_n)/i\omega_n$ where, as in all our previous equations, $\omega_\sigma = \omega_1 + \omega_2 + \cdots + \omega_n$. When $n=1$ we have to make the same substitutions, but also add on an additional isotropic contribution $-\lambda I_{\mu\alpha_1}/\varepsilon_0 V\omega_1^2$, where

$$\lambda = M^{(e)}\frac{e^2}{m} + \sum_k \frac{(Z_k e)^2}{M_k} \tag{4.130}$$

with $M^{(e)}$ denoting the number of electrons, and $\mathbf{I}$ denotes the second-rank unit tensor whose elements are $I_{\mu\alpha_1}=1$ when μ and α_1 label the same axis, otherwise $I_{\mu\alpha_1}=0$. Thus we arrive at the following alternative general formula for the nth-order susceptibility tensor:

$$\chi^{(n)}_{\mu\alpha_1\cdots\alpha_n}(-\omega_\sigma;\omega_1,\ldots,\omega_n) = -\delta_{n1}\frac{\lambda}{\varepsilon_0 V\omega_1^2}I_{\mu\alpha_1}$$
$$+\frac{1}{n!}\mathbf{S}\,(\varepsilon_0 V)^{-1}\hbar^{-n}(-i\omega_\sigma\omega_1\omega_2\cdots\omega_n)^{-1}$$
$$\times\int_{-\infty}^{0}\mathrm{d}\tau_1\int_{-\infty}^{\tau_1}\mathrm{d}\tau_2\cdots\int_{-\infty}^{\tau_{n-1}}\mathrm{d}\tau_n\,\mathrm{Tr}\{\rho_0[\cdots[[\Pi_\mu,\Pi_{\alpha_1}(\tau_1)],\Pi_{\alpha_2}(\tau_2)],\cdots$$
$$\cdots\Pi_{\alpha_n}(\tau_n)]\}\exp\left[-i\sum_{r=1}^{n}\omega_r\tau_r\right]. \qquad (4.131)$$

Equation (4.75) evaluates the trace in (4.28) on the assumption that the system is an assembly of independent, distinguishable molecules. When that assumption is dropped and the trace is simply evaluated in the many-particle energy representation, the result is (4.75) with the molecular density N replaced by V^{-1}, the matrix elements of $e\mathbf{r}$ replaced by matrix elements of $\mathbf{Q}$, and the state labels $a, b_1, \ldots, b_n$ interpreted as labels for many-particle energy eigenstates. Taking this many-particle formula, and introducing $\mathbf{\Pi}$ as outlined above, gives

$$\chi^{(n)}_{\mu\alpha_1\cdots\alpha_n}(-\omega_\sigma;\omega_1,\ldots,\omega_n) = -\delta_{n1}\frac{\lambda}{\varepsilon_0 V\omega_1^2}I_{\mu\alpha_1}$$
$$+\frac{1}{\varepsilon_0 V}\frac{1}{n!\hbar^n}(-i^{n+1}\omega_\sigma\omega_1\omega_2\cdots\omega_n)^{-1}\,\mathbf{S}_\mathrm{T}\sum_{ab_1\cdots b_n}\rho_0(a)$$
$$\times\left[\frac{\Pi^\mu_{ab_1}\,\Pi^{\alpha_1}_{b_1b_2}\cdots\Pi^{\alpha_{n-1}}_{b_{n-1}b_n}\,\Pi^{\alpha_n}_{b_na}}{(\Omega_{b_1a}-\omega_1-\cdots-\omega_n)(\Omega_{b_2a}-\omega_2-\cdots-\omega_n)\cdots(\Omega_{b_na}-\omega_n)}\right]. \qquad (4.132)$$

For a system of M independent, distinguishable molecules the reduction of (4.132) to monomolecular form follows the same lines as were used to go from (4.28) to (4.75). In the final formula we have to include the additional term exhibited in (4.132) when $n=1$, while the terms exhibited in (4.75) are modified by replacing the monomolecular matrix elements of $er_\mu, er_{\alpha_1}, \ldots, er_{\alpha_{n-1}}$ and er_{α_n} by the corresponding matrix elements of $\pi_\mu/(-i\omega_\sigma)$, $\pi_{\alpha_1}/(i\omega_1)$, ..., $\pi_{\alpha_{n-1}}/(i\omega_{n-1})$ and $\pi_{\alpha_n}/(i\omega_n)$ respectively, where the lower-case Greek letter $\boldsymbol{\pi}$ is used to denote the monomolecular current operator. Thus we find in place of (4.75) the alternative formula:

$$\chi^{(n)}_{\mu\alpha_1\cdots\alpha_n}(-\omega_\sigma;\omega_1,\ldots,\omega_n) = -\delta_{n1}\frac{\lambda}{\varepsilon_0 V\omega_1^2}I_{\mu\alpha_1}$$
$$+\frac{N}{\varepsilon_0}\frac{1}{n!\hbar^n}(-i^{n+1}\omega_\sigma\omega_1\omega_2\cdots\omega_n)^{-1}\,\mathbf{S}_\mathrm{T}\sum_{ab_1\cdots b_n}\rho_0(a)$$
$$\times\left[\frac{\pi^\mu_{ab_1}\,\pi^{\alpha_1}_{b_1b_2}\cdots\pi^{\alpha_{n-1}}_{b_{n-1}b_n}\,\pi^{\alpha_n}_{b_na}}{(\Omega_{b_1a}-\omega_1-\cdots-\omega_n)(\Omega_{b_2a}-\omega_2-\cdots-\omega_n)\cdots(\Omega_{b_na}-\omega_n)}\right], \qquad (4.133)$$

where N is the molecular density, π^{α}_{ab} denotes the (ab)th matrix element of π_{α}, and the state labels now refer to monomolecular states.

A more important case arises when there are $M^{(e)}$ independent electrons and the ion cores are fixed. To eliminate the contribution of the ion cores to the susceptibilities, we let the ion masses approach infinity in (4.130) so that only the electronic contribution $M^{(e)}e^2/m$ survives. More importantly, since electrons are Fermi particles, we can no longer use a representation in which the many-electron wave functions are simply products of one-electron wave functions. Instead we must use Slater determinants which are antisymmetrised combinations of product wave functions. Equation (4.41) remains valid where m is now an electron label. Moreover, the density operator ρ_0 continues to have the factorised Boltzmann form (4.32) which is appropriate for a macroscopic system of independent particles. Nevertheless, the final one-electron formula for F in (4.41) is different from (4.48) because we have to evaluate the many-particle trace in an antisymmetrised basis. In place of (4.41) we find the one-electron formula

$$F = \sum_a f_0(a) \int \mathrm{d}\varsigma \, u_a^* \, C_1 \, u_a \tag{4.134}$$

where, as previously, $\mathrm{d}\varsigma$ is a volume element in configuration space, the set of single-electron wave functions $\{u_a\}$ is an energy representation, and C_1 is the commutator for the electron labelled number 1. Finally,

$$f_0(a) = \{\exp[(\mathrm{E}_a - \mathrm{E}_\mathrm{F})/kT] + 1\}^{-1} \tag{4.135}$$

is the thermal average of the occupation number of the one-electron state a, given by the Fermi-Dirac function. In the context of semiconductor physics, the chemical potential E_F is often called the 'Fermi energy'.

By keeping these remarks in mind, and using (4.128) with the ionic term deleted, we now find in place of (4.75) the alternative formula:

$$\begin{aligned} \chi^{(n)}_{\mu\alpha_1\cdots\alpha_n}(-\omega_\sigma;\omega_1,\ldots,\omega_n) &= -\delta_{n1}\frac{N^{(e)}e^2}{\varepsilon_0 m\omega_1^2} I_{\mu\alpha_1} \\ &+ \frac{1}{\varepsilon_0 V}\frac{e^{n+1}}{n!\hbar^n m^{n+1}}(-i^{n+1}\omega_\sigma\omega_1\omega_2\cdots\omega_n)^{-1}\,\mathbf{S}_\mathrm{T}\sum_{ab_1\cdots b_n} f_0(a) \\ &\times\left[\frac{p^{\mu}_{ab_1}\,p^{\alpha_1}_{b_1b_2}\cdots p^{\alpha_{n-1}}_{b_{n-1}b_n}\,p^{\alpha_n}_{b_na}}{(\Omega_{b_1a}-\omega_1-\cdots-\omega_n)(\Omega_{b_2a}-\omega_2-\cdots-\omega_n)\cdots(\Omega_{b_na}-\omega_n)}\right], \end{aligned} \tag{4.136}$$

where $N^{(e)} = M^{(e)}/V$ is the electron density, and p^{α}_{ab} denotes the (ab)th matrix element of p_{α}. Similarly, we find that (4.76) takes the alternative form:

$$\chi^{(n)}(-\omega_\sigma;\omega_1,\dots,\omega_n) = -\delta_{n1}\frac{N^{(e)}e^2}{\varepsilon_0 m\omega_1^2}\mathbf{e}_\sigma^*\cdot\mathbf{e}_1$$

$$+\frac{1}{\varepsilon_0 V}\frac{e^{n+1}}{n!\hbar^n m^{n+1}}(-i^{n+1}\omega_\sigma\omega_1\omega_2\cdots\omega_n)^{-1}\,\mathbf{S}_\mathrm{T}\sum_{ab_1\cdots b_n} f_0(a)$$

$$\times\left[\frac{\mathbf{e}_\sigma^*\cdot\mathbf{p}_{ab_1}\,\mathbf{e}_1\cdot\mathbf{p}_{b_1b_2}\cdots\mathbf{e}_n\cdot\mathbf{p}_{b_na}}{(\Omega_{b_1a}-\omega_1-\cdots-\omega_n)(\Omega_{b_2a}-\omega_2-\cdots-\omega_n)\cdots(\Omega_{b_na}-\omega_n)}\right], \quad (4.137)$$

where $\mathbf{p}_{ab}$ denotes the (ab)th matrix element of the momentum operator $\mathbf{p}$. In these equations $f_0(a)$ is given by (4.135). To determine the chemical potential $\mathrm{E_F}$ we set $C_1=1$ in (4.41) and (4.134) so that $F=M^{(e)}$, and (4.134) then reduces to the following equation:

$$M^{(e)} = \sum_a f_0(a). \quad (4.138)$$

We note that when the electron density is so low that $f_0(a)$ is small for all a, the 1 in (4.135) may be ignored. As a confirmation of this we find from (4.138) and (4.135) that

$$\frac{f_0(a)}{V} = N^{(e)}\frac{\exp(-\mathrm{E}_a/kT)}{\sum_a\exp(-\mathrm{E}_a/kT)} = N^{(e)}\rho_0(a), \quad (4.139)$$

where $\rho_0(a)$ is a normalised Boltzmann distribution. Consequently, as expected, (4.136) assumes the form (4.133) which was derived for distinguishable molecules.

The resonant energy denominators involved in the alternative formulae are identical to those involved in the original formulae as described in §4.5. Consequently the rules governing the introduction of damping factors are also identical: the poles of the conductivity tensor must lie in the *lower* half of the complex-frequency plane (*cf.* (4.111) and (4.112)).

4.6.2 Derivation of the alternative formulae: Current density and the conductivity tensors

We may gain insight into the structure of the alternative susceptibility formulae by considering the commutator of $\mathbf{Q}$ in (4.1) with the field-free many-particle Hamiltonian H_0. Obviously $\mathbf{Q}$ commutes with the many-particle potential energy. Moreover, electron and ion core operators commute unless they are associated with the same particle. Finally, the fundamental commutator relations between the cartesian components of $\boldsymbol{\mu}$ and $\mathbf{p}$ for any particle: $[\mu_\alpha,p_\beta]=i\hbar\delta_{\alpha\beta}$ imply that $[\boldsymbol{\mu}_j,\mathbf{p}_j^2]=2i\hbar\mathbf{p}_j$ for electrons and $[\boldsymbol{\mu}_k,\mathbf{p}_k^2]=2i\hbar\mathbf{p}_k$ for ions. By using these results we find that

$$[\mathbf{Q}, H_0] = \left[-e\sum_j \boldsymbol{\mu}_j + e\sum_k Z_k \boldsymbol{\mu}_k \,,\ \frac{1}{2m}\sum_j \mathbf{p}_k + \sum_k \frac{1}{2M_k}\mathbf{p}_k^2 \right]$$

$$= \frac{-e}{2m}\sum_j [\boldsymbol{\mu}_j, \mathbf{p}_j^2] + e\sum_k \frac{Z_k}{2M_k}[\boldsymbol{\mu}_k, \mathbf{p}_k^2]$$

$$= i\hbar\,\mathbf{\Pi}. \tag{4.140}$$

We may put this result in a particularly illuminating form by pre-multiplying by $U_0(-t)$ and postmultiplying by $U_0(t) = \exp(-iH_0 t/\hbar)$:

$$\mathbf{\Pi}(t) = \mathrm{d}\mathbf{Q}(t)/\mathrm{d}t\,. \tag{4.141}$$

This equation may be used as the basis for the elimination of $\mathbf{Q}(t)$ in favour of $\mathbf{\Pi}(t)$ from our susceptibility formulae. This is an interesting exercise in operator algebra but it is, however, more instructive to proceed by an alternative route which we now describe in detail.

We have previously employed the polarisation $\mathbf{P}(t)$ to describe the response of the medium to the applied electric field. An equivalent description may be given in terms of the polarisation-current density $\mathbf{J}(t) = \partial\mathbf{P}(t)/\partial t$. By differentiating (2.1) we obtain the power series

$$\mathbf{J}(t) = \mathbf{J}^{(1)}(t) + \cdots + \mathbf{J}^{(n)}(t) + \cdots \tag{4.142}$$

where

$$\mathbf{J}^{(n)}(t) = \partial\mathbf{P}^{(n)}(t)/\partial t\,. \tag{4.143}$$

It follows from (2.40) that we may write

$$\mathbf{J}^{(n)}(t) = \int_{-\infty}^{+\infty} \mathrm{d}\omega_1 \cdots \int_{-\infty}^{+\infty} \mathrm{d}\omega_n\, \boldsymbol{\sigma}^{(n)}(-\omega_\sigma\,; \omega_1, ..., \omega_n)\,|$$

$$\mathbf{E}(\omega_1)\cdots\mathbf{E}(\omega_n)\exp(-i\omega_\sigma t), \tag{4.144}$$

where

$$\boldsymbol{\sigma}^{(n)}(-\omega_\sigma\,; \omega_1, ..., \omega_n) = -i\omega_\sigma \varepsilon_0\, \boldsymbol{\chi}^{(n)}(-\omega_\sigma\,; \omega_1, ..., \omega_n) \tag{4.145}$$

denotes the nth-order conductivity tensor.

To calculate $\boldsymbol{\sigma}^{(n)}(-\omega_\sigma\,; \omega_1, ..., \omega_n)$ directly we must evaluate $\mathbf{J}^{(n)}(t)$ by using the field-dependent Hamiltonian in a convenient gauge. In our previous work we have set the vector potential equal to zero and the scalar potential equal to $-\mathbf{E}(t)\cdot\boldsymbol{\mu}$. To calculate $\mathbf{J}^{(n)}(t)$ it is more convenient to use a gauge in which we set the scalar potential equal to zero and write the vector potential in the form:

$$\mathbf{A}(t) = -\int_{-\infty}^{t} \mathbf{E}(\tau)\,\mathrm{d}\tau$$

$$= \int_{-\infty}^{+\infty} \mathrm{d}\omega\, \mathbf{E}(\omega)\,(i\omega)^{-1}\exp(-i\omega t)\,, \tag{4.146}$$

where we have used the Fourier integral form (2.25) for $\mathbf{E}(t)$. These electromagnetic potentials also represent an electric field $\mathbf{E}(t)$ and zero magnetic field because we are ignoring the spatial dependence of $\mathbf{E}(t)$.

$\mathbf{A}(t)$ enters into the Hamiltonian only through the kinetic energy term, which becomes

$$\begin{aligned}\frac{1}{2m}\sum_j[\mathbf{p}_j + e\mathbf{A}(t)]^2 + \sum_k \frac{1}{2M_k}[\mathbf{p}_k - Z_k e\mathbf{A}(t)]^2 \\ = \frac{1}{2m}\sum_j \mathbf{p}_j^2 + \sum_k \frac{1}{2M_k}\mathbf{p}_k^2 - \mathbf{A}(t)\cdot\mathbf{\Pi},\end{aligned} \tag{4.147}$$

where $\mathbf{\Pi}$ is the current operator (4.128) in the absence of applied fields. We have dropped a term proportional to $\mathbf{A}^2(t)$ which has no physical effects. (Since $\mathbf{A}(t)$ is independent of $\boldsymbol{\mu}$ this term merely introduces a time-dependent phase factor into the wave function.) The field-dependent Hamiltonian (4.147) has the standard form considered in the general theory developed in §3.6: namely, $H = H_0 + H_{\mathrm{I}}(t)$, where the perturbation $H_{\mathrm{I}}(t)$ is $-\mathbf{A}(t)\cdot\mathbf{\Pi}$. The analysis given in that section therefore goes through unchanged to yield the power series (3.51) for the perturbed density operator. This is given at the end of Chapter 3 together with (3.82) and (3.76) which specify the nth term in the series with $H_{\mathrm{I}}(t) = -\mathbf{A}(t)\cdot\mathbf{\Pi}$.

It remains for us to use this series to calculate the macroscopic current density. Now, inspection of the electronic and ionic kinetic energies in (4.147) verifies that the velocity of the jth electron is $[\mathbf{p}_j + e\mathbf{A}(t)]/m$ and the velocity of the kth ion core is $[\mathbf{p}_k - Z_k e\mathbf{A}(t)]/M_k$ in the present gauge. It follows that the current operator now takes the form:

$$\begin{aligned}\mathbf{\Pi}' &= \sum_j \frac{-e}{m}[\mathbf{p}_j + e\mathbf{A}(t)] + \sum_k \frac{eZ_k}{M_k}[\mathbf{p}_k - Z_k e\mathbf{A}(t)] \\ &= \mathbf{\Pi} - \lambda\mathbf{A}(t),\end{aligned} \tag{4.148}$$

where λ is the parameter specified in (4.130). From (3.32), the macroscopic polarisation-current density is given by

$$\mathbf{J}(t) = V^{-1}\mathrm{Tr}[\rho(t)\,\mathbf{\Pi}']. \tag{4.149}$$

When $\rho(t)$ is substituted in the form (3.51) we arrive at the power series (4.142) for $\mathbf{J}(t)$, in which the nth term is given by (4.144) and the nth-order conductivity tensor is expressed in terms of the matrix elements of $\mathbf{\Pi}$.

The similarity of structure between the perturbation $-\mathbf{Q}\cdot\mathbf{E}(t)$ considered in §4.1 and the perturbation $-\mathbf{\Pi}\cdot\mathbf{A}(t)$ under discussion here means that the formula for the nth-order conductivity tensor has a similar structure to that of the nth-order susceptibility tensor. Matrix elements of $\mathbf{\Pi}$ replace matrix elements of $\mathbf{Q}$ and there are n additional factors $(-i\omega_1)^{-1}, (-i\omega_2)^{-1}, ..., (-i\omega_n)^{-1}$. The latter arise from the n Fourier transforms of $\mathbf{A}(t)$ which enter into the nth-order contribution to $\mathbf{J}(t)$ when (4.146) is used to express them in terms of the corresponding Fourier transforms of $\mathbf{E}(t)$. A final division by $-i\omega_\sigma$ in accordance with (4.145) yields (4.131) for the nth-order susceptibility tensor when $n \geq 2$.

The additional term which appears when $n = 1$ is simply V^{-1} times the Fourier transform of the second term in the last line of (4.148), which we have so far neglected, with $\mathbf{E}(\omega_1)$ factored out. Since $\mathbf{A}(t)$ is independent of $\boldsymbol{\mu}$ this term is not changed by the trace in (4.149) and contributes as it stands to the linear term in $\mathbf{J}(t)$.

4.7 Semiempirical approximations and models

In previous sections of this chapter (§4.1 and §4.6) we have derived fundamental formulae for the susceptibility tensors in terms of electric-dipole operators and their matrix elements (or, alternatively, current operators and their matrix elements). These basic formulae provide a route to understanding some of the most important and general topics in nonlinear optics; such as the rôle and classification of multi-photon processes, resonance enhancement, selection rules, spatial symmetry and geometrical properties. The application of the general formulae to media containing free charge-carriers is described in Chapter 8. Here we are concerned with nonlinear-optical effects in a system of bound charges, and we ask the question: can these quantal formulae be used to make an accurate *quantitative* prediction of the optical nonlinearity of a particular medium? Reliable quantitative calculations would be of great value if they could allow the enormous number of existing natural and synthetic materials to be assessed for their potential as useful nonlinear-optical media, without the need to embark on extensive time-consuming and expensive measurements; predictive calculations might also be used by synthetic chemists to assist in the design of new materials having useful nonlinear properties.

The simple answer to this question is that the general formulae can be used for calculations directly and without approximation, only if the set of wave functions and eigenenergies is known with sufficient accuracy. This is true only in the case of the very simplest atoms and molecules. The first-principles approach is to calculate the wave functions by solving the Schrödinger wave equation (3.3) exactly. This is possible only in the case of H and He atoms and a few very simple molecules such as H_2. In some cases the first-principles calculations can be tested against others which use experimentally-determined values of oscillator strengths and frequencies, and also against measurements of the optical properties themselves. Good agreement is found for the alkali-metal atoms which have simple hydrogen-like spectra due to the single valence electron; for example, calculated and measured values of the third-harmonic susceptibility $\chi^{(3)}(-3\omega;\omega,\omega,\omega)$ of alkali-metal vapours agree to within 15% in some cases (see §4.5.1). Similarly, the value of the Raman susceptibility of molecular hydrogen calculated using *ab initio* theoretical values for the dipole matrix elements is in good agreement with experimental measurements of the Raman cross-section (Pan *et al*, 1977). In the case of some other (but still relatively simple) atoms and molecules, a good deal of experimental data for transition frequencies and oscillator strengths may be found in the literature; these data may be used in direct application of the formulae for the susceptibilities. Examples of such materials are the rare gases (He, Ne, Ar, ...), alkaline earths (Ba, Mg, Sr, ...), and some small molecules (CO, HCl, HF, SF_6, ...) (Hanna *et al*, 1979).

However, for molecules more complex than these, it is impossible to produce a set of exact wave functions or obtain sufficient spectroscopic data, and it is therefore essential to devise some approximations. Over the years, the mechanisms of molecular optical nonlinearity have been explained with semiempirical models which range in complexity from naive simplicity to extensive many-electron computer calculations (Ducuing, 1977; Oudar, 1977; Williams, 1983; Chemla and Zyss, 1987). Here we outline a few approaches that are used; we have selected those that have proved particularly successful, or that readily offer a useful physical insight. Our account is necessarily very brief; further details may be found in the cited references. A comprehensive early account of alternative models is given by Flytzanis (1975), and several are discussed in detail in Chemla and Zyss (1987).

4.7.1 Resonant case

The greatest simplification of the general formulae for the susceptibilities can be made in the case when the nonlinear process of interest is resonantly enhanced. In this case it is usually possible to extract the significant resonant terms from the full expression for $\boldsymbol{\chi}^{(n)}$ and lump the contributions from all the other levels in a dispersionless background (which may indeed be negligible in the case of strong resonance enhancement). The resonant contribution may then be calculated using the best available wave functions for the resonant levels or experimental data for the selected transition frequencies and oscillator strengths. In the case when one pair of levels provide the dominant resonant contribution, the two-level model described in Chapter 6 is ideal.

4.7.2 Unsöld approximation and charge models

In the nonresonant case (when all the combinations of optical frequencies involved are far removed from molecular resonances), the hyperpolarisabilities are said to be 'integral properties' of the molecule which do not depend on the details of the energy structure but only on the chemical bonding (Hopfield, 1971). In other words, the hyperpolarisabilities depend on the detailed distribution of charges in the *ground* state only, together with statistical averages of the frequencies and oscillator strengths of transitions to the excited states. This implies that models which are able to estimate the general outline of the spectral density of states in complex molecules, together with a more exact picture of the ground-state charge distribution, should yield useful results. This argument may explain, to some extent, the success of the SOS approach (to be described in §4.7.4). We shall now substantiate the argument in a more direct way, and also obtain some physical insight, by applying the Unsöld approximation to the linear molecular polarisability $\boldsymbol{\alpha}$ and lowest-order hyperpolarisability $\boldsymbol{\beta}$. This approximation, proposed and applied to the model of linear molecular polarisability by Unsöld in 1927, was first generalised to nonlinear optics by Robinson (1967).

We begin by taking the expression for the linear molecular polarisability $\boldsymbol{\alpha}(-\omega;\omega)$, obtained from (4.65) and (4.107):

$$\alpha_{xx}(-\omega;\omega) = \frac{2}{\hbar}\sum_{b\neq g}\frac{|er^x_{gb}|^2\,\Omega_{bg}}{(\Omega_{bg}^2-\omega^2)}\,, \tag{4.150}$$

where the molecule is taken to be in the ground energy state labelled g (*i.e.*, $\rho_0(g)=1$), and the summation is over all molecular states except g. We assume that the optical frequency ω is in the 'transparency range' well below all electronic transitions from the ground state. According to the Unsöld approximation, we can then replace the transition frequencies Ω_{bg} which occur in the expression (4.150) for $\alpha_{xx}(-\omega;\omega)$ by a single mean value Ω_x outside the summation sign:

$$\alpha_{xx}(-\omega;\omega) = \frac{2\Omega_x}{\hbar(\Omega_x^2-\omega^2)}\sum_{b\neq g}|er^x_{gb}|^2\,. \tag{4.151}$$

It is convenient to introduce the centred coordinates $\hat{r}_\alpha = r_\alpha - \langle r_\alpha\rangle$, where the $\langle r_\alpha\rangle$ are mean electronic coordinates defined as $\langle r_\alpha\rangle = (\int d\varsigma\,\rho_e r_\alpha)/\int d\varsigma\,\rho_e$, and $\rho_e(\mathbf{r})$ is the electronic charge density (Chemla,

1980). Thus $e\hat{r}_\alpha$ represents the difference between the dipole moment er_α and its ground-state expectation value: $e\hat{r}_\alpha = er_\alpha - er^\alpha_{gg}$. Then the summation in (4.151) can be written as:

$$\sum_{b\neq g} |er^x_{gb}|^2 = \sum_b |er^x_{gb}|^2 - |er^x_{gg}|^2$$
$$= (|er_x|^2)_{gg} - |er^x_{gg}|^2 = (|e\hat{r}_x|^2)_{gg}. \tag{4.152}$$

These relations can be verified using the matrix-element algebra described in Chapter 3. The expression (4.151) then becomes:

$$\alpha_{xx}(-\omega;\omega) = \frac{2\Omega_x}{\hbar(\Omega_x^2-\omega^2)}(|e\hat{r}_x|^2)_{gg}. \tag{4.153}$$

We have thus used the closure property of the molecular eigenstates to replace the summation over matrix elements by an expression which only requires knowledge of the ground-state charge distribution for its evaluation. It only remains to assign a value to the mean transition frequency Ω_x. This can be done using the well-known Thomas-Kuhn sum rule for oscillator strengths, which states that

$$\sum_{b\neq g} \frac{2m}{e^2\hbar} |er^x_{gb}|^2\,\Omega_{bg} = 1. \tag{4.154}$$

Applying the Unsöld approximation to (4.154), and using (4.152), leads to

$$(\Omega_x)^{-1} = \frac{2m}{e^2\hbar}(|e\hat{r}_x|^2)_{gg}. \tag{4.155}$$

Thus, for low frequencies such that $\omega \ll \Omega_x$, the expression for the linear polarisability reduces to:

$$\alpha_{xx}(-\omega;\omega) = \frac{2}{\hbar\Omega_x}(|e\hat{r}_x|^2)_{gg}$$
$$= \frac{4m}{e^2\hbar^2}\left[(|e\hat{r}_x|^2)_{gg}\right]^2$$
$$= \frac{16\pi\varepsilon_0}{e^4}\frac{1}{a_0}\left[(|e\hat{r}_x|^2)_{gg}\right]^2, \tag{4.156}$$

where $a_0 = 4\pi\varepsilon_0\hbar^2/me^2$ is the Bohr radius.

The quadrupole moment $(|e\hat{r}_x|^2)_{gg}$ measures the physical extent of the ground-state electronic charge distribution. The part of the charge distribution which is well localised and tightly bound – such as the electronic orbitals close to the ion cores – has small ground-state moments and is therefore difficult to polarise, *i.e.*, its contribution to the polarisability is small. The extended valence-electron orbitals (such as σ and π orbitals), on the other hand, are much more readily polarisable, and these make the major contribution.

Robinson (1967) applied the same approach to the second-order molecular hyperpolarisability, and obtained the result:

$$\beta_{xyz}(-\omega_\sigma;\omega_1,\omega_2) = \frac{\Omega_x\Omega_y\Omega_z(\Omega_x+\Omega_y+\Omega_z)}{2\hbar^2(\Omega_x^2-\omega_\sigma^2)(\Omega_y^2-\omega_1^2)(\Omega_z^2-\omega_2^2)}\, e^3(\hat{r}_x\hat{r}_y\hat{r}_z)_{gg}\,, \tag{4.157}$$

where $e^3(\hat{r}_x\hat{r}_y\hat{r}_z)_{ab}$ denotes the (ab)th matrix element of $\widehat{er}_x\,\widehat{er}_y\,\widehat{er}_z$, and $\widehat{er}_\alpha = er_\alpha - er_{gg}^\alpha$ as before. We see therefore that, in the Unsöld approximation, the fine detail of the excited-state wave functions is superfluous. The dependence on the moment $(|e^{n+1}\hat{r}_{\alpha_1}\cdots\hat{r}_{\alpha_{n+1}}|^2)_{gg}$ provides important insight into the significance of the ground-state charge distribution in determining the magnitude of the hyperpolarisabilities. As in the case of the linear polarisability, the valence electrons make the major contribution.

For polyatomic molecules, there are various approximate methods for calculating the moments of the ground-state valence-electron charge distribution. For saturated molecules in which the electrons are localised in bonds, the moments can be estimated using the vector sum of the polarisabilities of the individual bonds, and the interaction between bonds is neglected. This is the 'bond additivity' scheme (Hermann and Ducuing, 1974; Ducuing, 1977). This scheme applies well in the calculation of the polarisability $\boldsymbol{\alpha}$ and hyperpolarisability $\boldsymbol{\gamma}$ of both centrosymmetric and noncentrosymmetric saturated molecules. Its success is based on the observation that the polarisabilities of σ bonds are almost the same in any molecule in which they occur. For the lowest-order hyperpolarisability $\boldsymbol{\beta}$, however, the method is less reliable because of the sensitivity to small deviations from centrosymmetry (see §5.3) (Ward and Miller, 1979). When strong delocalisation of the valence electrons occurs, as in conjugated molecules, the bond-additivity property can no longer be satisfied. For such systems one must adopt a description in which the electrons belong to the whole molecule rather than to particular bonds. One simple approach is to assume that the electrons are free to move in a potential box determined by the dimensions of the molecule, and the asymmetry is introduced by suitable choice of the shape of the well; this is the 'free-electron' scheme developed by Kuhn (1956) which is, however, suitable only for small molecules. A more accurate approach for larger molecules is to calculate the molecular orbitals using one of the established semiempirical methods (Pople and Beveridge, 1970).

There is great interest in conjugated organic molecules as nonlinear-optical media because, as suggested by expressions such as (4.157), the large delocalisation of the π electrons induces cooperative effects and can produce very large nonlinear coefficients. Usually non-substituted conjugated molecules are centrosymmetric, and therefore the lowest-order hyperpolarisability $\boldsymbol{\beta}$ is zero (see §5.3). However the π-electron system produces a very large polarisability $\boldsymbol{\alpha}$ and hyperpolarisability $\boldsymbol{\gamma}$ with a strong superlinear dependence on N, the number of conjugated electrons: $\alpha_{ij} \sim N^3$ and $\gamma_{ijkl} \sim N^5$ (Hermann and Ducuing, 1974; Chemla and Zyss, 1987). A nonzero second-order hyperpolarisability can be obtained by adding substituent groups to the conjugated molecule in order to distort the π-electron distribution. The magnitude and sign of the effects depend in a complex way on both the nature and the number of substituents as well as on the length of the conjugated system (Chemla and Zyss, 1987).

4.7.3 Miller's delta

A celebrated rule, proposed by Miller (1964), was one of the earliest attempts to predict the second-order nonlinear-optical behaviour of a wide range of crystalline materials in a consistent way. Miller defined a third-rank tensor $\boldsymbol{\Delta}$ through the relation:

$$\chi^{(2)}_{ijk}(-\omega_\sigma;\omega_1,\omega_2) = \Delta_{ijk}\, \chi^{(1)}_{ii}(-\omega_\sigma;\omega_\sigma)\, \chi^{(1)}_{jj}(-\omega_1;\omega_1)\, \chi^{(1)}_{kk}(-\omega_2;\omega_2), \tag{4.158}$$

and his conjecture was that Δ_{ijk} is a constant, at least for materials of a given symmetry. This applies to the case when all the optical frequencies lie in the transparency region of the crystal: well below the optical energy gap but well above lattice resonances. Miller's conjecture was based on the observation that, amongst the inorganic crystals whose $\boldsymbol{\chi}^{(1)}$ and $\boldsymbol{\chi}^{(2)}$ properties had been measured at that time, the variation in $\boldsymbol{\Delta}$ is much smaller than that in $\boldsymbol{\chi}^{(2)}$; whereas the measured nonlinear susceptibilities span over four orders of magnitude, the average value of Δ_{ijk} is 3×10^{-13} m V^{-1} with a standard deviation of only 1.9×10^{-13} m V^{-1} (Byer (1977) gives a useful compilation of measured values). With Δ_{ijk} equal to a constant value, (4.158) is known as Miller's rule.

A partial explanation for this relative constancy in the parameter Δ_{ijk} can be found by expressing (4.158), with the aid of (4.103), in terms of the molecular polarisability $\boldsymbol{\alpha}$ and hyperpolarisability $\boldsymbol{\beta}$, when it will be seen that the resulting relation:

$$\Delta_{ijk} = \frac{(\varepsilon_0/N)^2 <\beta_{ijk}(-\omega_\sigma;\omega_1,\omega_2)>}{<\alpha_{ii}(-\omega_\sigma;\omega_\sigma)><\alpha_{jj}(-\omega_1;\omega_1)><\alpha_{kk}(-\omega_2;\omega_2)>} \tag{4.159}$$

is independent of local-field factors. The relation (4.159) also reiterates the influence of the polarisability discussed above; the more polarisable the medium, the stronger is the nonlinearity. Ducuing and Flytzanis

(1968) related Miller's $\mathbf{\Delta}$ to the electronic dipole moment of the bonds in the crystal. The constancy of Δ_{ijk} can be attributed to the fact that σ bonds have similar polarisabilities in all materials. Miller's rule can also be deduced theoretically from the simple anharmonic oscillator model outlined in Chapter 1 (Garrett and Robinson, 1966). In the light of the above discussion of the Unsöld approximation, it can be appreciated that the anharmonic oscillator corresponds to a model in which a delta-function density of states is located at the resonant frequency of the oscillator.

Miller's rule has proved to be a simple and useful guide in the search for materials with large second-order nonlinearity. It is, however, no more than a guide since its great limitation is that it does not incorporate the symmetry properties of the medium. For example, centrosymmetric media, for which $\chi^{(2)}$ is identically zero (see §5.3), violate the rule. The variation of $\mathbf{\Delta}$ amongst organic crystals is also much greater; in particular, those having large polarisability due to delocalised electronic orbitals are intrinsic exceptions to Miller's rule. Some materials such as MNA (2-methyl-4-nitroaniline) and NPP (N-(4-nitrophenyl)-L-prolinol) have values of Δ_{ijk} which are $1-2$ orders greater than that for the typical inorganic nonlinear crystals. Perhaps the greatest achievement of Miller's conjecture has been that it stimulated the initial interest in organic materials for applications in electrooptics and nonlinear optics.

4.7.4 Sum over states

The most direct (and ambitious) application of the quantal formulae for the susceptibilities relies on established semiempirical techniques for computing molecular orbitals, from which values for transition frequencies and dipole-matrix elements are derived. This data is fed into the summations appearing in the general formulae for the susceptibilities. This is termed the sum-over-states (SOS) approach. It was pioneered by the work of Morrell and Albrecht (1979), and Garito and coworkers (Lalama and Garito, 1979; Teng and Garito, 1983a,b). Molecular-orbital calculations, which a decade ago would have seemed impossibly ambitious, are made possible by the increasing availability of powerful computers.

The procedure can be divided into three stages: (1) the calculation of the many-body electronic ground state of the molecular system; (2) accounting for correlations by configuration interaction calculations (Lowitz, 1967) to form the lowest-energy excited states and transition dipole moments of the molecule; and (3) feeding the transition frequencies and dipole moments into the formulae for the hyperpolarisabilities $\boldsymbol{\gamma}^{(n)}$.

The calculation (1) is usually accomplished using the Roothan variation of molecular-orbital Hartree-Fock theory (Roothan, 1951; Pople and Beveridge, 1970) and the LCAO (linear-combination-of-atomic-orbitals) approximation. The essence of the Hartree-Fock-Roothan model is that one considers the motion of a single electron in the electrostatic field of the nuclei and in the average Coulomb and exchange fields of all the other electrons. In the LCAO approximation, a molecular-orbital wave function ψ_i is expressed as

$$\psi_i = \sum_k C_{ki}\,\phi_k\,, \tag{4.160}$$

where ϕ_k are the valence atomic orbitals for the atoms of which the molecule is composed. The inner-shell electrons and nucleus of an atom are treated together as a core with some net charge. The coefficients C_{ki} of the atomic orbitals are calculated self-consistently using Roothan's variational method to obtain ψ_k and the corresponding one-electron energies. The unperturbed ground-state wave function ψ_g for the molecule, assumed to be a closed-shell system of $2l$ valence electrons, is expressed as a single antisymmetric product of l molecular-orbital functions ψ_i:

$$\psi_g = \det\{\psi_1\overline{\psi}_1 \cdots \psi_i\overline{\psi}_i \cdots \psi_l\overline{\psi}_l\}\,, \tag{4.161}$$

where $\psi_i, \overline{\psi}_i$ are the wave functions of the electrons with spin up and down respectively. These methods are by their nature semiempirical; the models contain input parameters which are adjusted for the best fit to experimental data. In most of the work reported to date this parameterisation is achieved using variations of the semiempirical CNDO (complete-neglect-of-differential-overlap) procedure which is well known in molecular-orbital theory (Pople and Beveridge, 1970).

Work by Pugh and coworkers (Docherty *et al*, 1985; Pugh and Morley, 1987) extends the SOS approach to a large set of organic molecules. The calculations are parameterised by comparison between calculated and experimental values of dipole moments and transition frequencies derived from optical absorption spectra; excellent correlation is obtained. Since these are the quantities that appear in the SOS in the formulae for the molecular hyperpolarisabilities $\boldsymbol{\gamma}^{(n)}$, it is reasonable to expect that the calculated values of $\boldsymbol{\gamma}^{(n)}$ will also be accurate. The correlations between the calculated values of the second-order hyperpolarisability $\boldsymbol{\beta}$ and those derived from a variety of experimental sources (Fig. 4.6) show that this is indeed true.

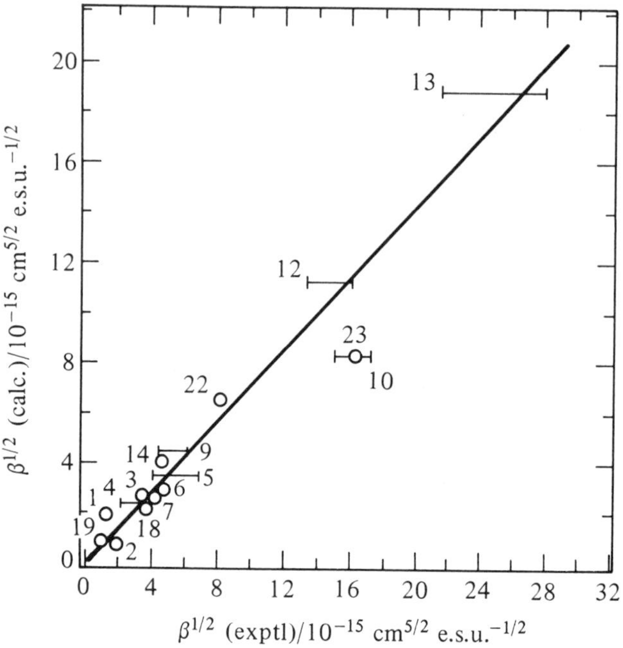

Fig. 4.6 Plot of calculated against experimental hyperpolarisabilities $\boldsymbol{\beta}$ for a range of organic molecules. For each molecule, β denotes the optimum coefficient of $\boldsymbol{\beta}$. The square-root plot is used merely for convenience in spanning the results. The error bars indicate the limits of the wide range of experimental β values often reported for the same molecule. The numbers 1–23 identify the chemical formulae of the molecules, which are listed in the cited work. (After Docherty *et al*, 1985.)

The SOS method has been used to scan through a large number of novel molecules, saving a great deal of chemical synthetic effort. It is found that, in general, larger molecules with extended conjugation show very large nonlinear coefficients. Thus work on attaining useful macroscopic nonlinear materials can concentrate on using these molecules. It should be borne in mind, however, that large molecular hyperpolarisability is only one of several requirements for a nonlinear material with practical applicability; other aspects such as an ability to grow crystals of sufficient size, good optical quality and low loss, resistance to optical damage, phase-matching ability (for parametric processes), chemical and environmental stability, and low toxicity, are all important considerations. The orientation of the molecules also plays a crucial rôle in determining the magnitude of the macroscopic second-order susceptibility. In particular, if the molecules are arranged in a centrosymmetric manner, $\boldsymbol{\chi}^{(2)}$ will be identically zero, regardless of the magnitude of $\boldsymbol{\beta}$ (see §5.3.2).

Docherty *et al* (1985) were able to study quite large organic molecules, such as 4-amino-4′-nitro-*trans*-stilbene, using calculations which incorporated as many as 100 states in the summation. The fact that such immense calculations have become possible recently – and yield useful results – is a tribute to the theoretical accuracy of the method and is due also to the increasing availability of substantial computer power. Further advances in treating large numbers of complex molecular systems can be expected in future.

4.7.5 Finite-field method

An alternative approach is the finite-field perturbation method, which was developed by Cohen and Roothan (1965) for the calculation of atomic polarisabilities, and subsequently extended by Zyss (1979) to calculate the hyperpolarisability $\boldsymbol{\beta}$ for several monosubstituted benzene derivatives. The principle of the method is to derive the various terms in the power series for the static induced molecular-dipole moment (4.83) in terms of derivatives with respect to the local field:

$$\begin{aligned} er_i &= er_i^{(0)} + \alpha_{ij}E_j + \tfrac{1}{2}\beta_{ijk}E_jE_k + \cdots \\ &= er_i^{(0)} + \left.\frac{\partial(er_i)}{\partial E_j}\right|_{\mathbf{E}=0} E_j + \tfrac{1}{2}\left.\frac{\partial^2(er_i)}{\partial E_j\,\partial E_k}\right|_{\mathbf{E}=0} E_jE_k + \cdots . \end{aligned} \qquad (4.162)$$

Thus if the molecular dipole moment can be calculated as a function of the local electric field $\mathbf{E}$, then β_{ijk} is obtained from the second derivative of this function at $\mathbf{E}=0$. The method followed by Zyss (1979) is to calculate semiempirical molecular orbitals using one of the established methods (the INDO approximation; Pople and Beveridge, 1970), and from this derive a diagonalised molecular Hamiltonian which is modified to include a static-field dipolar perturbation. A good general agreement with experimental trends is obtained (Zyss and Chemla, 1987).

5

Symmetry properties

In the previous chapter, explicit formulae for the nonlinear susceptibilities were derived. The susceptibilities exhibit various types of symmetry which are of fundamental importance in nonlinear optics: permutation symmetry, time-reversal symmetry, and symmetry in space. (Another kind of symmetry – the relationship between the real and imaginary parts of the susceptibilities – is described in Appendix 8.) The time-reversal and permutation symmetries are fundamental properties of the susceptibilities themselves, whereas the spatial symmetry of the susceptibility tensors reflects the structural properties of the nonlinear medium. All of these have important practical implications. In this chapter we outline the essential features and some practical consequences.

5.1 Permutation symmetry

The permutation-symmetry properties of the nonlinear susceptibilities have already been encountered in earlier chapters. Intrinsic permutation symmetry, first described in §§2.1 and 2.2, implies that the nth-order susceptibility $\chi^{(n)}_{\mu\alpha_1\cdots\alpha_n}(-\omega_\sigma;\omega_1,...,\omega_n)$ is invariant under all $n!$ permutations of the pairs $(\alpha_1,\omega_1), (\alpha_2,\omega_2),..., (\alpha_n,\omega_n)$. Intrinsic permutation symmetry is a fundamental property of the nonlinear susceptibilities which arises from the principles of time invariance and causality, and which applies universally. In some circumstances a susceptibility may also possess a more general property, overall permutation symmetry, in which the susceptibility is invariant when the permutation includes the additional pair $(\mu,-\omega_\sigma)$; *i.e.,* the nth-order susceptibility is invariant under all $(n+1)!$ permutations of the pairs $(\mu,-\omega_\sigma), (\alpha_1,\omega_1),..., (\alpha_n,\omega_n)$. As indicated in §4.3, the overall permutation-symmetry property is an approximation which is valid only when all of the optical frequencies which occur in the formula for the susceptibility are far removed from the transition frequencies of the nonlinear medium, so that the medium is transparent at all the relevant frequencies.

The overall permutation-symmetry property was introduced in §4.3 mainly because it allows the various formulae for the susceptibilities to be written in a compact form. Here we consider some of its important physical consequences.

5.1.1 Physical consequences of overall permutation symmetry

The first point to notice is that overall permutation symmetry implies that the low-frequency limits of the susceptibility tensors, when they exist, will be symmetrical in all their subscripts. Consider, for example, the expression (4.71) for the second-order susceptibility:

$$\chi^{(2)}_{\mu\alpha\beta}(-\omega_\sigma;\omega_1,\omega_2)=\frac{Ne^3}{\varepsilon_0\,2\hbar^2}\,\mathbf{S}_\mathrm{T}\sum_{abc}\rho_0(a)\left[\frac{r^\mu_{ab}\,r^\alpha_{bc}\,r^\beta_{ca}}{(\Omega_{ba}-\omega_1-\omega_2)\,(\Omega_{ca}-\omega_2)}\right],\tag{5.1}$$

which is invariant under the six permutations of the pairs $(\mu,-\omega_\sigma)$, (α,ω_1), (β,ω_2). It is evident that, provided all the optical frequencies which appear in the denominator terms are much smaller than any of the transition frequencies Ω_{ba} and Ω_{ca}, the frequencies $-\omega_\sigma$, ω_1 and ω_2 alone may be freely permuted without significantly changing the susceptibility. In other words, the dispersion of the medium at the relevant optical frequencies is negligible. This would be the case, for example, if the nonlinear polarisation were predominantly electronic in origin and the optical frequencies all fell within the transparent spectral region of the medium. It follows that the susceptibility is then invariant under all permutations of the subscripts μ, α and β. A similar symmetry property obtains for each of the higher-order susceptibilities in the low-frequency limit. This property came to be known as 'Kleinman symmetry', after Kleinman (1962) who originally pointed it out for $\boldsymbol{\chi}^{(2)}$ and discussed its significance as a tool for deciding whether ionic or electronic processes make the dominant contribution to the tensor. An application of the Kleinman-symmetry property to reduce the number of independent elements of the second-order susceptibility tensor is discussed in §5.3.

Perhaps the most interesting consequence of overall permutation symmetry is the Manley-Rowe power relations (Manley and Rowe, 1956; Haus, 1958) which describe the exchange of power between electromagnetic waves in a purely reactive nonlinear medium. To show how these relations arise, let us suppose that the medium is subjected initially to two simple harmonic fields with nonzero incommensurable frequencies ω' and ω''. The nonlinearity of the constitutive relation will result in the generation of, in general, all possible combination frequencies denoted as follows:

$$\omega_{jk} = j\,\omega' + k\,\omega'', \tag{5.2}$$

where j and k are arbitrary integers (positive, negative, or zero) and the ω_{jk} are positive or zero. Thus the total electric field has the form of (2.49); *i.e.*,

$$\mathbf{E}(t) = \tfrac{1}{2}\sum_{jk\geq 0}\left[\mathbf{E}_{\omega_{jk}}\exp(-i\omega_{jk}t) + \mathbf{E}_{-\omega_{jk}}\exp(i\omega_{jk}t)\right]. \tag{5.3}$$

In deriving the Manley-Rowe relations we shall, for simplicity, concentrate our attention on the second-order polarisation given by the expression (2.52) with $n=2$, namely:

$$\mathbf{P}^{(2)}(t) = \tfrac{1}{2}\sum_{jk\geq 0}\left[\mathbf{P}^{(2)}_{\omega_{jk}}\exp(-i\omega_{jk}t) + \mathbf{P}^{(2)}_{-\omega_{jk}}\exp(i\omega_{jk}t)\right]. \tag{5.4}$$

From (2.55) the amplitude is given by

$$\mathbf{P}^{(2)}_{\omega_{jk}} = \varepsilon_0 \sum_{pqrs} K(-\omega_{jk};\omega_{pq},\omega_{rs})\,\boldsymbol{\chi}^{(2)}(-\omega_{jk};\omega_{pq},\omega_{rs})\,|\,\mathbf{E}_{\omega_{pq}}\,\mathbf{E}_{\omega_{rs}} \times \delta(j,p+r)\,\delta(k,q+s). \tag{5.5}$$

The summation in (5.5) is over all integers p, q, r and s, and is equivalent to the summation term Σ_ω which appears in (2.55). The last two factors in the summand are Kronecker deltas, *e.g.*, $\delta(j,p+r)=1$ if $j=p+r$, and $=0$ otherwise; these ensure that the only nonzero terms in the summand are those for which $\omega_{pq}+\omega_{rs}=\omega_{jk}$. Since the input frequencies ω' and ω'' were taken to be incommensurable, $j\,\omega' \neq k\,\omega''$ for all j and k. Thus, by reference to Table 2.1, we can set $K(-\omega_{jk};\omega_{pq},\omega_{rs})=1$ for all values of j and k. (The choice of incommensurable input frequencies is made for simplicity; it is not essential to the final result.)

We now consider the flow of power from the electric field to the medium. First, consider the linear case. When monochromatic radiation of frequency ω_i passes through a medium (which for the moment we suppose to have a negligible nonlinear response), it experiences absorption and a change of phase velocity due to the imaginary and real parts of the dielectric constant, respectively (see §6.3.3). The real power input at ω_i to unit volume of the medium at any point, W_i, is given by

$$\begin{aligned} W_i &= \left\langle \mathbf{E}(t)\cdot\frac{\mathrm{d}}{\mathrm{d}t}\mathbf{P}(t)\right\rangle \\ &= \tfrac{1}{2}\,\omega_i\,\mathrm{Re}(i\,\mathbf{E}_{\omega_i}\cdot\mathbf{P}^{(1)}_{\omega_i}{}^*) = \tfrac{1}{2}\,\omega_i\,\mathrm{Im}(\mathbf{E}_{\omega_i}{}^*\cdot\mathbf{P}^{(1)}_{\omega_i}), \end{aligned} \tag{5.6}$$

as shown in texts on electromagnetic theory, and where the angle brackets here denote a cycle average. This description may now be generalised to include optical nonlinearity. Thus, the power input to unit volume of the medium at the frequency ω_{jk} through the second-order polarisation is given by:

$$
\begin{aligned}
W_{jk} &= \langle \tfrac{1}{2}[\mathbf{E}_{\omega_{jk}}\exp(-i\omega_{jk}t) + \mathbf{E}_{\omega_{jk}}{}^{*}\exp(i\omega_{jk}t)] \\
&\qquad\qquad \cdot \tfrac{1}{2}\frac{\mathrm{d}}{\mathrm{d}t}[\mathbf{P}^{(2)}_{\omega_{jk}}\exp(-i\omega_{jk}t) + \mathbf{P}^{(2)}_{\omega_{jk}}{}^{*}\exp(i\omega_{jk}t)] \rangle \\
&= \tfrac{1}{2}\,\omega_{jk}\,\mathrm{Re}(i\,\mathbf{E}_{\omega_{jk}}\cdot\mathbf{P}^{(2)}_{\omega_{jk}}{}^{*}) \\
&= \tfrac{1}{2}\,\omega_{jk}\,\mathrm{Re}[i\,(E_{\omega_{jk}})_{\mu}\,(P^{(2)}_{-\omega_{jk}})_{\mu}]\,.
\end{aligned} \tag{5.7}
$$

In the final line of (5.7), the repeated label μ is understood to be summed over the cartesian coordinate axes x, y and z. Then, with (5.5), we find

$$
\sum_{jk}\frac{j}{\omega_{jk}}W_{jk} = \tfrac{1}{2}\varepsilon_0\,\mathrm{Re}\Big\{i\sum_{jkpqrs} j\,\chi^{(2)}_{\mu\alpha\beta}(\omega_{jk};\omega_{pq},\omega_{rs})(E_{\omega_{jk}})_{\mu}(E_{\omega_{pq}})_{\alpha}(E_{\omega_{rs}})_{\beta}
\times\delta(-j,p+r)\,\delta_{\mathrm{K}}(-k,q+s)\Big\}. \tag{5.8}
$$

It is not difficult to show that the right-hand side of this equation vanishes in consequence of overall permutation symmetry, as follows. By making the interchange of summation variables $(j, k, \mu)\leftrightarrow(p, q, \alpha)$ on the right-hand side of (5.8) we obtain

$$
\begin{aligned}
\sum_{jk}\frac{j}{\omega_{jk}}W_{jk}
&= \tfrac{1}{2}\varepsilon_0\,\mathrm{Re}\Big\{i\sum_{jkpqrs} p\,\chi^{(2)}_{\alpha\mu\beta}(\omega_{pq};\omega_{jk},\omega_{rs})\,(E_{\omega_{pq}})_{\alpha}(E_{\omega_{jk}})_{\mu}(E_{\omega_{rs}})_{\beta} \\
&\qquad\qquad \times\delta(-p,j+r)\,\delta(-q,k+s)\Big\} \\
&= \tfrac{1}{2}\varepsilon_0\,\mathrm{Re}\Big\{i\sum_{jkpqrs} p\,\chi^{(2)}_{\alpha\mu\beta}(\omega_{pq};\omega_{jk},\omega_{rs})\,(E_{\omega_{jk}})_{\mu}(E_{\omega_{pq}})_{\alpha}(E_{\omega_{rs}})_{\beta} \\
&\qquad\qquad \times\delta(-j,p+r)\,\delta(-k,q+s)\Big\}.
\end{aligned} \tag{5.9}
$$

In the second line we have interchanged the order of the factors $(E_{\omega_{jk}})_{\mu}$ and $(E_{\omega_{pq}})_{\alpha}$, and used the identities

$$
\begin{aligned}
\delta(-p,j+r) &= \delta(-j,p+r) \\
\delta(-q,k+s) &= \delta(-k,q+s)\,.
\end{aligned} \tag{5.10}
$$

We now observe that the only nonzero terms in the sum in (5.9) are those for which $-j=p+r$ and $-k=q+s$, so that $\omega_{jk} = -(\omega_{pq}+\omega_{rs})$. It follows from overall permutation symmetry that the element of $\boldsymbol{\chi}^{(2)}$, which appears in (5.9), is identical to that which appears in (5.8). Hence, the final result of making the interchange of summation variables $(j, k, \mu)\leftrightarrow(p, q, \alpha)$ in (5.8) is merely to replace the initial factor j in the summand by p. Similarly, the interchange of summation variables $(j, k, \mu)\leftrightarrow(r, s, \beta)$ merely replaces the initial factor j in the summand by r. Finally, by adding together these three alternative expressions for the right-hand side of (5.8) and dividing by 3, we find that the initial factor j in the summand is replaced by $\frac{1}{3}(j+p+r)$. It follows that the right-hand side of (5.8) vanishes because

$$\tfrac{1}{3}(j+p+r)\,\delta(-j, p+r) = 0 \tag{5.11}$$

for all values of j, p and r. Thus we obtain the relation

$$\sum_{jk} \frac{j}{\omega_{jk}} W_{jk} = 0. \tag{5.12}$$

A similar discussion yields the relation

$$\sum_{jk} \frac{k}{\omega_{jk}} W_{jk} = 0. \tag{5.13}$$

In these equations W_{jk} denotes the power input to unit volume of the medium at the frequency $|\omega_{jk}|$ through the second-order polarisation. However, the relations depend only on the overall permutation-symmetry property which holds good in all orders. Consequently, (5.12) and (5.13) remain valid when W_{jk} is identified with the power input density at the frequency $|\omega_{jk}|$ through the nth-order polarisation, where n is arbitrary. Finally, by summing the relations over all orders we see that W_{jk} may in fact be identified with the *total* power input density at the frequency $|\omega_{jk}|$. With this interpretation of W_{jk}, (5.12) and (5.13) are the Manley-Rowe power relations. They are usually presented in a slightly different form by making use of the fact that $W_{-j\,-k} = W_{jk}$. Thus we have:

$$\sum_{j=1}^{\infty} \sum_{k=-\infty}^{\infty} \frac{j}{\omega_{jk}} W_{jk} = \sum_{k=1}^{\infty} \sum_{j=-\infty}^{\infty} \frac{k}{\omega_{jk}} W_{jk} = 0 \tag{5.14}$$

where W_{jk} is the total power input density at the frequency $|\omega_{jk}|$. When (5.12) is multiplied by ω' and added to (5.13) multiplied by ω'' we obtain the simple result: $\sum_{jk} W_{jk} = 0$; *i.e.*, there is *no* real net power flow from the total electric field to the medium. The flow of power from the fields to the medium produces a *virtual* excitation which is reversible (see §4.3.3). In other words, the medium is purely reactive; it can be considered to act only as a catalyst to the nonlinear process, mediating the flow of power between the various optical fields.

Manley-Rowe relations, such as (5.12) and (5.13), may be used to describe the flow of power between monochromatic fields which are coupled through one specific order of nonlinearity; indeed, this is their most common application. For example, let us consider the second-order process of sum-frequency generation: $\omega_1 + \omega_2 \rightarrow \omega_3$. In the notation of (5.2), ω_3 is equivalent to ω_{11}, and $\omega_1 \equiv \omega' = \omega_{10}$, $\omega_2 \equiv \omega'' = \omega_{01}$. If we again use W_i to denote the power input to unit volume of the medium at the frequency ω_i, then from (5.12) and (5.13) we obtain the relation:

$$W_1/\omega_1 = W_2/\omega_2 = -W_3/\omega_3 . \tag{5.15a}$$

Since $\omega_1 + \omega_2 = \omega_3$, we find the result:

$$-W_3 = W_1 + W_2 , \tag{5.15b}$$

i.e., the power generated at ω_3 is equal in magnitude to the total power lost by the fields at ω_1 and ω_2. As a further example, consider the second-order process of difference-frequency generation: $\omega_1 - \omega_2 \rightarrow \omega_3$. Similar to (5.15), for this process we obtain the relation:

$$W_1/\omega_1 = -W_2/\omega_2 = -W_3/\omega_3 . \tag{5.16}$$

Thus, in this case, the source field at ω_1 loses power not only to the generated frequency ω_3, but *also* to the source field ω_2. In other words, if the difference frequency ω_3 is generated using the two input fields ω_1 and ω_2, then *both* ω_2 and ω_3 gain in power. This property is exploited in parametric amplifiers and oscillators (see §7.2.2).

The above derivation of the Manley-Rowe relations was made assuming classical fields. The relations can, however, be given a quantum-mechanical interpretation in terms of photon numbers. This interpretation is made by dividing relations such as (5.15a) and (5.16) by Planck's constant $\hbar$. Now, $W_i/\hbar\omega_i$ is equal to the number of photons of frequency ω_i which are annihilated per unit volume in unit time during the nonlinear optical interaction (a negative value of W_i implies photon creation). Thus, for the second-order sum-frequency process described by (5.15a), the number of photons created at the sum frequency ω_3 is equal to the number of photons annihilated at each of the two source frequencies ω_1 and ω_2. Similarly, for the difference-frequency process described by (5.16), the number of photons annihilated at the source frequency ω_1 is equal to the number of photons created at each of the frequencies ω_2 and ω_3. This latter process has sometimes been described as 'photon splitting'. In such nonlinear processes, the total *number* of photons is not conserved.

By way of contrast, let us consider stimulated Raman scattering, a nonlinear process whose governing susceptibility does not exhibit overall permutation symmetry (*cf.* §4.5.2); therefore the Manley-Rowe relations do not apply. The pump (ω_P) and Stokes (ω_S) frequencies are related by $\omega_P - \omega_S \simeq \Omega_{fg}$, where Ω_{fg} denotes a transition frequency of the medium. Here the total photon energy is clearly *not* conserved; there is instead a net flow of energy from the pump field to the medium. For each pump photon that is annihilated, a Stokes photon is created and the medium simultaneously receives a quantum of energy $\hbar(\omega_P - \omega_S) \simeq \hbar\Omega_{fg}$. In this case, the total *number* of photons *is* conserved:

$$W_S/\omega_S = -W_P/\omega_P\,. \tag{5.17}$$

Historically, those nonlinear optical processes for which the Manley-Rowe relations are valid (*i.e.,* those involving purely reactive, or lossless systems) were termed 'parametric' processes. According to this definition, stimulated Raman scattering, for example, would not be termed a parametric process, and so neither would the process of two-photon-resonant third-harmonic generation depicted in Fig. 4.1. Unfortunately, with the passage of time, the meaning of the term 'parametric' has altered and is now much less precise. It is often used to refer broadly to those processes whose efficiencies, in a travelling-wave interaction, depend on phase-matching (see §7.2). Conversely, those that do not depend on phase-matching are sometimes termed 'nonparametric' (§7.3). The process of two-photon-resonant third-harmonic generation (Fig. 4.1) is commonly cited as a parametric process because of its dependence on phase-matching, despite the fact that the resonant susceptibility which governs the process does not exhibit overall permutation symmetry. Stimulated Raman scattering in a molecular gas, on the other hand, should clearly be classified as nonparametric; its susceptibility does not have overall permutation symmetry, and the travelling-wave interaction does not depend on phase-matching. There are anomalies in this classification, however, which are mentioned in §7.3.

5.1.2 Contracted susceptibility tensors for second-order processes

We now consider in more detail the permutation-symmetry properties of the second-order susceptibility tensors, and also describe a useful contracted notation for these tensors.

For the process of second-harmonic generation, it is a common practice to write the induced nonlinear polarisation in terms of a third-rank tensor **d**, which is usually defined by

$$P_{2\omega}^{(2)} = \varepsilon_0\, \mathbf{d}(-2\omega;\omega,\omega) : \mathbf{E}_\omega \mathbf{E}_\omega. \tag{5.18}$$

If we compare this with the earlier expression (2.55), and noting from Table 2.1 that $K(-2\omega;\omega,\omega) = \frac{1}{2}$, we find the relation

$$\mathbf{d}(-2\omega;\omega,\omega) = \tfrac{1}{2}\,\boldsymbol{\chi}^{(2)}(-2\omega;\omega,\omega). \tag{5.19}$$

(There are several other definitions for the tensor **d** given in the literature; the one given here is the most common one.) The intrinsic permutation symmetry of the tensor $\chi^{(2)}_{\mu\alpha\beta}(-\omega_\sigma;\omega_1,\omega_2)$ allows the pairs of indices (α,ω_1), (β,ω_2) to be freely interchanged. It follows that for the special case of second-harmonic generation $(\omega_1=\omega_2=\omega)$, the susceptibility is invariant to the permutation of α and β. (This symmetry is identical to that of the piezoelectric tensors (Nye, 1959).) The third-rank tensor $d_{\mu\alpha\beta}(-2\omega;\omega,\omega)$ can therefore be represented in contracted form as a 3x6 matrix $d_{\mu m}(-2\omega;\omega,\omega)$, defined as follows:

$$(P^{(2)}_{2\omega})_\mu = \varepsilon_0 \sum_{\alpha\beta} d_{\mu m}(-2\omega;\omega,\omega)\,(E_\omega)_\alpha\,(E_\omega)_\beta\,, \tag{5.20}$$

where the suffix m takes the values 1−6 with the following correspondence to the pairs of cartesian-axis labels $\alpha\beta$:

m	1	2	3	4	5	6
$\alpha\beta$	xx	yy	zz	zy yz	zx xz	xy yx

(5.21)

In writing (5.20) and (5.21), we recognise that the order in which the field components appear is irrelevant (so that, for example, $(E_\omega)_y(E_\omega)_z + (E_\omega)_z(E_\omega)_y = 2(E_\omega)_y(E_\omega)_z$). It is customary for the cartesian label μ on $d_{\mu m}$ to be assigned the values 1, 2 and 3, representing the axes x, y and z, respectively. The notation is also simplified by omitting the frequency arguments on $d_{\mu m}$. We can therefore write (5.25) and (5.26) in matrix form, as follows:

$$\begin{bmatrix}(P^{(2)}_{2\omega})_x\\(P^{(2)}_{2\omega})_y\\(P^{(2)}_{2\omega})_z\end{bmatrix} = \varepsilon_0 \begin{bmatrix} d_{11} & d_{12} & d_{13} & d_{14} & d_{15} & d_{16}\\ d_{21} & d_{22} & d_{23} & d_{24} & d_{25} & d_{26}\\ d_{31} & d_{32} & d_{33} & d_{34} & d_{35} & d_{36}\end{bmatrix} \begin{bmatrix}(E_\omega)_x^2\\(E_\omega)_y^2\\(E_\omega)_z^2\\2(E_\omega)_y(E_\omega)_z\\2(E_\omega)_x(E_\omega)_z\\2(E_\omega)_x(E_\omega)_y\end{bmatrix}. \tag{5.22}$$

The effect of intrinsic permutation symmetry is therefore to allow the $3^3 = 27$ elements of the tensor $d_{\mu\alpha\beta}$ to be reduced to 18 elements in the matrix $d_{\mu m}$. Later in this chapter (§5.3), the form of the matrix $d_{\mu m}$ for the various crystal classes is given, and it is shown that because of spatial symmetry, there may be fewer than 18 nonzero independent elements. It is also shown that Kleinman symmetry, when valid, can result in a yet further reduction in the number of nonzero independent elements of $d_{\mu m}$. The matrix notation $d_{\mu m}$ has the advantage of being more compact than the tensor notation; however, it should be noted that in spite of their appearance with two suffixes, the matrix elements $d_{\mu m}$ do *not* transform like the components of a second-rank tensor.

If Kleinman symmetry holds (*i.e.*, all the field frequencies are in the transparent spectral region, far removed from resonances of the nonlinear medium), the *d*-coefficient notation for second-harmonic generation may be extended to other second-order processes. The previous relation (5.19) can be modified to give a more general definition:

$$\mathbf{d}(-\omega_\sigma;\omega_1,\omega_2) = \tfrac{1}{2}\,\boldsymbol{\chi}^{(2)}(-\omega_\sigma;\omega_1,\omega_2). \tag{5.23}$$

Thus the **d**-tensor, like the $\boldsymbol{\chi}^{(2)}$-tensor defined in Chapter 2, is a smoothly continuous function as the various frequency arguments tend to zero or become equal (*e.g.*, $\omega_1 \to \omega_2$). From (2.55), the induced second-order polarisation in the nondegenerate case ($\omega_1 \neq \omega_2$) is given, in **d**-tensor notation, by

$$\mathbf{P}^{(2)}_{\omega_\sigma} = 2\varepsilon_0\, K(-\omega_\sigma;\omega_1,\omega_2)\,\mathbf{d}(-\omega_\sigma;\omega_1,\omega_2) : \mathbf{E}_{\omega_1}\,\mathbf{E}_{\omega_2}. \tag{5.24}$$

The intrinsic permutation symmetry of the susceptibility tensor allows the pairs (α,ω_1) and (β,ω_2) to be interchanged, thus: $d^{(2)}_{\mu\alpha\beta}(-\omega_\sigma;\omega_1,\omega_2) = d^{(2)}_{\mu\beta\alpha}(-\omega_\sigma;\omega_2,\omega_1)$. If Kleinman symmetry holds, this allows the frequencies ω_1 and ω_2 to be interchanged: $d^{(2)}_{\mu\beta\alpha}(-\omega_\sigma;\omega_2,\omega_1) = d^{(2)}_{\mu\beta\alpha}(-\omega_\sigma;\omega_1,\omega_2)$, and so we obtain the relationship: $d^{(2)}_{\mu\alpha\beta}(-\omega_\sigma;\omega_1,\omega_2) = d^{(2)}_{\mu\beta\alpha}(-\omega_\sigma;\omega_1,\omega_2)$. Thus, using the contracted suffix notation defined by (5.21), we arrive at the expression

$$(P^{(2)}_{\omega_\sigma})_\mu = 2\varepsilon_0 \sum_{\alpha\beta} d_{\mu m}(-\omega_\sigma;\omega_1,\omega_2)\,(E_{\omega_1})_\alpha\,(E_{\omega_2})_\beta\,, \tag{5.25}$$

or, in matrix form:

$$\begin{bmatrix} (P^{(2)}_{\omega_\sigma})_x \\ (P^{(2)}_{\omega_\sigma})_y \\ (P^{(2)}_{\omega_\sigma})_z \end{bmatrix} = 2\,\varepsilon_0\, K(-\omega_\sigma;\omega_1,\omega_2) \times \begin{bmatrix} d_{11} & \cdots & d_{16} \\ d_{21} & \cdots & d_{26} \\ d_{31} & \cdots & d_{36} \end{bmatrix} \begin{bmatrix} (E_{\omega_1})_x(E_{\omega_2})_x \\ (E_{\omega_1})_y(E_{\omega_2})_y \\ (E_{\omega_1})_z(E_{\omega_2})_z \\ (E_{\omega_1})_y(E_{\omega_2})_z + (E_{\omega_2})_y(E_{\omega_1})_z \\ (E_{\omega_1})_x(E_{\omega_2})_z + (E_{\omega_2})_x(E_{\omega_1})_z \\ (E_{\omega_1})_x(E_{\omega_2})_y + (E_{\omega_2})_x(E_{\omega_1})_y \end{bmatrix} \tag{5.26}$$

Here, as in (5.22), the frequency arguments have been omitted from the coefficients $d_{\mu m}$; this reflects the fact that in the transparent spectral region in which Kleinman symmetry is assumed to hold, the frequency dispersion of the **d**-tensors can be considered to be small.

Finally, by expressing the monochromatic fields $\mathbf{E}_{\omega_j}$ in terms of a unit polarisation vector $\mathbf{e}_j$ and a scalar amplitude E_j, as in (2.57): $\mathbf{E}_{\omega_j} = \mathbf{e}_j E_j$, the polarisation $\mathbf{P}^{(2)}_{\omega_\sigma}$ with unit polarisation vector $\mathbf{e}_\sigma$ can be written in the simple scalar form:

$$P^{(2)}_\sigma = 2\varepsilon_0 K(-\omega_\sigma;\omega_1,\omega_2)\, d_{\text{eff}} E_1 E_2. \tag{5.27}$$

Here d_{eff}, the effective d-coefficient, is a scalar parameter defined by

$$d_{\text{eff}} = \sum_{\mu\alpha\beta} d_{\mu m}\,(e_\sigma{}^*)_\mu\,(e_1)_\alpha\,(e_2)_\beta = \tfrac{1}{2}\chi^{(2)}(-\omega_\sigma;\omega_1,\omega_2), \tag{5.28}$$

with $\chi^{(2)}$ defined as in (2.58). Equation (5.28) may be used together with the results of §5.3 to derive expressions for d_{eff} in the different crystal classes, for various configurations of the polarisation directions of the input fields; expressions obtained for the uniaxial crystal classes are listed in Appendix 4.

5.2 Time-reversal symmetry

In this section we consider another symmetry property of the susceptibility tensors, which arises when the Hamiltonian is invariant under time-reversal. We begin by discussing briefly what is meant by time-reversal invariance. For simplicity, the discussion will be confined to a system of spinless particles. The time-reversal symmetry properties of the susceptibility tensors are not affected when spin is taken into account, but a more sophisticated treatment is necessary to derive them (*e.g.*, Wigner, 1959; Butcher and McLean, 1963, 1964).

In classical mechanics, time-reversal means changing the sense in which time is measured (*i.e.*, replacing t by $-t$). We find two important classes of classical dynamical variables which are distinguished by their behaviour under time-reversal. First, there are the dynamical variables invariant under time-reversal, *e.g.*, position coordinates, functions of position coordinates, and even functions of the momenta such as the kinetic energy. Second, there are the dynamical variables which change sign under time-reversal, *e.g.*, the momenta, odd functions of the momenta, and the angular momenta. When we make the transition to quantum mechanics we find that dynamical variables in the first class are represented by real operators in the Schrödinger picture, while those in the second class are represented by pure imaginary operators. Thus the operation of time-reversal is equivalent, in the quantum-mechanical formalism for a system of spinless particles, to complex conjugation. Operators which are real in the Schrödinger picture are invariant under time-reversal.

To derive the time-reversal symmetry properties of the susceptibility tensors we use the formula (4.75) for $\boldsymbol{\chi}^{(n)}$. The matrix elements of the dipole-moment operator which appear there are given by

$$er^{\alpha}_{ab} = \int d\varsigma\, u_a^*(\boldsymbol{\Theta})\, er_\alpha\, u_b(\boldsymbol{\Theta})\,, \tag{5.29}$$

where, as in §4.2, $\boldsymbol{\Theta}$ is a vector of the molecular configuration coordinates, the wave functions $u_i(\boldsymbol{\Theta})$ are eigenfunctions of the molecular Hamiltonian H_0, *i.e.*,

$$H_0 u_i(\boldsymbol{\Theta}) = E_i u_i(\boldsymbol{\Theta}), \tag{5.30}$$

and

$$er_\alpha = -e\sum_j (\mu_j)_\alpha + e\sum_k Z_k (\mu_k)_\alpha\,. \tag{5.31}$$

In (5.31) the summation is over all the charged particles (electrons and ion cores) in the molecule, and the symbols have the meanings defined previously in (4.1), *i.e.*, the position vector of the jth electron (charge $-e$) is denoted by $\boldsymbol{\mu}_j$, and the position vector of the kth nucleus of charge $Z_k e$ is denoted by $\boldsymbol{\mu}_k$. Now, when the particles are spinless and the molecule is not subjected to a d.c. magnetic field, the Hamiltonian H_0 consists of a sum of kinetic energy terms and an interaction potential energy which is a real function of the vector coordinates $\boldsymbol{\mu}_j$ and $\boldsymbol{\mu}_k$. Hence H_0 is invariant under time-reversal. It follows from (5.30) that the energy eigenfunctions $u_i(\boldsymbol{\Theta})$ may be chosen to be real. With this choice of energy eigenfunctions, we see from (5.29) that the matrix elements er^{α}_{ab} are all real, since, from (5.31), er_α is obviously real. The time-reversal symmetry property of $\boldsymbol{\chi}^{(n)}$ follows from this fact.

Inspecting (4.75), we see that the Boltzmann factor $\rho_0(a) = \eta\exp(-E_a/kT)$ and the product of dipole matrix elements in the numerator are both real. Moreover, the molecular transition frequencies Ω_{ba} which appear in the denominator are also real. Consequently, when we take the complex conjugate of (4.75), the only result is to replace the frequencies $\omega_1, ..., \omega_n$ by their complex conjugates. Thus we have the time-reversal-symmetry property:

$$[\boldsymbol{\chi}^{(n)}(-\omega_\sigma;\omega_1, ..., \omega_n)]^* = \boldsymbol{\chi}^{(n)}(-\omega_\sigma{}^*;\omega_1{}^*, ..., \omega_n{}^*). \tag{5.32}$$

We may put this relation into a simpler form by combining it with the reality condition (2.43):

$$[\boldsymbol{\chi}^{(n)}(-\omega_\sigma;\omega_1, ..., \omega_n)]^* = \boldsymbol{\chi}^{(n)}(\omega_\sigma{}^*;-\omega_1{}^*, ..., -\omega_n{}^*). \tag{5.33}$$

In §2.2 this relation was derived from the fact that $\mathbf{P}^{(n)}(t)$ and $\mathbf{E}(t)$ are both real vectors. We note, in passing, that (5.33) may also be obtained from the explicit formula (4.75) by using the Hermitian character of the matrix elements of $\mathbf{r}$. From (5.32) and (5.33) we see that

$$\boldsymbol{\chi}^{(n)}(\omega_\sigma; -\omega_1, \ldots, -\omega_n) = \boldsymbol{\chi}^{(n)}(-\omega_\sigma; \omega_1, \ldots, \omega_n), \tag{5.34}$$

i.e., $\boldsymbol{\chi}^{(n)}$ is invariant when all the frequencies $\omega_\sigma, \omega_1, \ldots, \omega_n$ are negated.

The most important consequence of time-reversal symmetry is the fact that the first-order susceptibility is symmetrical in its subscripts. By setting $n = 1$ in (5.34) we have

$$\chi^{(1)}_{\mu\alpha}(-\omega; \omega) = \chi^{(1)}_{\mu\alpha}(\omega; -\omega). \tag{5.35}$$

Now, the overall permutation symmetry of $\boldsymbol{\chi}^{(1)}$ implies that

$$\chi^{(1)}_{\mu\alpha}(\omega; -\omega) = \chi^{(1)}_{\alpha\mu}(-\omega; \omega), \tag{5.36}$$

and so, by combining (5.35) and (5.36), we have

$$\chi^{(1)}_{\mu\alpha}(-\omega; \omega) = \chi^{(1)}_{\alpha\mu}(-\omega; \omega); \tag{5.37}$$

i.e., $\boldsymbol{\chi}^{(1)}$ is a symmetrical tensor. The same result may be derived from general energy considerations (*e.g.*, Nye, 1959). We shall make use of the fact that $\boldsymbol{\chi}^{(1)}$ is symmetrical in our discussion of the spatial symmetry properties of the susceptibilities in the following section. The relation (5.37) is also the basis of the reciprocity theorems in electromagnetic theory (*e.g.*, Landau and Lifshitz, 1960).

Another consequence of time-reversal symmetry is obvious by inspection of (5.32): when the frequencies $\omega_\sigma, \omega_1, \ldots, \omega_n$ are real and avoid the singularities of $\boldsymbol{\chi}^{(n)}$, the nth-order susceptibility tensor is real. In deriving the time-reversal-symmetry properties of the susceptibility tensors, starting from (4.75), we have assumed implicitly the simultaneous validity of overall permutation symmetry. The time-reversal symmetry of the susceptibilities, like overall permutation symmetry, breaks down unless all the optical frequencies which occur in the explicit formulae for the susceptibilities are sufficiently far removed from all the transition frequencies of the medium to allow the linewidths or damping factors to be neglected. However, it is important to notice that (5.37), which involves both time-reversal and overall permutation symmetries in its derivation, is valid for *all* real frequencies. The reason is that (5.37) involves the same frequency ω on both sides. We saw in Chapter 2 that real frequencies should always be approached from the upper half of the complex frequency plane. There is no problem in bringing ω down from above to any point on the real axis in (5.37). In (5.34), on the other hand, the frequencies on one side are minus those on the other and we cannot bring *all* the frequencies involved down to points on the real axis from above. Consequently time-reversal symmetry in the form (5.34) is valid for real frequencies only if we keep well away from resonances so that the direction in which the frequencies approach the real axis is immaterial. Similar remarks apply to overall permutation symmetry in the form described in §5.2. By inspection of the explicit formulae, for example (4.112), it can be seen that the regime in which these symmetry properties remain valid is precisely the regime in which the susceptibilities can be taken to be real.

5.3 Spatial symmetry

In the previous sections it has been shown that the susceptibility tensors $\boldsymbol{\chi}^{(n)}$ are subject to certain internal symmetry restrictions: intrinsic and overall permutation symmetry, and time-reversal symmetry. In addition to these, the susceptibility tensors are subject to restrictions imposed by the spatial symmetry of the nonlinear medium which they describe. Depending on the symmetry properties of the medium, the effect of these restrictions may be to cause some of the tensor coefficients to vanish, and some of the nonvanishing coefficients may be related so that the number of independent coefficients is thus reduced. The spatial symmetry of the susceptibility tensors is an essential feature in many spectroscopic applications, and is also an important factor in the design of nonlinear-optical devices. This section is concerned mainly with the problem of expressing the susceptibility tensors, for a given nonlinear medium of known symmetry, in a form in which all the vanishing tensor elements are identified and all symmetry relationships between the nonvanishing tensor elements are stated explicitly. There are several methods for reducing the tensors, of varying complexity and generality, and in the following pages we shall briefly outline some of these approaches, giving examples. In the final sub-section we shall make the connection between the general discussion of spatial symmetry properties and the explicit formulae derived in Chapter 4 for the nonlinear susceptibility of a molecular gas.

We should note here that a full description of crystal symmetry groups is given in the *International Tables for X-Ray Crystallography* (Henry and Lonsdale, 1952). The text by Nye (1959) is a very readable account of the influence of spatial symmetry on the tensorial properties of crystals. There are several more advanced texts on the group-theoretical treatment of tensor symmetries, including those by Wigner (1959) and Tinkham (1964). Applications are emphasised in the books by Lomont (1959), Lax (1974), and in the introductory text by Burns (1977). A survey of techniques with many practical examples relating to the photoelastic and electrooptic properties of crystals is given by Narasimhamurty (1981).

The mathematical complexities of some of the topics that we touch on here can be quite considerable, especially those towards the end of the section; our aim, however, is only to sketch the underlying principles and illustrate these with some examples of practical relevance in nonlinear optics.

5.3.1 The transformation laws and Neumann's principle

The guiding principle is a fundamental postulate, known as Neumann's principle, which applies to all of the physical properties of a system that exhibits spatial symmetry (a gas, liquid, crystal or individual molecule). The principle states that any physical property must include all the point-symmetry operations characteristic of the system, *i.e.*, all of the symmetry elements of the corresponding point group. In applying this principle to the susceptibilities (or molecular hyperpolarisabilities), each component of the tensor $\boldsymbol{\chi}^{(n)}$ (or $\boldsymbol{\gamma}^{(n)}$) is to be regarded as a distinct physical property of the system. Thus Neumann's principle requires that the components of the susceptibility (or hyperpolarisability) tensor will remain invariant under any transformation of coordinates that is governed by a valid symmetry operation for the medium. (Although in the following pages the discussion is in terms of the susceptibilities $\boldsymbol{\chi}^{(n)}$, it applies equally to the hyperpolarisabilities $\boldsymbol{\gamma}^{(n)}$.)

We begin, therefore, by describing the coordinate transformation laws for the susceptibility tensors. The susceptibility tensors transform between different coordinate systems according to the usual transformation laws for polar tensors, which are easily derived. We consider two coordinate systems which are related to one another by a proper or improper rotation. (Proper rotations leave a right-handed coordinate system right-handed; improper rotations change a right-handed coordinate system to a left-handed coordinate system – they involve an inversion operation.) The coordinates of a particular point will be denoted by x_α in one system and by x'_α in the other. Here α labels the coordinate axes and takes the values 1, 2 and 3. In our previous work we have used x, y and z to label the coordinate axes, but this notation is inconvenient when two coordinate systems are involved. The coordinates of the point in the two systems of coordinates are related by the linear transformation

$$\mathbf{x}' = \mathbf{R}\mathbf{x}, \tag{5.38}$$

where $\mathbf{R}$ is a 3×3 matrix which defines the transformation, while $\mathbf{x}'$ and $\mathbf{x}$ are column matrices with elements x'_α and x_α, respectively. The characteristic property of rotations, both proper and improper, is that they preserve lengths (the length L of a vector $\mathbf{v}$ which is represented by the 3×1 matrix $\mathbf{x}$ is given by $L^2 = \mathbf{v}\cdot\mathbf{v} = \mathbf{x}^{\mathrm{T}}\mathbf{x}$, where the superscript ^{T}

denotes the transpose). We must have, therefore,

$$\mathbf{x}'^{\mathrm{T}}\mathbf{x}' = \mathbf{x}^{\mathrm{T}}\mathbf{R}^{\mathrm{T}}\mathbf{R}\mathbf{x} = \mathbf{x}^{\mathrm{T}}\mathbf{x} \tag{5.39}$$

for all $\mathbf{x}$; here we have used the matrix identity $(\mathbf{R}\mathbf{x})^{\mathrm{T}} = \mathbf{x}^{\mathrm{T}}\mathbf{R}^{\mathrm{T}}$. It follows that $\mathbf{R}^{\mathrm{T}}\mathbf{R}$ is the unit matrix, *i.e.*, the inverse of $\mathbf{R}$ is equal to its transpose. In suffix notation, (5.38) reads

$$x'_{\alpha} = R_{\alpha\beta}x_{\beta}\,, \tag{5.40}$$

while the inverse transformation is

$$x_{\beta} = (R_{\beta\alpha})^{-1}x'_{\alpha} = R_{\alpha\beta}\,x'_{\alpha}\,. \tag{5.41}$$

The electric field $\mathbf{E}(t)$ and the polarisation $\mathbf{P}(t)$ are both polar vectors which transform in the same way as the coordinates; so too are the Fourier transforms of these vectors. Thus we have the transformation laws (written in the form that we shall use them):

$$\begin{aligned} P'_{\mu}(t) &= R_{\mu u}\,P_u(t) \\ E_a(\omega) &= R_{\alpha a}\,E'_{\alpha}(\omega). \end{aligned} \tag{5.42}$$

The transformation laws for the susceptibility tensors follow from (5.42). As a specific example we consider the first-order (linear) susceptibility tensor. From the definition (2.33), the first-order polarisation expressed in the unprimed coordinate system is given by

$$P_u^{(1)}(t) = \varepsilon_0 \int_{-\infty}^{+\infty} \mathrm{d}\omega\, \chi_{ua}(1)(-\omega;\omega)\,E_a(\omega)\exp(-i\omega t). \tag{5.43}$$

Hence, from (5.42) we find in the primed coordinate system:

$$P_{\mu}^{(1)\prime}(t) = \varepsilon_0 \int_{-\infty}^{+\infty} \mathrm{d}\omega\, \chi_{\mu\alpha}^{(1)\prime}(-\omega;\omega)\,E_{\alpha}(\omega)\exp(-i\omega t), \tag{5.44}$$

where

$$\chi_{\mu\alpha}^{(1)\prime}(-\omega;\omega) = R_{\mu u}\,R_{\alpha a}\,\chi_{ua}^{(1)}(-\omega;\omega). \tag{5.45}$$

This is the usual transformation law for polar tensors of rank 2.

Similarly, we find that the transformation law for $\boldsymbol{\chi}^{(2)}(-\omega_{\sigma};\omega_1,\omega_2)$ is:

$$\chi_{\mu\alpha\beta}^{(2)\prime}(-\omega_{\sigma};\omega_1,\omega_2) = R_{\mu u}\,R_{\alpha a}\,R_{\beta b}\,\chi_{uab}^{(2)}(-\omega_{\sigma};\omega_1,\omega_2), \tag{5.46}$$

and the transformation law for the nth-order susceptibility tensor is:

$$\begin{aligned} &\chi_{\mu\alpha_1\cdots\alpha_n}^{(n)\prime}(-\omega_{\sigma};\omega_1,\ldots,\omega_n) = \\ &\qquad R_{\mu u}\,R_{\alpha_1 a_1}\cdots R_{\alpha_n a_n}\,\chi_{ua_1\cdots a_n}^{(n)}(-\omega_{\sigma};\omega_1,\ldots,\omega_n). \end{aligned} \tag{5.47}$$

The elements of the transformation matrix $\mathbf{R}$ are given by the direction cosines $R_{\alpha a} = \cos\theta_{\alpha a}$, where $\theta_{\alpha a}$ is the angle between the α and a axes.

A parallel argument in terms of the electric field $\mathbf{E}(t)$ and the molecular dipole moment $e\langle\mathbf{r}\rangle$ gives the transformation law for the hyperpolarisability tensor $\boldsymbol{\gamma}^{(n)}$ which is identical to (5.47). As an example of its application, the law is used in §4.4.3 to express an orientation average.

5.3.2 Reduction of the susceptibility tensors: The analytical and direct-inspection methods

According to Neumann's principle, the elements of a susceptibility tensor taken with respect to two coordinate systems, which are related by one of the symmetry operations of the medium, must be identical. Thus, when the rotation defined by the transformation matrix **R** in the transformation law (5.47) belongs to the point group of the medium, the susceptibility tensors must be identical in the two coordinate systems, *i.e.*,

$$\chi^{(n)\prime}_{\mu\alpha_1\cdots\alpha_n}(-\omega_\sigma;\omega_1,\dots,\omega_n) = \chi^{(n)}_{\mu\alpha_1\cdots\alpha_n}(-\omega_\sigma;\omega_1,\dots,\omega_n). \tag{5.48}$$

The transformation law (5.47) taken together with (5.48) thus imposes restrictions on the elements of the susceptibility tensors. By considering the point-group symmetry associated with the medium, the various relationships which exist among the elements of the susceptibility tensors can be established. For example, by considering the point groups for the 32 crystal classes, the relationships for the susceptibility tensors of any crystalline solid can be found. From these relationships the nonzero independent elements can be extracted. This is the analytical method.

For crystals not belonging to the hexagonal class, the simpler but equivalent method of direct inspection can be used (Nye, 1959). For these crystals the different symmetry operations (inversion, rotation about axes, reflection, *etc.*) when applied to (5.47) result in a reshuffling only of the indices and/or sign changes, so that (5.48) becomes

$$\chi^{(n)}_{\mu\alpha_1\cdots\alpha_n}(-\omega_\sigma;\omega_1,\dots,\omega_n) = \varepsilon_{\mathrm{T}}\,\chi^{(n)}_{\{\mu\alpha_1\cdots\alpha_n\}}(-\omega_\sigma;\omega_1,\dots,\omega_n), \tag{5.49}$$

where $\varepsilon_{\mathrm{T}} = \pm 1$, and the subscript symbol $\{\mu\alpha_1\cdots\alpha_n\}$ denotes a single permutation of $\mu\alpha_1\cdots\alpha_n$. Hence the nonzero ($\varepsilon_{\mathrm{T}} = +1$) and independent elements can be found by direct inspection of (5.49).

We now give a particularly simple example of the direct-inspection method, which leads to an important selection rule. The inversion operation is $R_{\alpha\beta} = -\delta_{\alpha\beta}$, where $\delta_{\alpha\beta}$ is the Kronecker delta. When this is a symmetry transformation for the nonlinear medium, (5.47) and (5.48) yield:

$$\chi_{\mu\alpha_1\cdots\alpha_n}(-\omega_\sigma;\omega_1,\dots,\omega_n) = (-1)^{n+1}\chi_{\mu\alpha_1\cdots\alpha_n}(-\omega_\sigma;\omega_1,\dots,\omega_n). \tag{5.50}$$

Clearly, from this relation it follows that $\boldsymbol{\chi}^{(n)}$ vanishes when n is even. Thus we arrive at the important result that all the even-order susceptibility tensors vanish in a medium with inversion symmetry. For example, a second-order process such as second-harmonic generation is thus forbidden in materials with a centre of inversion (for example, a gas which is isotropic). The lowest-order nonvanishing nonlinear polarisation in such a medium is $\mathbf{P}^{(3)}(t)$. Among the 32 crystal classes, there are 21 which lack a centre of inversion (the classes concerned are identified later), and second-order and higher even-order nonlinear effects can occur only in these crystal classes. The relation (5.50) is an example of a selection rule in the electric-dipole approximation. Further selection rules will become apparent when the detailed results for isotropic media and the crystal classes are given below. (It should be reiterated here that throughout this book (except for Appendix 9) we confine attention to the susceptibilities in the electric-dipole approximation. When a similar treatment is applied to the multipole susceptibilities considered in Appendix 9, it is found that the selection rules are different; for example, the second-order susceptibility $\boldsymbol{\chi}^{(2)}$ which incorporates electric-quadrupolar interactions is nonzero for media possessing inversion symmetry.)

As a further demonstration of the operation of the direct-inspection method, we turn now to a detailed discussion of the forms of the first- and second-order susceptibility tensors for crystals belonging to the crystal class $\bar{4}2\text{m}$ in the tetragonal system; examples of crystals in this class are KDP and ADP (potassium and ammonium dihydrogen phosphate respectively), materials which are widely used in nonlinear optics for applications such as second-harmonic generation. The crystal class symbol $\bar{4}2\text{m}$ indicates that the crystal has an inversion tetrad axis ($\bar{4}$) with two orthogonal diad axes (2) perpendicular to it. There are also two mirror planes (m) containing the inversion tetrad axis inclined at 45° to the diad axes. The conventional orientation of the reference axes $Ox_1x_2x_3$ is with Ox_3 along the inversion tetrad axis and Ox_1 and Ox_2 along the diad axes. The symmetry elements and the reference axes are indicated schematically, according to the usual conventions (*e.g.*, Nye, 1959), in Fig. 5.1.

When the reference axes are subjected to one of the symmetry operations in the point group $\bar{4}2\text{m}$ to yield new axes $Ox_1'x_2'x_3'$, the coordinate transformations take a very simple form. The 180° rotation associated with the x_1 (diad) axis yields the symmetry transformation:

$$\begin{aligned} x_1' &= x_1 \\ x_2' &= -x_2 \\ x_3' &= -x_3 \, . \end{aligned} \tag{5.51}$$

The 180° rotation associated with the x_2 axis yields a symmetry

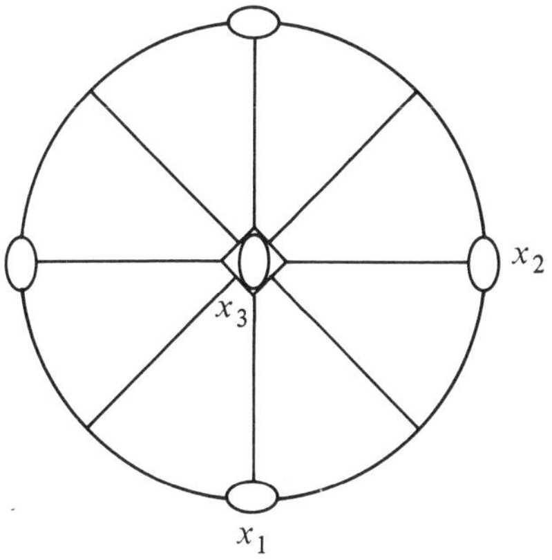

Fig. 5.1 Crystal stereogram showing the symmetry elements and reference axes for the crystal class $\bar{4}2m$.

transformation which is derived from (5.51) by permuting the subscripts according to (123)→(231). The 90° rotation followed by inversion associated with the x_3 (inversion tetrad) axis yields the symmetry transformation:

$$\begin{aligned} x_1' &= -x_2 \\ x_2' &= x_1 \\ x_3' &= -x_3 . \end{aligned} \tag{5.52}$$

We notice that, when this transformation is applied twice, (5.51) is reproduced with the subscripts permuted according to (123)→(312). Thus the x_3 axis is also a diad axis. Finally, the reflection operation associated with the mirror plane between the x_1 and x_2 axes yields the symmetry transformation:

$$\begin{aligned} x_1' &= x_2 \\ x_2' &= x_1 \\ x_3' &= x_3 . \end{aligned} \tag{5.53}$$

All the symmetry restrictions imposed on the first- and second-order susceptibility tensors can be obtained by considering the three diad transformations derived from (5.51) by cyclic permutation of the subscripts and the mirror transformation (5.53). The other transformations associated with the point group $\bar{4}2m$ yield no new information.

The symmetry transformations with which we are concerned are of a simple type. Each coordinate of a particular point in the rotated coordinate system is equal to plus or minus some coordinate in the reference coordinate system. It is therefore convenient to replace the actual transformation equations by a rule which (1) determines how the subscript on a coordinate in the rotated coordinate system has to be altered so as to give the subscript on the related coordinate in the reference coordinate system, and (2) also determines the sign which must be put in front of it. As an example we consider the inverse tetrad transformation (5.52); the corresponding rule is $1 \rightarrow -2$, $2 \rightarrow 1$, $3 \rightarrow -3$ where the arrow may be read as 'goes to'. The advantage of the rule over the actual transformation equations which it represents is that it works for tensors of arbitrary rank, assuming that the minus signs are to be multiplied together and put in front of the related element in the reference coordinate system. Take, for example, the 113 element of $\boldsymbol{\chi}^{(2)}(-\omega_\sigma;\omega_1,\omega_2)$ in the rotated coordinate system derived by means of the inverse tetrad operation. According to our rule, this must be equal to $(-1)(-1)(-1)$ times the 223 element of $\boldsymbol{\chi}^{(2)}(-\omega_\sigma;\omega_1,\omega_2)$ in the original coordinate system, *i.e.*, in an obvious shorthand notation: $113 \rightarrow -223$. It is easy to verify that this is so. The matrix of the transformation is

$$\mathbf{R} = \begin{bmatrix} 0 & -1 & 0 \\ 1 & 0 & 0 \\ 0 & 0 & -1 \end{bmatrix}. \tag{5.54}$$

By setting $\mu=\alpha=1$ and $\beta=3$ in the transformation law (5.46) for the second-order susceptibility, we obtain

$$\begin{aligned} \chi^{(2)\prime}_{113}(-\omega_\sigma;\omega_1,\omega_2) &= R_{1u}R_{1a}R_{3b}\,\chi^{(2)}_{uab}(-\omega_\sigma;\omega_1,\omega_2) \\ &= (-1)(-1)(-1)\,\chi^{(2)}_{223}(-\omega_\sigma;\omega_1,\omega_2). \end{aligned} \tag{5.55}$$

The same considerations apply to all the transformations with which we are concerned here because, like (5.54), the rows and columns in their matrices consist of two zeros and ± 1.

Let us now consider what restrictions are placed on $\boldsymbol{\chi}^{(1)}(-\omega;\omega)$ by the symmetry elements in the $\bar{4}2m$ point group. The diad transformation (5.51) takes the form $1 \rightarrow 1$, $2 \rightarrow -2$, $3 \rightarrow -3$. Hence $12 \rightarrow -12$ and $13 \rightarrow -13$. However, since we are dealing with a symmetry transformation, $\boldsymbol{\chi}^{(1)}(-\omega;\omega)$ must be identical in both coordinate systems. Hence $\chi^{(1)}_{12}(-\omega;\omega)$ and $\chi^{(1)}_{13}(-\omega;\omega)$ both vanish. By considering the diad transformations associated with the remaining two coordinate axes we find that all the off-diagonal elements of $\boldsymbol{\chi}^{(1)}(-\omega;\omega)$ vanish in the reference coordinate system. Finally, the mirror transformation (5.53) takes the form $1 \rightarrow 2$, $2 \rightarrow 1$, $3 \rightarrow 3$. Hence $11 \rightarrow 22$, and so the 11 and 22 diagonal elements of $\boldsymbol{\chi}^{(1)}(-\omega;\omega)$ are equal in the reference coordinate system. This exhausts the symmetry restrictions on $\boldsymbol{\chi}^{(1)}(-\omega;\omega)$.

We turn now to a discussion of the symmetry restrictions placed on $\boldsymbol{\chi}^{(2)}(-\omega_\sigma;\omega_1,\omega_2)$ by the symmetry elements in the $\bar{4}2m$ point group. By considering the three diad transformations we find that any element of $\boldsymbol{\chi}^{(2)}(-\omega_\sigma;\omega_1,\omega_2)$ with either two or three identical subscripts must vanish. For example, the diad transformation (5.52) yields the results $222 \rightarrow -222$ and $112 \rightarrow -112$. Consequently $\chi^{(2)}_{222}(-\omega_\sigma;\omega_1,\omega_2)$ and $\chi^{(2)}_{112}(-\omega_\sigma;\omega_1,\omega_2)$ must vanish. We conclude that the $\mu\alpha\beta$ element of $\boldsymbol{\chi}^{(2)}(-\omega_\sigma;\omega_1,\omega_2)$ in the reference coordinate system can only be nonzero if μ, α and β are all different. Consideration of the mirror transformation (5.53) shows that these six elements are equal in pairs which differ only by the interchange of the suffixes 1 and 2. Thus we have finally:

$$\begin{aligned}\chi^{(2)}_{132}(-\omega_\sigma;\omega_1,\omega_2) &= \chi^{(2)}_{231}(-\omega_\sigma;\omega_1,\omega_2)\\ \chi^{(2)}_{213}(-\omega_\sigma;\omega_1,\omega_2) &= \chi^{(2)}_{123}(-\omega_\sigma;\omega_1,\omega_2)\\ \chi^{(2)}_{312}(-\omega_\sigma;\omega_1,\omega_2) &= \chi^{(2)}_{321}(-\omega_\sigma;\omega_1,\omega_2),\end{aligned} \tag{5.56}$$

and this exhausts the symmetry relations between the nonvanishing elements of $\boldsymbol{\chi}^{(2)}(-\omega_\sigma;\omega_1,\omega_2)$ in the case when $\omega_1 \neq \omega_2$. There are thus three independent nonvanishing elements. As we have remarked earlier in §5.1.2, in the case that $\omega_1 = \omega_2 = \omega$, $\boldsymbol{\chi}^{(2)}(-\omega_\sigma;\omega_1,\omega_2)$ is symmetrical in its last two subscripts. So $\chi^{(2)}_{132}(-2\omega;\omega,\omega) = \chi^{(2)}_{123}(-2\omega;\omega,\omega)$, and the relations (5.56) are therefore replaced by:

$$\begin{aligned}&\chi^{(2)}_{132}(-2\omega;\omega,\omega) = \chi^{(2)}_{231}(-2\omega;\omega,\omega) = \chi^{(2)}_{213}(-2\omega;\omega,\omega) = \chi^{(2)}_{123}(-2\omega;\omega,\omega)\\ &\chi^{(2)}_{312}(-2\omega;\omega,\omega) = \chi^{(2)}_{321}(-2\omega;\omega,\omega).\end{aligned} \tag{5.57}$$

There are then only two independent nonvanishing elements, and the form of $\boldsymbol{\chi}^{(2)}(-2\omega;\omega,\omega)$ is the same as that of the piezoelectric tensor.

The discussion of the symmetry of the susceptibility tensors proceeds along the same lines in all the crystal classes except those belonging to the hexagonal system. The sixfold rotation characteristic of the hexagonal system leads to transformation matrices with two non-vanishing elements in each row, and the transformation equations can no longer be reduced to a simple rule. In this case we must have recourse to the full analytical method described earlier, which follows the same general principle but is merely more complicated in practice. (For further worked examples of tensorial properties in several crystal classes, see Narasimhamurty (1981).)

As a final example in this sub-section, we consider the symmetry properties of the third-order susceptibility $\boldsymbol{\chi}^{(3)}(-\omega_\sigma;\omega_1,\omega_2,\omega_3)$ in an isotropic medium. In such media, namely, in gases, liquids and amorphous solids, $\boldsymbol{\chi}^{(3)}$ is the lowest-order nonlinear susceptibility with nonvanishing components. Because of the isotropy, $\boldsymbol{\chi}^{(3)}$ is invariant under any rotation operation (inversion, reflection, rotation, *etc.*). We first consider the coordinate transformation produced by a reflection in the *yz* plane, which takes the form $1 \rightarrow -1$, $2 \rightarrow 2$, $3 \rightarrow 3$. It follows therefore that all components of $\boldsymbol{\chi}^{(3)}$ with an odd number of *x* indices must vanish. Similarly, by considering reflections in the *xy* and *xz* planes, all tensor components with an odd number of y or z indices must be zero, so that of the 81 components of the general fourth-rank tensor, only 21 remain. Next, we consider the coordinate transformation produced by a 90° rotation about the z axis, which takes the form $1\rightarrow2$, $2\rightarrow-1$, $3\rightarrow3$. In the case of the transformation of the tensor $\boldsymbol{\chi}^{(3)}$, the minus sign in $2\rightarrow-1$ has no effect since the number of x and y indices of all the non-vanishing components is even. Hence we have the relations $1111\rightarrow2222$, $1122\rightarrow2211$, $1212\rightarrow2121$ and $1221\rightarrow2112$. Equivalent relations are obtained for 90° rotations about the x and y axes; taken altogether they are:

$$\begin{aligned}&1111\rightarrow2222\rightarrow3333\,,\\&2233\rightarrow3322\rightarrow3311\rightarrow1133\rightarrow1122\rightarrow2211\,,\\&2323\rightarrow3232\rightarrow3131\rightarrow1313\rightarrow1212\rightarrow2121\,,\\&2332\rightarrow3223\rightarrow3113\rightarrow1331\rightarrow1221\rightarrow2112\,.\end{aligned}\tag{5.58}$$

The final step is to consider the invariance of $\boldsymbol{\chi}^{(3)}$ to a rotation through an arbitrary angle θ about an axis, say the z axis. The matrix for such a coordinate transformation is:

$$\mathbf{R}=\begin{bmatrix}\cos\theta & \sin\theta & 0\\ -\sin\theta & \cos\theta & 0\\ 0 & 0 & 1\end{bmatrix}.\tag{5.59}$$

Setting $\mu=\alpha=\beta=\gamma=1$ in the transformation law (5.47) for $n=3$, together with (5.48), gives

$$\chi^{(3)}_{xxxx}=(\cos^4\theta+\sin^4\theta)\,\chi^{(3)}_{xxxx}+2\cos^2\theta\sin^2\theta\,(\chi^{(3)}_{xxyy}+\chi^{(3)}_{xyxy}+\chi^{(3)}_{xyyx}).\tag{5.60}$$

It is then straightforward to show that, for arbitrary values of θ, (5.60) is valid only if

$$\chi^{(3)}_{xxxx}=\chi^{(3)}_{xxyy}+\chi^{(3)}_{xyxy}+\chi^{(3)}_{xyyx}\,.\tag{5.61}$$

From this relation, together with (5.58), we see that there are at last only three independent tensor elements of $\boldsymbol{\chi}^{(3)}$ in isotropic media, namely: $\chi^{(3)}_{xxyy}$, $\chi^{(3)}_{xyxy}$ and $\chi^{(3)}_{xyyx}$, or their equivalents. If, in addition, Kleinman symmetry holds, then

$$\chi^{(3)}_{xxyy}=\chi^{(3)}_{xyxy}=\chi^{(3)}_{xyyx}=\tfrac{1}{3}\,\chi^{(3)}_{xxxx}\,,\tag{5.62}$$

and thus only one nonzero independent element remains finally.

5.3.3 Group-theoretical methods

For those point groups which contain three- and higher-fold axes, and particularly for tensors of higher rank, the direct-inspection and analytical methods become impractical. Then a more unified and convenient way for reducing the tensors within a chosen frame is that based on group-theoretical considerations. Fumi (1952) and Lomont

(1959) describe a group-theoretical procedure that can be used to determine the number of independent components of the susceptibility $\boldsymbol{\chi}^{(n)}$. The vanishing components may then be identified using the method of invariants, described by Erdös (1964), or a related group-theoretical method due to Lax (1974). Most recently, sequences of crystallographic and limit group generators were used to compute numerically the non-vanishing components of susceptibility tensors up to third order (Shang and Hsu, 1987), and by this means errors in the original tables of Butcher (1965) were detected. The tables of susceptibility tensors given in Appendix 3 take account of these corrections.

5.3.4 Results for isotropic media and the 32 crystal classes

The tables given in Appendices 3 and 4 contain the results obtained for the first-, second- and third-order susceptibility tensors, for all the 32 crystal classes and also for isotropic media. In the tables the reference axes, denoted *Oxyz*, are the principal crystallographic axes, whose relation to the symmetry elements of the crystal conforms with the IRE convention.† In Table A3.1 we have used the fact (proved in §5.2) that $\boldsymbol{\chi}^{(1)}(-\omega;\omega)$ is a symmetric tensor in any medium. Moreover, the crystal classes have been grouped together into crystal systems because the form of $\boldsymbol{\chi}^{(1)}(-\omega;\omega)$ is the same for each member of a particular system. Only those crystal classes for which $\boldsymbol{\chi}^{(2)}(-\omega_\sigma;\omega_1,\omega_2)$ is not identically zero are listed in Table A3.2, *i.e.*, the 21 classes whose point groups do not contain the inversion operation (non-centrosymmetric point groups). Table A4.1 lists the matrices of second-order coefficients $d_{\mu m}$, as defined in §5.1.2, for each of the crystal classes which appear in Table A3.2. The entries for $d_{\mu m}$ given in Table A4.1 were derived from the elements of $\boldsymbol{\chi}^{(2)}(-\omega_\sigma;\omega_1,\omega_2)$ listed in Table A3.2, for the case $\omega_1=\omega_2=\omega$, and by permuting the last two subscripts (as described in §5.1.2). Also listed in Table A4.1 are supplementary relations between the second-order coefficients which apply in the case that Kleinman symmetry holds. In presenting the results for the third-order susceptibility tensors in Table A3.3, the 32 crystal classes need be considered only in 11 distinct groups. This stems from a general property which applies to all susceptibility tensors of odd order (*i.e.*, even rank); namely, in the transformation of even-rank tensors, as given by the transformation law (5.47), the matrix $[R_{\alpha\beta}]$ occurs an even number of times so that corresponding proper and improper rotations, which differ only by an inversion operation, have the same effect. A particular

† *Standards of Piezoelectric Crystals,* Proc. IRE **37**, 1378 (1949)

example of this occurred in §5.3.2 when we derived the symmetry restrictions on $\boldsymbol{\chi}^{(3)}$ in an isotropic medium.

We notice an interesting result which is apparent from the tabulated data. In its linear response to an electromagnetic field, a material belonging to the cubic system is indistinguishable from an isotropic medium. However, this is no longer the case when nonlinear terms are included in the response. Certain classes of the cubic system (432, $\bar{4}$3m and 23) have nonzero second-order susceptibility tensors, whereas this tensor vanishes for an isotropic medium.

Many polarisation selection rules are readily obtainable from the tabulated relationships between tensor components. We illustrate this by one example: we show that a circularly-polarised beam cannot generate a third-harmonic polarisation in an isotropic medium. Let us suppose that the incident monochromatic field $\mathbf{E}_\omega$ has components in the x and y directions, as follows:

$$\mathbf{E}_\omega = E_x \, \mathbf{x} + E_y \, \mathbf{y} \,, \tag{5.63}$$

where $\mathbf{x}$ and $\mathbf{y}$ are unit polarisation vectors. Then, from (2.54) and Table A3.3, the scalar amplitude of the x-component of the third-harmonic polarisation in an isotropic medium is given by

$$\begin{aligned} (P^{(3)}_{3\omega})_x &= \frac{\varepsilon_0}{4} \left[\chi^{(3)}_{xxxx} E_x^3 + (\chi^{(3)}_{xxyy} + \chi^{(3)}_{xyxy} + \chi^{(3)}_{xyyx}) E_x E_y^2 \right] \\ &= \frac{\varepsilon_0}{4} (\chi^{(3)}_{xxyy} + \chi^{(3)}_{xyxy} + \chi^{(3)}_{xyyx}) E_x (E_x^2 + E_y^2) \,, \end{aligned} \tag{5.64}$$

where for brevity we have omitted the frequency arguments on $\boldsymbol{\chi}^{(3)}(-3\omega; \omega, \omega, \omega)$. Now for a circularly-polarised beam, $E_y = i \, E_x$, and it follows therefore from (5.64) that $P_x = 0$, and similarly $P_y = 0$, *i.e.*, the third-harmonic polarisation vanishes. The same result can be deduced from a quantum-optics approach by considering the conservation of angular momentum associated with the photons (*e.g.*, Bey and Rabin, 1967; Schubert and Wilhelmi, 1986).

The practical importance of polarisation-direction selection rules is that they allow spurious or competing nonlinear processes to be suppressed and also information on the structure of a nonlinear medium to be obtained by isolating specific tensor components. Several polarisation-direction selection rules of particular interest for nonlinear-spectroscopic applications are tabulated in the book by Levenson and Kano (1988).

Although our discussion of the symmetry properties has been couched in terms of the electric-dipole susceptibility tensors, in fact the restrictions placed on a polar tensor by the spatial symmetry of the medium are independent of the particular physical property which the tensor describes. Hence Tables A3.1–A3.3 describe the forms taken by the conductivity tensors (§4.6) or any other polar-property tensors of rank 2, 3 and 4, respectively.

5.3.5 Free atoms and rotating molecules

We now extend this discussion of the spatial symmetry properties of the susceptibility tensors by making the connection with the explicit formulae for the nonlinear susceptibilities of a molecular gas, which are derived in §§4.2 and 4.3. We shall show that formulae such as (4.75) for the general nth-order susceptibility, or the alternative form (4.133), do indeed transform as tensors according to Neumann's principle (§5.3.2). We shall also be finally able to drop the unrealistic assumption made in §§4.2 and 4.3 that the molecules in the gas are identically oriented. The following discussion serves as a brief introduction to more advanced concepts. The treatment below does in fact apply to the description of any physical tensor property, such as the susceptibility, obtainable in the form of (3.32): $<O> = \mathrm{Tr}(\rho O)$ (Yuratich, 1976).

It is helpful first to consider briefly some basic ideas from the theory of group representations. For simplicity we shall introduce only those few definitions and properties that are directly relevant here; more advanced discussions may be found in several texts on the physical applications of group theory, including those cited in the introduction to this section.

The mathematical properties of a group are possessed by the collection of point-symmetry operations characteristic of any molecule or crystal, and these operations may thus define a point group. As we have already seen, one (but not unique) method of representing the effects of symmetry operations is to attach to the molecule or crystal a convenient set of orthogonal axes (the 'basis' vectors), and to express each symmetry operation as a square matrix, such as (5.54). The set of such matrices forms a particular 'representation' of the point group.

As an alternative to using vectors as the basis for symmetry group representations, we are free to use certain mathematical functions, as we now describe. For each rotation R belonging to the group of symmetry operations, there is a corresponding rotation operator O_R. Consider then a set of n linearly independent functions $\psi_1, \ldots, \psi_n$ such that O_R will transform any ψ_i into a linear combination of the functions. By $O_R \psi_i$ we mean 'transform the coordinates and express the contours of the function that are fixed in space in terms of the new coordinates' – note that nothing is done to the functions, they are only expressed in terms of new coordinates. Thus we can write:

$$O_R \psi_i = \sum_j \psi_j \Gamma_{ji}(R)\,, \tag{5.65}$$

and the matrices $\Gamma_{ji}(R)$ are said to form a representation of the group, for which the functions ψ_i form the basis. In our earlier work when the basis functions were unit cartesian vectors, we used (5.65) implicitly in the form of the transformation laws for the susceptibility tensors.

There is an important theorem that relates group theory to quantum mechanics, and that provides the vital link between our quantum-mechanical susceptibility formulae and the concepts of spatial symmetry. The theorem, which is proved in several texts on physical applications of group theory (*e.g.*, Burns, 1977, pp. 131–2), states that eigenfunctions that share the same eigenvalue form a basis of a representation. Thus we may use as a basis the set of wave functions which describe the degenerate states of a particular molecular energy level. We shall therefore now suppose that Γ^a is a representation of the symmetry group G of the molecular Hamiltonian. The molecular states $\psi_{a\alpha}$ form the basis, where α denotes the quantum numbers which distinguish degenerate states of energy E_a, and l_a is the dimension of the representation (equal to the total number of degenerate states of energy E_a). Following the notation of (5.65), the matrices that comprise the representation will be denoted, for example, by $\Gamma^a_{\alpha\alpha'}$.

There is one last theorem to which we shall need to refer; this states that if the basis functions are indeed orthonormal (which is the case for the energy eigenfunctions – as described previously in §3.4), then the representation is 'unitary', a condition that can be expressed as

$$\sum_\alpha \Gamma^a_{\alpha\alpha'}(R)^* \,\Gamma^a_{\alpha\alpha''}(R) = \delta_{\alpha'\alpha''}\,. \tag{5.66}$$

The proof of this theorem rests on the fact that for two functions ψ and ϕ,

$$\int \mathrm{d}\varsigma\, (O_R\psi)^*\, (O_R\phi) = \int \mathrm{d}\varsigma\, \psi^*\phi\,. \tag{5.67}$$

This follows from the meaning of ψ and $O_R\psi$; both of these functions have the same *value* at the same position in space, and it is only the position that is expressed in different coordinates. Therefore for the orthonormal basis functions $\psi_{\alpha\alpha'}$ and $\psi_{\alpha\alpha''}$ we have

$$\begin{aligned}
\delta_{\alpha'\alpha''} &= \int \mathrm{d}\varsigma\, \psi_{a\alpha'}{}^*\, \psi_{a\alpha''} \\
&= \int \mathrm{d}\varsigma\, (O_R\psi_{a\alpha'}{}^*)(O_R\psi_{a\alpha''}) \\
&= \sum_{\alpha\alpha'''} \Gamma^a_{\alpha\alpha'}(R)^*\, \Gamma^a_{\alpha'''\alpha''}(R) \int \mathrm{d}\varsigma\, \psi_{a\alpha}{}^*\, \psi_{a\alpha'''} \\
&= \sum_\alpha \Gamma^a_{\alpha\alpha'}(R)^*\, \Gamma^a_{\alpha\alpha''}(R)\,,
\end{aligned} \tag{5.68}$$

where the first and last steps follow from the definition of orthonormality (3.4), and the second step follows from (5.65). This completes the proof.

Now let us move a step closer to the susceptibility formulae by considering the expression:

$$T = \sum_{\alpha} \int \mathrm{d}\varsigma \, \psi_{a\alpha}^{*} \hat{T} \, \psi_{a\alpha} \, , \tag{5.69}$$

where T is a number which results from applying an operator $\hat{T}$. Under a positive rotation R belonging to the symmetry group G, the new value of T which is 'measured' by $\hat{T}$ is

$$\begin{aligned} T' &= \sum_{\alpha} \int \mathrm{d}\varsigma \, (O_R \psi_{a\alpha})^{*} \hat{T} \, (O_R \psi_{a\alpha}) \\ &= \sum_{\alpha\alpha'\alpha''} \Gamma^{a}_{\alpha\alpha'}(R)^{*} \, \Gamma^{a}_{\alpha\alpha''}(R) \int \mathrm{d}\varsigma \, \psi_{a\alpha'}^{*} \, \hat{T} \, \psi_{a\alpha''} \\ &= T \, , \end{aligned} \tag{5.70}$$

where in the last step we have made use of (5.66). Thus we see that, by virtue of the sum over degenerate states, the number T is invariant for all the symmetry operations of G. Next we consider the components of a vector $\mathbf{T} = \sum_{\mu} T_{\mu} \boldsymbol{\mu}$ where $\boldsymbol{\mu}$ takes the values of the unit cartesian vectors **x**, **y** and **z**. We suppose that the component T_{μ} is 'measured' by an operator $\hat{T}_{\mu}$ in (5.69). Under a positive rotation R of the molecule, represented by the rotation matrix **R**, the new component measured by $\hat{T}_{\mu}$ is

$$\begin{aligned} T_{\mu}' &= \sum_{\alpha} \int \mathrm{d}\varsigma \, (O_R \psi_{a\alpha})^{*} \hat{T}_{\mu} (O_R \psi_{a\alpha}) \\ &= \sum_{\alpha} \int \mathrm{d}\varsigma \, \psi_{a\alpha}^{*} \, O_R \hat{T}_{\mu} O_R^{-1} \psi_{a\alpha} \\ &= \sum_{\alpha} \int \mathrm{d}\varsigma \, \psi_{a\alpha}^{*} \Big(\sum_{u} R_{\mu u} \hat{T}_{u} \Big) \psi_{a\alpha} \\ &= R_{\mu u} T_{u} \, , \end{aligned} \tag{5.71}$$

where the second step follows from recognising that the measurement made by $\hat{T}_{\mu}$ after rotation of the basis functions is entirely equivalent to leaving the basis functions alone but using instead the rotated operator $\hat{T}_{\mu}' = O_R \hat{T}_{\mu} O_R^{-1}$. In the third step in (5.71) we have assumed that $\hat{T}_{\mu}$ is a vector operator. It follows that, for T_{μ} to behave like a vector component, $\hat{T}_{\mu}$ must be a vector operator. By extension, the $(n+1)$-rank tensor components $T^{(n)}_{\mu\alpha_1 \cdots \alpha_n}$ require a tensor operator in (5.71) of the form $\hat{T}^{(n)}_{\mu\alpha_1 \cdots \alpha_n}$. The sum over degenerate states in (5.71) is irrelevant to this result, but is vital for (5.70).

We are now in a position to consider the symmetry properties of the susceptibility formulae of Chapter 4. Let us consider the operator:

$$\sum_b (\Omega_{ba} - \omega')^{-1} \left[\psi_b \int \mathrm{d}\varsigma\, \psi_b{}^*\right] \tag{5.72}$$

which appears numerous times in the hyperpolarisability formulae, where ω' represents some general combination of the field frequencies ω_σ, $\omega_1, \ldots, \omega_n$. We recognise, from (3.25), that the expression in square brackets in (5.72) is the projection operator $P(\psi_b)$, and it arises from terms such as

$$\cdots e\mathbf{r}_{ab}\, e\mathbf{r}_{bc} \cdots \;=\; \cdots \int \mathrm{d}\varsigma\, \psi_a{}^* e\mathbf{r}\, \psi_b \int \mathrm{d}\varsigma\, \psi_b{}^* e\mathbf{r}\, \psi_c \cdots \tag{5.73}$$

which appear in the numerators of the formulae for the hyperpolarisabilities. From the definition (3.36) for the unperturbed equilibrium Hamiltonian H_0, (5.72) is equivalent to

$$\sum_b (H_0 - \mathrm{E}_a - \omega')^{-1} P(\psi_b) = (H_0 - \mathrm{E}_a - \omega')^{-1}, \tag{5.74}$$

since $\sum_b P(\psi_b) = 1$. The expression (5.72) is therefore invariant under all rotations belonging to the molecular point group. Hence, so far as rotation properties are concerned, the nth-order molecular hyperpolarisability $\boldsymbol{\gamma}^{(n)}$ (a tensor of rank $n+1$) is found to be a collection of terms of the form

$$\gamma^{(n)}_{\mu\alpha_1 \cdots \alpha_n} \rightarrow \sum_{a\alpha} \rho_0(a\alpha) \int \mathrm{d}\varsigma\, \psi_{a\alpha}{}^*\, er_\mu er_{\alpha_1} \cdots er_{\alpha_n} \psi_{a\alpha}\,. \tag{5.75}$$

We can neglect here the permutation operators $\mathbf{S}$ and $\mathbf{S}_\mathrm{T}$, since the permutations commute with rotations. The expression (5.75) satisfies the requirements of (5.71) so that, as expected, our hyperpolarisability formula will transform as a tensor. It is a polar tensor since $e\mathbf{r} \rightarrow -e\mathbf{r}$ under inversion. Furthermore, the hyperpolarisability tensor will be invariant under any rotation R belonging to the symmetry group G if it can be cast in the form of (5.69), implying that in (5.75):

$$\sum_{a\alpha} \rho_0(a\alpha) \cdots \;=\; \sum_a \overline{\rho_0(a)}\, l_a^{-1} \sum_\alpha \cdots \tag{5.76}$$

where $\overline{\rho_0(a)}$ denotes the *total* fraction of population in the energy level E_a and, as before, l_a is the number of degenerate sub-states. Thus it is necessary that each degenerate sub-state be equally populated in order that the tensor components be invariant under all rotations R belonging to G. This can be appreciated physically by considering the simplest example of a single free atom; if, for example, some magnetic sub-states of the atom were preferentially occupied, then under the equilibrium conditions to which ρ_0 applies, the atom would be 'oriented', *i.e.,* its symmetry would be lower than G.

We now finally consider the macroscopic susceptibility $\boldsymbol{\chi}^{(n)}$, which is defined by (4.104) and (4.105) in terms of an orientation average of the molecular hyperpolarisabilities $\boldsymbol{\gamma}^{(n)}$. As mentioned in §4.4.3, the general calculation of the orientation average in the nonlinear case $n \geq 2$ becomes large. However, in the special case of a medium consisting of freely-rotating molecules we are saved from this labour because the point group is itself the full rotation group. Since, as we have just shown, the formula for the molecular hyperpolarisability $\boldsymbol{\gamma}^{(n)}$ is invariant to any rotation of the point group, it follows that in this special case $\boldsymbol{\chi}^{(n)}$ must *already* be orientation-averaged. This justifies the simplifying assumption made in our treatment of the susceptibilities of a molecular gas in §§4.2 and 4.3, where the orientation average is neglected.

As implied in the discussion above, a complete description of the spatial properties of the hyperpolarisabilities of rotating molecules can be made in terms of angular-momentum states (Condon and Shortley, 1957; Wigner, 1959; Shore and Menzel, 1968; Brink and Satchler, 1971). Yuratich and Hanna (1976,1977), Cotter and Hanna (1976) have described the application of irreducible-tensor methods, Racah angular-momentum algebra and the Wigner-Eckart theorem, to the atomic and molecular hyperpolarisabilities. In this way, the key spectroscopic parameters (selection rules, the angular dependence on polarisation vectors and the summation over intermediate degeneracies) can all be given in compact and useful forms. Hellwarth (1977) used an elegant spherical-tensor approach to describe the spatial symmetry in the important practical case of the third-order susceptibility $\boldsymbol{\chi}^{(3)}$ of isotropic materials in the situation when the Born-Oppenheimer approximation applies (*i.e.*, when electronic and vibrational-rotational nuclear motions can be treated independently; see §4.5.4).

5.3.6 Microscopic and macroscopic symmetries

An important application of irreducible-tensor decomposition techniques in nonlinear optics is to relate microscopic and macroscopic symmetries in condensed matter. This approach is generally more practical than the lengthy evaluation of the geometrical formulae given in §4.4.3. Irreducible-tensor methods have long been used in atomic and molecular physics, but their use in the description of condensed matter is a more recent development (a general review of irreducible-tensor methods in solid-state nonlinear optics is given by Jerphagnon *et al* (1978)). Nevertheless, the exact calculation of the susceptibility of a crystal is a formidable task, that requires some approximations to be tractable. Most often the system is decomposed into an assembly of building blocks whose structure is simpler and easier to describe (Zyss and Oudar, 1982; Chemla and Zyss, 1987).

6

Resonant nonlinearities

In earlier chapters we have considered in detail the susceptibility formalism of nonlinear optics, which is perhaps the most familiar approach and has a wide range of application. Starting from the constitutive relations of Chapter 2, the susceptibility formalism is quite general. In many practical applications, a particular phenomenon can be described accurately by a single order of nonlinearity, and the susceptibility then provides a useful and convenient description. However, this is not always the case. Some of the most interesting phenomena in nonlinear optics involve close resonance with the transition frequencies of the medium, and perhaps also the use of very intense optical fields. As remarked in §4.5, in these circumstances the resonant susceptibilities display mathematical divergences which are clearly unphysical. Strictly, these divergences occur only because higher-order nonlinearities have been neglected. Successive orders of nonlinearity take into account such effects as saturation, power broadening and level shifts (optical Stark effect). For intense fields or very close resonance, the contributions from several orders of nonlinearity may be comparable in magnitude. Therefore, despite its generality, the susceptibility formalism does not necessarily provide the most *practical* approach for the description of resonant processes.

This chapter is concerned with the problem of deriving alternative and more manageable descriptions of resonant nonlinear processes, illustrated by examples. For much of the chapter, the nonlinear medium is treated as a two-level system (described in §6.2). The essential point about this simple model is that it incorporates saturation and other dynamic properties, and thus it overcomes the major failings of the small-perturbation analysis (Chapter 4) in the limit of strong interactions and resonances. Despite its great simplicity, this model has been very successfully applied to a wide range of interesting and complex problems in nonlinear optics. By modelling a resonant system in this way to describe a particular nonlinear process, the functional form of the polarisation in terms of the field is often found to be the same as a conventional single-order susceptibility, but with field-dependent corrections. This gives rise to the idea of a field-dependent susceptibility, and in §6.3 this is illustrated by the example of a resonant intensity-dependent refractive index. In §§6.4 and 6.5, we describe the development of the dynamic two-level atom model to treat some more complex situations that arise in nonlinear optics: the case of more than one driving field, and multi-photon resonances. To illustrate the latter, and following on from the discussion in §4.5, two-photon-resonant effects are considered in §6.5 – stimulated Raman scattering is given as a particular example. Throughout the chapter, an important objective is to make the link between the different approaches: the classical model of light scattering described in Chapter 1, the susceptibilities of Chapter 4, and the Bloch-equation approach introduced here. Finally, it turns out that interesting and important phenomena can occur when the nonlinear medium is excited with intense short pulses of duration comparable to the material relaxation times. The different regimes of interest for time-dependent processes are summarised in the last section of the chapter.

6.1 Quasi-monochromatic fields

We begin this discussion of resonant nonlinear processes by reexamining the idea of adiabatic fields, introduced in §2.4, in the light of the explicit susceptibility formulae derived in Chapter 4. For clarity, the discussion here is mainly in terms of the linear optical response, but the generalisation to the nonlinear case is shown to be straightforward.

Throughout this chapter, we are concerned with applied fields that comprise a superposition of quasi-monochromatic waves (2.60):

$$\mathbf{E}(t) = \tfrac{1}{2}\sum_{\omega'\geq 0}\Big[\mathbf{E}_{\omega'}(t)\exp(-i\omega' t) + \mathbf{E}_{\omega'}{}^{*}(t)\exp(i\omega' t)\Big]. \tag{6.1}$$

Each wave amplitude $\mathbf{E}_{\omega'}(t)$ may be expressed in terms of its Fourier transform (2.25), which we denote by $\mathbf{E}_{\omega'}(\omega)$:

$$\mathbf{E}_{\omega'}(t) = \int_{-\infty}^{+\infty} \mathrm{d}\omega\, \mathbf{E}_{\omega'}(\omega-\omega')\exp[-i(\omega-\omega')t], \tag{6.2}$$

and it is assumed that for a quasi-monochromatic wave the transform is sharply peaked around $\omega = \omega'$.

In the early sections of the present chapter, we consider the simple case when the applied field consists of a single quasi-monochromatic wave, with envelope function $\mathbf{E}_{\omega_1}(t)$. In this case, when (6.2) is substituted into (2.33) together with (2.61), we obtain the following expression for the Fourier amplitude of the linear polarisation wave as a function of time:

$$\mathbf{P}_{\omega_1}^{(1)}(t) = \varepsilon_0 \int_{-\infty}^{+\infty} \mathrm{d}\omega\, \boldsymbol{\chi}^{(1)}(-\omega;\omega)\cdot \mathbf{E}_{\omega_1}(\omega-\omega_1)\exp[-i(\omega-\omega_1)t]. \quad (6.3)$$

Let us now consider in more detail the meaning of the adiabatic and transient limits for the applied field, which were introduced in §2.4. We follow here some simple derivations given by Hanna *et al* (1979); the results so obtained are equivalent to those derived from first principles by Puell and Vidal (1976).

6.1.1 Adiabatic response

Let us assume, at first, that ω_1 is far removed from the transition frequencies of the medium, so that the susceptibility $\boldsymbol{\chi}^{(1)}(-\omega;\omega)$ is a slowly-varying function of frequency when $\omega=\omega_1$ (this is unnecessarily restrictive, as will become clear shortly). The susceptibility may then be expanded in a Taylor series about ω_1, so that (6.3) becomes

$$P_1^{(1)}(t) = \varepsilon_0 \Bigg\{ \chi^{(1)}(-\omega_1;\omega_1) \int_{-\infty}^{+\infty} \mathrm{d}\omega\, E_1(\omega-\omega_1)\exp[-i(\omega-\omega_1)t]$$
$$+ \frac{\mathrm{d}}{\mathrm{d}\omega}\chi^{(1)}(-\omega;\omega)\Big|_{\omega_1} \int_{-\infty}^{+\infty} \mathrm{d}\omega\,(\omega-\omega_1)\,E_1(\omega-\omega_1)\exp[-i(\omega-\omega_1)t]$$
$$+ \cdots \Bigg\}, \quad (6.4)$$

where we use the scalar notation defined in §2.3.4. From the definition of the Fourier transform (2.25) we see that the first integral in (6.4) is simply $E_1(t)$, and the second is $i\,\mathrm{d}E_1(t)/\mathrm{d}t$. Thus

$$P_1^{(1)}(t) = \varepsilon_0 \left\{ \chi^{(1)}(-\omega_1;\omega_1)E_1(t) + i\frac{\mathrm{d}}{\mathrm{d}\omega}\chi^{(1)}(-\omega;\omega)\Big|_{\omega_1} \frac{\mathrm{d}}{\mathrm{d}t}E_1(t) + \cdots \right\}, \quad (6.5)$$

where the series continues with higher-order derivatives of $\chi^{(1)}$ and $E_1(t)$. Now, from (2.70) we have, in the adiabatic limit,

$$P_1^{(1)}(t) = \varepsilon_0 \chi^{(1)}(-\omega_1;\omega_1)\,E_1(t), \quad (6.6)$$

which we recognise as the first term in the series (6.5). Therefore, a definition of the adiabatically-applied field, more precise than that given in §2.4, is that the second term in (6.5) should be negligible, *i.e.*,

$$\left| [\chi^{(1)}(-\omega_1;\omega_1)]^{-1} \frac{\mathrm{d}}{\mathrm{d}\omega}\chi^{(1)}(-\omega;\omega)\Big|_{\omega_1} [E_1(t)]^{-1}\frac{\mathrm{d}}{\mathrm{d}t}E_1(t) \right| \ll 1. \quad (6.7)$$

We now examine what the condition (6.7) implies in terms of the explicit form of the linear susceptibility derived in Chapter 4. From

(4.111),

$$\chi^{(1)}(-\omega;\omega) = \frac{N}{\varepsilon_0}\frac{e^2}{\hbar}\sum_b |\mathbf{e}_\omega\cdot\mathbf{r}_{ba}|^2\left\{(\Omega_{ba}-\omega-i\Gamma)^{-1}+(\Omega_{ba}+\omega+i\Gamma)^{-1}\right\} \tag{6.8}$$

where it is assumed that $\rho_0(a)=1$, and thus:

$$\frac{\mathrm{d}}{\mathrm{d}\omega}\chi^{(1)}(-\omega;\omega)\bigg|_{\omega_1} = \frac{N}{\varepsilon_0}\frac{e^2}{\hbar}\sum_b |\mathbf{e}_{\omega_1}\cdot\mathbf{r}_{ba}|^2 \times\left\{(\Omega_{ba}-\omega_1-i\Gamma)^{-2} - (\Omega_{ba}+\omega_1+i\Gamma)^{-2}\right\}. \tag{6.9}$$

Therefore, as ω_1 approaches a particular resonance so that $\Delta\equiv \Omega_{ba}-\omega_1 \ll \Omega_{ba}+\omega_1$, the adiabatic limit (6.7) becomes

$$\left|(\Delta-i\Gamma)^{-1}E_1(t)^{-1}\frac{\mathrm{d}}{\mathrm{d}t}E_1(t)\right| \ll 1. \tag{6.10}$$

If we introduce a time scale τ_c such that the characteristic rate of change of the field envelope can be denoted by $E_1(t)/\tau_c \sim \mathrm{d}E_1(t)/\mathrm{d}t$, then (6.10) becomes $|\Delta-i\Gamma|\tau_c \gg 1$. For 'smooth' pulses, τ_c might be the pulse length or rise time. The pulses obtained from practical laser sources often have considerable structure, and a more realistic estimate of τ_c might be obtained from $\tau_c\sim 1/\delta\omega$, where $\delta\omega$ is the optical bandwidth; (6.10) thus states that the frequency spread of the pulse should not overlap the molecular-transition frequency. Another way of expressing the adiabatic condition, which was stated without justification in §2.4, is to require the time scale of changes in the driving field, τ_c, to be much longer than the impulse response time of the polarisation, which for radiation of frequency ω_1 is characterised by $|\Delta-i\Gamma|^{-1}$ (*cf.* (6.14)–(6.16)).

As an example from nonlinear optics, consider third-harmonic generation. The algebra is only a little more complicated than in the linear case. Starting from (2.48) it is found readily that, to first order in $\mathrm{d}E_1(t)/\mathrm{d}t$,

$$P_3^{(3)}(t) = \tfrac{1}{4}\varepsilon_0\left\{\chi^{(3)}(-3\omega_1;\omega_1,\omega_1,\omega_1) + 3i\frac{\mathrm{d}}{\mathrm{d}\omega}\chi^{(3)}\bigg|_{\omega_1}[E_1(t)]^{-1}\frac{\mathrm{d}}{\mathrm{d}t}E_1(t)\right\}[E_1(t)]^3 \tag{6.11}$$

where $\omega_3=3\omega_1$ and

$$\frac{\mathrm{d}}{\mathrm{d}\omega}\chi^{(3)}\bigg|_{\omega_1} \equiv \frac{\mathrm{d}}{\mathrm{d}\omega}\chi^{(3)}(-2\omega_1-\omega;\omega_1,\omega_1,\omega)\bigg|_{\omega_1}. \tag{6.12}$$

These equations may be compared with (6.5). The susceptibility in (6.12) may be written from (4.112). When the derivative is evaluated, the condition (6.10) is again found, where $\Delta - i\Gamma$ is now the smallest of any of the denominator terms in $\chi^{(3)}(-3\omega;\omega,\omega,\omega)$. For example, for the two-photon-resonant process depicted in Fig. 4.1, we have $|\Omega_{kg} - 2\omega - i\Gamma|\tau_c >> 1$. When this condition is satisfied, the incident field can be taken to be adiabatic and we can use the adiabatic formula (2.69), *i.e.*, retain only the first part of (6.11). Similar conditions apply for all nonlinear processes. For those processes, such as third-harmonic generation, in which a new frequency is generated, a further requirement is that the generated field should also satisfy the adiabatic condition – in some cases this is the most stringent condition that must be satisfied.

Another factor to be considered is the effect of the fields on the medium, as manifested by higher-order nonlinearities. This is the subject of later sections of this chapter; however we note here that the adiabatic-following regime (to be discussed later) is described by the adiabatic terms (*e.g.*, the first parts of (6.5) and (6.11)) in each order of nonlinearity, and that corrections may be obtained by including the derivative terms.

6.1.2 Adiabatic condition violated

Now consider what happens to the linear polarisation in the limiting case when ω_1 is tuned close to resonance with a transition. Although it is possible to continue to work in the frequency domain, it is easier and more instructive in this case to transform to the time domain and use the envelope response function introduced in §2.4. We recall, from (2.66), that the linear susceptibility $\boldsymbol{\chi}^{(1)}$ may be related to the first-order envelope response function $\boldsymbol{\Phi}^{(1)}(t)$. By defining a scalar $\Phi^{(1)}(t)$ in the manner analogous to the definition (2.58) for $\chi^{(1)}(-\omega;\omega)$, the relation may be written:

$$\chi^{(1)}(-\omega;\omega) = \int_{-\infty}^{+\infty} \mathrm{d}\tau\, \Phi^{(1)}(\tau)\,. \qquad (6.13)$$

In the limit when the applied quasi-monochromatic field is in close resonance with a particular single transition Ω_{ba} such that $|\Delta| << \Omega_{ba} + \omega_1$, it is straightforward to show from (6.8) and (6.13) that

$$\Phi^{(1)}(\tau) = \frac{iNe^2}{\varepsilon_0 \hbar} |\mathbf{e}_1 \cdot \mathbf{r}_{ba}|^2 \exp[-(i\Delta + \Gamma)\tau] \quad \text{for } \tau \geq 0$$

$$= 0 \quad \text{for } \tau < 0\,, \qquad (6.14)$$

where the second line follows from the causality condition. Thus, as is well known, the exponential decay of the polarisation wave in the time domain is associated with a Lorentzian spectral line shape. Now, from (2.65), the linear polarisation is expressed in terms of the envelope response function as

$$P_1^{(1)}(t) = \varepsilon_0 \int_{-\infty}^{+\infty} d\tau \, \Phi^{(1)}(t-\tau) \, E_1(\tau) \, , \tag{6.15}$$

and substituting (6.14) into (6.15) gives

$$P_1^{(1)}(t) = \frac{iNe^2}{\hbar} |\mathbf{e}_1 \cdot \mathbf{r}_{ba}|^2 \int_{-\infty}^{t} d\tau \, E_1(\tau) \exp[-(i\Delta+\Gamma)(t-\tau)] \, . \tag{6.16}$$

A series expansion of the integral in (6.16) may be obtained by integration by parts:

$$P_1^{(1)}(t) = \frac{iNe^2}{\hbar} |\mathbf{e}_1 \cdot \mathbf{r}_{ba}|^2 \left\{ \int_{-\infty}^{t} d\tau \, E_1(\tau) + (i\Delta+\Gamma) \int_{-\infty}^{t} d\tau \int_{-\infty}^{\tau} d\tau' \, E_1(\tau') + \cdots \right\} . \tag{6.17}$$

The first integral in (6.17) is closely related to the 'field area' which, as described in the next section, is a key parameter in the theory of coherent resonant interactions.

The two expansions (6.5) and (6.17) are complementary: since the field area is proportional to $\tau_c E_1$, it follows that the first term in (6.17) is a good approximation only if $|\Delta - i\Gamma| \tau_c << 1$, which is just the converse of (6.10). In the extreme situation of exact resonance and no damping, the adiabatic condition can never be fulfilled, whereas (6.17) terminates at the first term. The limit $|\Delta - i\Gamma| \tau_c << 1$ defines the coherent-transient regime referred to in §2.4.3 (see also §§6.3 and 6.6). As described in §6.5.2, the detuning factor Δ takes a more general definition in the case of a multi-photon resonance.

6.1.3 Equivalent susceptibility in the transient regime

Lastly, we consider the situation when the quasi-monochromatic wave $E_1(t)$ acting on the medium consists of a short pulse, incident at time $t \simeq 0$, whose duration τ_P is very much less than the polarisation dephasing time ($\tau_P << |\Delta - i\Gamma|^{-1}$). From (6.16) it follows that the polarisation at subsequent times t is given by

$$P_1^{(1)}(t) = \frac{iNe^2}{\hbar} |\mathbf{e}_1 \cdot \mathbf{r}_{ba}|^2 E_1(0) \tau_P \exp[-(i\Delta+\Gamma)t] \, , \tag{6.18}$$

where the pulse duration τ_P is defined by writing the total field area as $\int_{-\infty}^{+\infty} dt \, E_1(t) = E_1(0) \tau_P$ and it is assumed that the peak of the pulse occurs at time $t = 0$. It follows that for pulsed excitation in the coherent-transient regime, the response of the medium depends on the field *area*, rather than the instantaneous field as in the adiabatic or steady-state regimes. It is also clear from (6.18) that, in the coherent-transient regime, the polarisation set up by the pulsed optical field exists for times longer than the pulse duration.

It is sometimes considered convenient to be able to describe the response in the coherent-transient regime in terms of an 'equivalent' susceptibility $\chi^{(1)}_{\text{equ}}$. This is defined by writing (6.18) in a form that is superficially similar to the adiabatic equation (6.6), *i.e.*,

$$P_1^{(1)}(t) = \varepsilon_0 \chi^{(1)}_{\text{equ}}(-\omega_1;\omega_1) E_1(0), \qquad (\tau_{\text{P}} < t << |i\Delta+\Gamma|^{-1}). \quad (6.19)$$

By comparison between (6.8) and (6.18) it follows that the true and 'equivalent' susceptibilities are related as

$$\chi^{(1)}_{\text{equ}}(-\omega_1;\omega_1) = \chi^{(1)}(-\omega_1;\omega_1)\tau_{\text{P}}(i\Delta+\Gamma). \quad (6.20)$$

Here we have assumed that the second term in braces in (6.8) is negligible in comparison with the first; this is the rotating-wave approximation (§6.2.3). The relation (6.20) implies that, in certain situations in the transient regime, it is possible to write an expression for the polarisation in the familiar form of the adiabatic response, except that the true susceptibility is reduced in the ratio of the pulse duration to the polarisation dephasing time. Relations similar to (6.20) can be found for the nonlinear susceptibilities. This rule is quite often applied in a rough-and-ready fashion to a variety of systems – but sometimes also when it is hardly justified. In particular, it is incorrect to apply the rule to those systems in which the relaxation processes are themselves dependent on the optical fluence or intensity.

6.2 Resonant two-level systems

The description of resonant optical processes is greatly simplified by restricting attention to the dominant resonant transition. Probably the most widely used description is based on a model of a two-level atom. As already mentioned in the introduction to the chapter, the two-level model, despite its simplicity, can be usefully applied to many complex systems – a good example is the bandgap-resonant optical nonlinearity of semiconductors discussed in Chapter 8. In this section we review the dynamic two-level model, which is applied in later sections to problems in nonlinear optics. An excellent detailed account of optical resonance and two-level atoms is given by Allen and Eberly (1975).

6.2.1 The two-level atom

We begin by considering the two-level system which consists of a single isolated atom with only two energy eigenstates, of energy E_a and E_b (with $\mathrm{E}_b > \mathrm{E}_a$). To describe this system we can make use of a number of expressions presented in Chapter 3.

At some instant of time ($t=0$), it is determined that the atom is in the lower unperturbed energy level E_a, whose energy eigenstate u_a is determined by the eigenvalue equation (3.36):

$$H_0 u_a = \mathrm{E}_a u_a . \tag{6.21}$$

The state ψ of the system evolves according to the Schrödinger equation (3.3):

$$i\hbar \, \mathrm{d}\psi/\mathrm{d}t = H\psi . \tag{6.22}$$

Without any perturbation being given to the system (*i.e.*, $H=H_0$), the wave function which describes the evolving state of the system for times $t \geq 0$ is given by (3.67) and (3.73), which may be combined and written as:

$$\psi = \exp(-i\mathrm{E}_a t/\hbar)\, u_a . \tag{6.23}$$

Here an arbitrary phase factor (the phase at time $t=0$) has been set equal to zero. Alternatively, if the system is first determined to be in the upper energy eigenstate u_b, then the subsequent unperturbed evolution of the wave function (for $t \geq 0$) is given by

$$\psi = \exp(-i\mathrm{E}_b t/\hbar)\, u_b . \tag{6.24}$$

The more general case is when the wave function for the system is a coherent superposition of the two states (6.23) and (6.24):

$$\psi = a \exp(-i\mathrm{E}_a t/\hbar) u_a + b \exp(-i\mathrm{E}_b t/\hbar) u_b , \tag{6.25}$$

where the state is specified by the complex numbers a and b, and again an overall arbitrary phase factor has been set equal to unity. The probability that the system is to be found in either the lower or upper energy level is $|a|^2$ or $|b|^2$ respectively, and $|a|^2+|b|^2=1$.

Let us now consider the system comprising an *ensemble* of many two-level atoms which are assumed to be identical and noninteracting. We wish to determine the evolution of this ensemble of atoms under the influence of an applied optical field. The time evolution of the state of a single atom is described conveniently by the Schrödinger equation (6.22). However, when considering an ensemble of a large number of entities, it is more appropriate to consider the time evolution of the density operator ρ for the ensemble as a whole, as described in §3.4. The equation of motion for ρ is given by (3.34):

$$i\hbar\, \mathrm{d}\rho/\mathrm{d}t = [H, \rho]. \tag{6.26}$$

From the definition (3.33), we can write the density operator for the ensemble as

$$\rho = \sum_{\psi} p_{\psi} P(\psi) \tag{6.27}$$

where, from (3.25), the projection operator is

$$P(\psi) = \psi \int d\varsigma\, \psi^*, \tag{6.28}$$

and p_{ψ} is the probability that the two-level system is in a particular superposition state ψ. The density operator can be specified in terms of a 2×2 matrix $[\rho_{ij}]$ whose (ij)th element is defined by (3.7):

$$\rho_{ij} = \int d\varsigma\, u_i^* \rho\, u_j\,. \tag{6.29}$$

From (6.27) – (6.29) we obtain

$$\begin{bmatrix} \rho_{aa} & \rho_{ab} \\ \rho_{ba} & \rho_{bb} \end{bmatrix} = \sum_{\psi} p_{\psi} \begin{bmatrix} |a|^2 & ab^* \exp(i\Omega_{ba}t) \\ a^*b \exp(-i\Omega_{ba}t) & |b|^2 \end{bmatrix} \tag{6.30}$$

where $\Omega_{ba} = (\mathrm{E}_b - \mathrm{E}_a)/\hbar$. When evaluating the various integrals (6.29) we have made repeated use of the fact that the states u_a and u_b are orthonormal, as described by (3.4). Notice that $\rho_{ba} = \rho_{ab}^*$, as expected for an Hermitian operator, and that $\rho_{aa} + \rho_{bb} = 1$. The diagonal elements of ρ represent the probabilities of finding the system in either one of the two energy eigenstates, whilst the off-diagonal elements are a measure of the coherence intrinsic to a superposition state.

If it were known with complete certainty that all the two-level systems in the ensemble were in the same state ψ, *i.e.*, the values of a and b were the same for each system, then quite simply $p_{\psi} = 1$ in (6.30). The ensemble is then said to be completely coherent. It is possible in very special circumstances to achieve such a coherent ensemble, and this is the basis of some specialised techniques in optical coherent-transient spectroscopy (Levenson and Kano, 1988; Brewer, 1977). More generally, the ensemble contains a distribution of quantum states ψ, and the probability p_{ψ} is described by a statistical probability distribution. A simple example is when the probabilities, $|a|^2$ and $|b|^2$, of finding a particular two-level system in the eigenstate u_a or u_b, respectively, are the same for all systems (*i.e.*, $|a|^2$ and $|b|^2$ have constant values), but the phase angle $\phi = \arg(a) - \arg(b)$ of the coherent superposition is randomly distributed between 0 and 2π. In that case $\sum_{\psi} p_{\psi} \rightarrow \int_0^{2\pi} \mathrm{d}\phi/2\pi$ in (6.30), and since $\int_0^{2\pi} \mathrm{d}\phi \exp(i\phi) = 0$, the resulting density matrix is

$$\begin{bmatrix} \rho_{aa} & \rho_{ab} \\ \rho_{ba} & \rho_{bb} \end{bmatrix} = \begin{bmatrix} |a|^2 & 0 \\ 0 & |b|^2 \end{bmatrix}. \tag{6.31}$$

The fact that the off-diagonal elements are zero indicates a total absence of coherence in the ensemble. It is interesting to notice, however, that the density matrix (6.31) remains a useful description of the totally-incoherent ensemble, even though there is no wave function that can describe such a case.

6.2.2 Equations of motion for the density matrix elements

Consider now the Hamiltonian for the two-level system. Similar to (3.49), we can separate the total Hamiltonian into different parts:

$$H = H_0 + H_{\mathrm{I}}(t) + H_{\mathrm{R}}, \tag{6.32}$$

where H_0 is the Hamiltonian describing the system when external forces are absent, and is given by

$$H_0 = \begin{bmatrix} \mathrm{E}_a & 0 \\ 0 & \mathrm{E}_b \end{bmatrix}. \tag{6.33}$$

The perturbation Hamiltonian $H_{\mathrm{I}}(t)$ is the operator representing the energy of the interaction between the two-level system and the applied electric field $\mathbf{E}(t)$. The last part of (6.32), H_{R}, does not appear in the previous work and it is introduced here to represent the various relaxation processes that act to bring the ensemble back to its thermal-equilibrium condition. For resonant phenomena we must take into account the various decay times required for the medium to relax after excitation. These relaxation processes and the form of H_{R} are discussed in detail shortly.

We briefly mention here that an alternative approach is to consider the eigenstates of the system comprising the two-level atom and the optical field taken together. The effect of this radiation-matter coupling is that the equilibrium superposition state (6.25) evolves into states of the so-called 'dressed atom', *i.e.*, the atom that has been modified or 'dressed' by the field. Quantities such as ρ_{aa} and ρ_{bb} should then more properly be described as probability densities for the dressed states. This 'dressed-atom' approach properly requires a quantum treatment of the optical field, which is beyond the scope of this book (see Cohen-Tannoudji (1977), Feneuille (1977)).

To proceed, we now consider the situation of interest here in which the two energy levels are coupled by an electric-dipole transition. The dipole moment of the two-level system is denoted by $e\mathbf{r}$, and its energy in the applied optical field is $-e\mathbf{r}\cdot\mathbf{E}(t)$. The perturbation

Hamiltonian is then given by

$$H_{\mathrm{I}}(t) = \begin{bmatrix} \delta \mathrm{E}_a & -e\mathbf{r}_{ab}\cdot\mathbf{E}(t) \\ -e\mathbf{r}_{ba}\cdot\mathbf{E}(t) & \delta \mathrm{E}_b \end{bmatrix}. \tag{6.34}$$

The off-diagonal matrix elements of $H_{\mathrm{I}}(t)$ induce transitions between the coupled states, as may be seen by inserting the Hamiltonian (6.32) into the equation of motion (6.26). The result is the following set of equations:

$$(i\hbar)\,\mathrm{d}\rho_{aa}/\mathrm{d}t = -e(\rho_{ba}\mathbf{r}_{ab} - \rho_{ab}\mathbf{r}_{ba})\cdot\mathbf{E}(t) + [H_{\mathrm{R}}, \rho]_{aa} \tag{6.35a}$$

$$(i\hbar)\,\mathrm{d}\rho_{ab}/\mathrm{d}t = -\rho_{ab}(\hbar\Omega_{ba} + \delta\mathrm{E}_b - \delta\mathrm{E}_a) - (\rho_{bb} - \rho_{aa})e\mathbf{r}_{ab}\cdot\mathbf{E}(t) + [H_{\mathrm{R}}, \rho]_{ab} \tag{6.35b}$$

$$(i\hbar)\,\mathrm{d}\rho_{bb}/\mathrm{d}t = e(\rho_{ba}\mathbf{r}_{ab} - \rho_{ab}\mathbf{r}_{ba})\cdot\mathbf{E}(t) + [H_{\mathrm{R}}, \rho]_{bb}\,. \tag{6.35c}$$

The fourth equation of motion, for $\mathrm{d}\rho_{ba}/\mathrm{d}t$, is superfluous since we already know that $\rho_{ba} = \rho_{ab}{}^*$. These equations of motion for the density matrix elements are sometimes called the 'master equations'.

The diagonal matrix elements of $H_{\mathrm{I}}(t)$ in (6.34) correspond to optically-induced Stark shifts of the energy levels E_a and E_b which appear in the diagonal positions of the matrix H_0. For simplicity, we neglect optical Stark shifts in the following discussion. Although these shifts can be significant in high-resolution laser spectroscopy, it is a satisfactory approximation to neglect them for many practical applications in nonlinear optics. The optical Stark effect is considered in a later section (§6.3.3).

The field incident on the ensemble is now taken to be a single quasi-monochromatic wave of frequency ω (as depicted in Fig. 6.1). The field is of the form (6.1):

$$\mathbf{E}(t) = \tfrac{1}{2}\Big[\mathbf{E}_\omega(t)\exp(-i\omega t) + \mathbf{E}_\omega{}^*(t)\exp(i\omega t)\Big] \tag{6.36}$$

and, again using the scalar notation of (2.57), we shall write $\mathbf{E}_\omega(t) = \mathbf{e}E(t)$. The strength of the resonant interaction between the field and the two-level ensemble is described by the parameter $\beta(t)$:

$$\beta(t) = e\mathbf{r}_{ab}\cdot\mathbf{e}E(t)/\hbar\,, \tag{6.37}$$

which is known as the 'on-resonance Rabi flopping frequency' or more often simply as the 'Rabi frequency'. We are free to choose the states u_a and u_b so that (in the absence of a magnetic field) the matrix element $e\mathbf{r}_{ab}$ is real and positive, and we shall assume that this is the case. We shall also make the assumption that the applied field $\mathbf{E}(t)$ is linearly polarised, and that the envelope function $E(t)$ is real and positive. None of these are essential assumptions, but are made just for simplicity: the

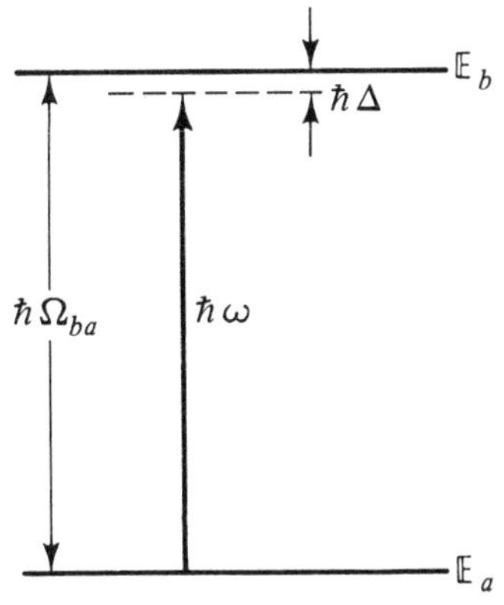

Fig. 6.1 Energy diagram for the resonant interaction between the two-level system and a quasi-monochromatic optical field. The transition energy is $\hbar\Omega_{ba} = \mathbb{E}_b - \mathbb{E}_a$, and the optical-frequency detuning is $\Delta = \Omega_{ba} - \omega$.

result is that the on-resonance Rabi frequency $\beta(t)$ is also real and positive. Introducing (6.36) into the master equations (6.35), we obtain

$$\mathrm{d}\rho_{aa}/\mathrm{d}t = i(\rho_{ba} - \rho_{ab})\beta(t)\cos\omega t - i\hbar^{-1}[H_\mathrm{R}, \rho]_{aa} \tag{6.38a}$$

$$\mathrm{d}\rho_{ab}/\mathrm{d}t = i\rho_{ab}\Omega_{ba} + i(\rho_{bb} - \rho_{aa})\beta(t)\cos\omega t - i\hbar^{-1}[H_\mathrm{R}, \rho]_{ba} \tag{6.38b}$$

$$\mathrm{d}\rho_{bb}/\mathrm{d}t = -i(\rho_{ba} - \rho_{ab})\beta(t)\cos\omega t - i\hbar^{-1}[H_\mathrm{R}, \rho]_{bb} . \tag{6.38c}$$

6.2.3 Rotating-wave approximation

A very useful simplification of (6.38) can be made by means of the 'rotating-wave approximation', in which terms that oscillate at the frequency $\omega + \Omega_{ba}$ are neglected. This approximation can be better understood and justified if we examine the time dependence in (6.38); this is made more explicit by substituting values for the density-operator matrix elements from (6.30). Thus, for example, (6.38a) becomes:

$$\begin{aligned}\mathrm{d}\rho_{aa}/\mathrm{d}t &= \mathrm{d}(aa^*)/\mathrm{d}t \\ &= \tfrac{1}{2} i\Big\{a^*b\Big[\exp(-i\Delta t) + \exp(-i\{\omega+\Omega_{ba}\}t)\Big] \\ &\quad - ab^*\Big[\exp(i\Delta t) + \exp(i\{\omega+\Omega_{ba}\}t)\Big]\Big\}\beta(t) - i\hbar^{-1}[H_\mathrm{R}, \rho]_{aa} ,\end{aligned} \tag{6.39}$$

where $\Delta = \Omega_{ba} - \omega$. When equations such as (6.39) are integrated, terms with denominators Δ and $\omega + \Omega_{ba}$ are obtained. Under near-resonance conditions, $\omega + \Omega_{ba} \gg \Delta$, and therefore the terms with the denominator $\omega + \Omega_{ba}$ can be safely neglected for the great majority of situations. (We should note, however, that the rotating-wave approximation thus neglects the small component of the total polarisation which occurs at the sum frequency $\omega+\Omega_{ba}$, and which may have useful spectroscopic applications.) By making the substitution

$$\rho_{ab} = \rho^{\Omega}_{ab} \exp[i(\Omega_{ba}-\Delta)t] \tag{6.40}$$

(remembering that $\rho_{ba}=\rho_{ab}{}^*$), and applying the rotating-wave approximation to the master equations (6.38), the result is

$$\mathrm{d}\rho_{aa}/\mathrm{d}t = \tfrac{1}{2}i(\rho^{\Omega}_{ba}-\rho^{\Omega}_{ab})\beta(t) - i\hbar^{-1}[H_{\mathrm{R}},\rho]_{aa} \tag{6.41a}$$

$$\mathrm{d}\rho^{\Omega}_{ab}/\mathrm{d}t = i\Delta\rho^{\Omega}_{ab} + \tfrac{1}{2}i(\rho_{bb}-\rho_{aa})\beta(t) - i\hbar^{-1}[H_{\mathrm{R}},\rho]_{ab}\exp[-i(\Omega_{ba}-\Delta)t] \tag{6.41b}$$

$$\mathrm{d}\rho_{bb}/\mathrm{d}t = -\tfrac{1}{2}i(\rho^{\Omega}_{ba}-\rho^{\Omega}_{ab})\beta(t) - i\hbar^{-1}[H_{\mathrm{R}},\rho]_{bb}\,. \tag{6.41c}$$

(We should mention here that for the special case of circularly-polarised light, the rotating-wave approximation is unnecessary since terms containing the frequency factor $\omega+\Omega_{ba}$ vanish. In terms of the vector model to be described below, for circularly-polarised light there is no counter-rotating component in the rotating frame.) Finally, a further simplifying transformation can be made by introducing the variables:

$$\begin{aligned} u &= \rho^{\Omega}_{ba} + \rho^{\Omega}_{ab} \\ v &= i(\rho^{\Omega}_{ba} - \rho^{\Omega}_{ab}) \\ w &= \rho_{bb} - \rho_{aa}\,. \end{aligned} \tag{6.42}$$

The equations of motion in the rotating-wave approximation, (6.41), then become

$$\mathrm{d}u/\mathrm{d}t = -\Delta v - i\hbar^{-1}\Big\{[H_{\mathrm{R}},\rho]_{ab}\exp(-i\{\Omega_{ba}-\Delta\}t) + [H_{\mathrm{R}},\rho]_{ba}\exp(i\{\Omega_{ba}-\Delta\}t)\Big\} \tag{6.43a}$$

$$\mathrm{d}v/\mathrm{d}t = \Delta u + \beta(t)w - \hbar^{-1}\Big\{[H_{\mathrm{R}},\rho]_{ab}\exp(-i\{\Omega_{ba}-\Delta\}t) - [H_{\mathrm{R}},\rho]_{ba}\exp(i\{\Omega_{ba}-\Delta\}t)\Big\} \tag{6.43b}$$

$$\mathrm{d}w/\mathrm{d}t = -\beta(t)v - i\hbar^{-1}\Big\{[H_{\mathrm{R}},\rho]_{bb} - [H_{\mathrm{R}},\rho]_{aa}\Big\}. \tag{6.43c}$$

These equations can be given a simple graphical interpretation in the case that relaxation can be neglected ($H_{\mathrm{R}}\rightarrow 0$), as discussed in §6.2.7.

6.2.4 The macroscopic polarisation

Having made these successive transformations, and before proceeding further, it is helpful to gain some appreciation of the physical significance of the parameters u, v and w. This is straightforward in the case of w; it is the fractional difference in populations for the two energy levels, often called the 'population inversion'. In other words, $w\hbar\Omega_{ba}/2$ is the expectation value of the energy of the two-level system. The case $w<0$ corresponds to an absorbing medium, whereas $w>0$ corresponds to an inverted medium exhibiting gain. The physical significance of the quantities u and v is less obvious, but they can in fact be simply related to the macroscopic polarisation, as we now show. The polarisation $\mathbf{P}(t)$ of the ensemble which contains N two-level atoms per unit volume is given by

$$\mathbf{P}(t) = N\langle e\mathbf{r}\rangle = N\,\mathrm{Tr}(\rho\, e\mathbf{r})\,, \tag{6.44}$$

cf. (4.3). The second step in (6.44) follows from (3.32).† Evaluating the trace for the two-level system, we have

$$\mathbf{P}(t) = Ne(\rho_{ab}\mathbf{r}_{ba} + \rho_{ba}\mathbf{r}_{ab})\,, \tag{6.45}$$

and from the definitions (6.42) we obtain the relations

$$\begin{aligned} u - iv &= 2\rho_{ba}\exp(i\omega t) \\ u + iv &= 2\rho_{ab}\exp(-i\omega t)\,. \end{aligned} \tag{6.46}$$

Combining (6.44)–(6.46) the result is

$$\mathbf{P}(t) = \frac{N}{2}\Big[(u-iv)e\mathbf{r}_{ab}\exp(-i\omega t) + (u+iv)e\mathbf{r}_{ba}\exp(i\omega t)\Big]\,, \tag{6.47}$$

and, noting once again that $\mathbf{r}_{ba}=\mathbf{r}_{ab}{}^{*}$, it follows that the scalar wave amplitude for the polarisation at frequency ω is given by

$$P(t) = Ne\mathbf{r}_{ab}\cdot\mathbf{e}(u-iv)\,. \tag{6.48}$$

Thus we see that u and $-v$ are the components of the ensemble-averaged microscopic polarisation, in units of the two-level transition moment $e\mathbf{r}_{ab}\cdot\mathbf{e}$, which are in-phase and in-quadrature with the field $\mathbf{E}(t)$. This interpretation is confirmed by (6.43c), which shows that v is the component effective in coupling to the field to produce energy changes. That is, v is the absorptive component of the polarisation, whilst u is the dispersive component. Taken together, u and v are a measure of the coherence of the ensemble. As remarked earlier, if the ensemble lacks coherence then the off-diagonal matrix elements of the density operator are zero, *i.e.*, $u=v=0$, and so the macroscopic polarisation vanishes.

† The ensemble average of spatial orientations of the two-level systems is implicit in the choice of orthonormal basis states (§5.3).

6.2.5 Relaxation

We now return to consider the Hamiltonian H_R, which represents the myriad of relaxation processes that tend to restore the ensemble to thermal equilibrium. Because of the likely complexity of these processes, it is usual to treat them in a simple phenomenological fashion. There are interactions, such as collisions in a gas or phonon scattering in a solid, which can disturb the phase of the dipole oscillations of the two-level system without affecting its energy. Therefore the inversion w can decay at a different rate from u and v, and it is customary, following the work of Bloch (1946), to assign different relaxation times. The processes – such as spontaneous emission – that tend to restore the population in the excited level E_b back to the ground level E_a can be described by the relaxation-operator matrix elements:

$$\begin{aligned} [H_R, \rho]_{bb} &= -i\hbar\rho_{bb}/T_b \\ [H_R, \rho]_{aa} &= i\hbar\rho_{bb}/T_b = i\hbar(1-\rho_{bb})/T_b \, , \end{aligned} \tag{6.49}$$

where T_b is the lifetime of the upper level, and the ground state for the two-level system is assumed to have an infinite lifetime. The lifetime of the dipole moment itself is described by a different relaxation time, conventionally denoted by T_2, and the corresponding off-diagonal elements of the relaxation operator are

$$\begin{aligned} [H_R, \rho]_{ab} &= -i\hbar\rho_{ab}/T_2 \\ [H_R, \rho]_{ba} &= -i\hbar\rho_{ba}/T_2 \, . \end{aligned} \tag{6.50}$$

Thus T_2 is the lifetime of the coherent superposition state ψ in (6.25), and is often referred to as the 'dephasing time' for the dipole oscillator. For historical reasons connected with the first application of the equations of motion (6.43) in nuclear magnetic resonance, T_2 is also sometimes termed the 'transverse' relaxation time.

In realistic situations the two-level model is only an approximation, and the conservation relation $\rho_{aa}+\rho_{bb}=1$ may be violated whenever the optically-coupled states u_a and u_b can interact with a reservoir of other eigenstates of similar energy. A way of treating this situation is to allow each of the density matrix elements ρ_{aa} and ρ_{bb} to relax towards the thermal equilibrium values $\rho_0(a)$ and $\rho_0(b)$, respectively, at their own rates. Thus

$$\begin{aligned} [H_R, \rho]_{bb} &= i\hbar(\rho_0(b)-\rho_{bb})/T_b \\ [H_R, \rho]_{aa} &= i\hbar(\rho_0(a)-\rho_{aa})/T_a \, . \end{aligned} \tag{6.51}$$

A common approximation, however, is to assume that the states u_a and u_b are sufficiently similar that their lifetimes can be considered equal, *i.e.*, $T_a=T_b=T_1$, and again for historical reasons, T_1 is conventionally termed the 'longitudinal' relaxation time.

We should, at this point, comment that throughout the discussion we have assumed for simplicity that all the two-level systems in the ensemble share the same resonant frequency, *i.e.*, the ensemble is homogeneously broadened. The dipole-moment lifetime T_2 due to incoherent interactions that affect all the two-level systems homogeneously (such as collisions, radiative decay and coupling between rotational, vibrational and electronic interactions) should be distinguished from the lifetime due to inhomogeneous effects, which conventionally is denoted by T_2*. In gases the main origin of an inhomogeneous lifetime is Doppler broadening, whilst in solids the effect is principally due to local variations in the crystal field. These effects produce a random distribution of resonant frequencies Ω_{ba} in the ensemble, and the resultant dephasing of the dipole moments of the individual two-level systems leads to damping of the *macroscopic* polarisation, even in the case that homogeneous damping (*i.e.*, damping of each of the *individual* dipole moments) is negligible ($T_2 \to \infty$). In general the presence of significant inhomogeneous broadening greatly complicates the analysis of optical resonance, but leads to the possibility of unique coherent-ensemble phenomena, such as photon echoes. These topics are beyond the scope of this book, but for very clear accounts the reader is referred to Allen and Eberly (1975), Brewer (1977) and Levenson and Kano (1988). Some aspects of inhomogeneous broadening are discussed in later sections (especially §6.3.4).

In the T_1, T_2 approximation, the equations of motion (6.43) for the variables u, v and w become

$$du/dt = -\Delta v - u/T_2 \qquad (6.52a)$$
$$dv/dt = \Delta u + \beta(t)w - v/T_2 \qquad (6.52b)$$
$$dw/dt = -\beta(t)v - (w - w_0)/T_1 , \qquad (6.52c)$$

where $w_0 = \rho_0(b) - \rho_0(a)$ is the equilibrium inversion with no optical field applied. Expressed in this form (6.52) are commonly known as 'Bloch equations'.

6.2.6 Steady-state and rate-equation regimes

Particularly simple solutions of the Bloch equations (6.52) can be found in the 'steady-state' regime. A steady-state response is obtained whenever $\tau_c >> T_1, T_2$, where τ_c is the characteristic time scale for fluctuations of the envelope of the driving field (introduced in §6.1). The steady-state solutions of (6.52), to which we refer later, are obtained by setting $du/dt \to 0$, $dv/dt \to 0$, and $dw/dt \to 0$, and the results are

$$u - iv = \frac{-\beta w}{\Delta - i/T_2} \tag{6.53a}$$

$$w = \frac{w_0\,[1+(\Delta T_2)^2]}{1+(\Delta T_2)^2+\beta^2 T_1 T_2}\,. \tag{6.53b}$$

There are two interesting points about these expressions that should be made. The first is the intensity-dependent saturation of the population inversion, which is indicated by the term β^2 in the denominator of (6.53b). Regardless of the thermal-equilibrium value w_0, in the limit of strong fields the population difference is driven to zero. The second point of interest concerns the relaxation parameter in the denominator in (6.53a). From (6.48), we know that the polarisation $P(t)$ is proportional to $u - i\,v$, and for steady-state excitation, this is given by (6.53a). In the weak-field limit $w \simeq w_0$, and in that case the relaxation of the polarisation $P(t)$ is determined by the denominator term $1/T_2$ in (6.53a). Thus it follows that the homogeneous damping terms Γ, which in §4.5 were introduced into the resonant denominators of $\chi^{(n)}$, should be identified as $\Gamma \equiv 1/T_2$, where T_2 is the dephasing time for the resonant transition.

We should mention briefly another regime of interest which occurs when dephasing is rapid ($T_2 \ll T_1$), and the characteristic time for fluctuations of the driving field τ_c is long compared to T_2 but comparable to T_1. The 'rate-equation' approximation to the Bloch equations can then be made, in which u and v are assumed to come to equilibrium with w very rapidly. Setting $du/dt \to 0$ and $dv/dt \to 0$ in (6.52), we find that u and v are again given in terms of w by (6.53a), but w is now found directly from the rate equation (6.52c). In the rate-equation regime, it is not possible to produce a population inversion of the two-level system by optical pumping (*i.e.*, $w > 0$ is impossible). This can be achieved, however, in the 'coherent-transient regime' ($\tau_c \lesssim T_1, T_2$), as described below.

The various time regimes of interest in nonlinear optics are summarised in §6.6.

6.2.7 The vector model

In the general case, for arbitrary values of T_1, T_2 and τ_c, the solutions of the Bloch equations (6.52) can assume great complexity. A useful observation, however, is that the Bloch equations are identical in form to those that govern spin precession in magnetic resonance. Consequently, much of the theory which was developed in earlier years to analyse problems in nuclear magnetic-resonance spectroscopy, can be carried over to study optical resonance phenomena. In particular, the equations of motion (6.52) can be rewritten as a single vector-precession equation. This is particularly illuminating since it allows optical resonance problems to be visualised using a simple geometrical picture. This picture, often referred to as the 'vector model', was devised by Bloch (1946) and Feynman *et al* (1957) – Allen and Eberly (1975) give further historical references.

The stratagem is to define a pair of vectors $\mathbf{R}=(u, v, w)$ and $\mathbf{\Omega}=(-\beta(t), 0, \Delta)$. The vectors are defined in an abstract cartesian coordinate frame whose axes we denote by the unit vectors $\hat{\mathbf{u}}$, $\hat{\mathbf{v}}$ and $\hat{\mathbf{w}}$. It is then straightforward to verify that the equations of motion (6.52) can be expressed in the form

$$\mathrm{d}\mathbf{R}/\mathrm{d}t = \mathbf{\Omega}\times\mathbf{R} + (\text{relaxation terms})\,. \tag{6.54}$$

Equation (6.54) indicates that the vector $\mathbf{R}$ precesses around $\mathbf{\Omega}$ exactly in the manner of the precession of the angular-momentum vector of a top spinning in a gravitational field. The vector $\mathbf{\Omega}$ is known as the 'pseudo-field vector' or, by reference to the spinning-top analogy, as the 'torque vector'. The vector $\mathbf{R}$ is known as the 'rotating-frame Bloch-Feynman vector', or usually more simply as the 'Bloch vector'. In the vector model the field-matter interaction is represented by the vector $\mathbf{\Omega}$, and $\mathbf{R}$ describes the material response. The coordinate frame of reference $(\hat{\mathbf{u}}, \hat{\mathbf{v}}, \hat{\mathbf{w}})$ is termed the 'rotating frame', because the pseudo-field vector $\mathbf{\Omega}$ is stationary in that frame with respect to the fast oscillations of the field $\mathbf{E}(t)$ at the optical frequency ω. The vector $\mathbf{\Omega}$ varies only in response to *changes* in the amplitude, phase or frequency of the driving field. When the field is removed ($\beta=0$), the subsequent material response is determined by the relaxation terms in (6.54).

We now assume, for the moment, that the relaxation terms are small enough to be neglected; this is the 'coherent-transient regime'. It is easy to show that in this case a constant of motion for the Bloch vector $\mathbf{R}$ is $u^2+v^2+w^2=1$. This relation is verified by multiplying the three equations (6.52a) – (6.52c) by u, v and w respectively, adding the resultant expressions and integrating to obtain the relation $u^2+v^2+w^2=k$. The constant k is 1 because of the normalisation condition $\mathrm{Tr}(\rho)=1$. In this case, therefore, $\mathbf{R}$ is a unit vector whose tip moves on the surface of the unit sphere.

When the ensemble is in thermal equilibrium, the Bloch vector points downwards along the $-\hat{\mathbf{w}}$ direction (as shown in Fig. 6.2(a)) indicating that the majority of the ensemble is in the lower energy state and that there is no coherence in the properties of the two-level systems. If an optical field with amplitude $E(t)$ is applied at time $t=0$ (*i.e.*, $E(t)>0$

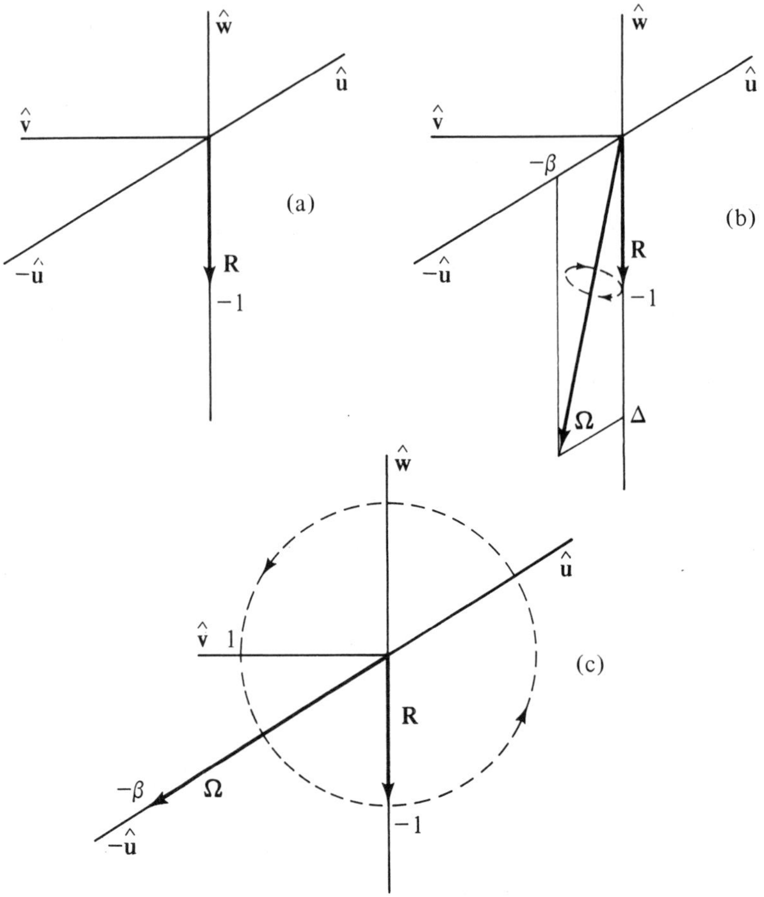

Fig. 6.2 Vector model for resonant two-level optical interactions in the absence of relaxation. (a) In thermal equilibrium the Bloch vector **R** points downwards along the $-\hat{\mathbf{w}}$ direction. If the field is suddenly applied to the system, the Bloch vector begins to precess around the pseudo-field vector $\mathbf{\Omega}=(-\beta, 0, \Delta)$. The tip of **R** describes a circle in a plane perpendicular to $\mathbf{\Omega}$. In (b) this is shown for the case when the field is detuned from the two-level resonance by an amount $-\Delta$. In case (c) the field is in exact resonance so that $\mathbf{\Omega}$ lies along the $\hat{\mathbf{u}}$ direction and the precession of **R** describes a unit circle in the $\hat{\mathbf{v}}-\hat{\mathbf{w}}$ plane.

for $t>0$), the pseudo-field vector $\mathbf{\Omega}$ begins to move in the $\hat{\mathbf{u}}-\hat{\mathbf{w}}$ plane ($\mathbf{\Omega}$ is always confined to that plane in the case that $\beta(t)$ is real, which we have assumed). The vector equation (6.54) tells us that the Bloch vector **R** will precess around $\mathbf{\Omega}$ (examples are shown in Figs. 6.2(b) and (c). The tip of the **R** vector moves in a circle which is in a plane perpendicular to $\mathbf{\Omega}$. The angular frequency Ω of the precession is given by

$$\Omega(t) = \mathrm{sgn}(\Delta)\,|\mathbf{\Omega}| = \mathrm{sgn}(\Delta)\sqrt{\beta^2(t) + \Delta^2}\ , \tag{6.55}$$

and $|\boldsymbol{\Omega}|$ is known as the 'off-resonance Rabi flopping frequency' or sometimes as the 'total Rabi frequency'. In the rotating frame, the angular frequency Ω takes the sign of the detuning factor Δ, as indicated in (6.55).

This vector picture may be most easily visualised in the simple case that the field $E(t)$ takes the form of a rectangular pulse which is applied *suddenly* at time $t=0$. The amplitude of the pulse is taken to be E_{P}, and its duration is τ_{P}. (This is called the 'Rabi case', by analogy with the simplest situation in nuclear magnetic resonance in which a magnetic field of fixed amplitude is simply switched on and off.) Since we are neglecting relaxation processes for the moment, we must assume that τ_{P} is much shorter than any of the characteristic relaxation times. Then, for times $0<t<\tau_{\mathrm{P}}$, the pseudo-field vector $\boldsymbol{\Omega}$ takes up a fixed position in the $\hat{\mathbf{u}}-\hat{\mathbf{w}}$ plane, as shown in Fig. 6.2(b) (where it has been assumed, for definiteness, that the detuning Δ is negative). The Bloch vector $\mathbf{R}$ precesses around $\boldsymbol{\Omega}$, the tip of $\mathbf{R}$ following a circle in the plane perpendicular to $\boldsymbol{\Omega}$. The angular frequency of precession Ω is given by (6.55), where the Rabi frequency β for the constant-amplitude field E_{P} is given, from (6.37), by $\beta=e\mathbf{r}_{ab}\cdot\mathbf{e}E_{\mathrm{P}}/\hbar$. As shown in Fig. 6.2(b), the line of $\mathbf{R}$ traces out a cone, whose apex half-angle is $\tan^{-1}(\beta/\Delta)$. Thus, for large detuning of the field from the two-level transition (large Δ), or for weak fields, the cone is narrow indicating that the ensemble is only weakly perturbed from its thermal-equilibrium state. If, however, the field is brought into closer resonance ($\Delta\to0$), the Bloch vector $\mathbf{R}$ sweeps out a larger cone and the inversion w may depart significantly from its thermal-equilibrium value.

In the limit in which the rectangular-pulsed field is brought into exact resonance ($\Delta=0$), the pseudo-field vector $\boldsymbol{\Omega}$ is constrained to lie along the $\hat{\mathbf{u}}$ direction, and therefore the Bloch vector $\mathbf{R}$ must precess in the $\hat{\mathbf{v}}-\hat{\mathbf{w}}$ plane (as shown in Fig. 6.2(c)). The angular frequency of the precession of $\mathbf{R}$ is then equal to the on-resonance Rabi frequency β. The angle through which the Bloch vector $\mathbf{R}$ will have rotated by the end of the pulse ($t=\tau_{\mathrm{P}}$) is therefore equal to $\beta\tau_{\mathrm{P}}$, and this quantity is known as the 'pulse area'. A case of special interest is the pulse with area π, which causes the Bloch vector $\mathbf{R}$ for the ensemble, initially in thermal equilibrium, to rotate until it points upwards in the $\hat{\mathbf{w}}$ direction, indicating that all of the two-level systems are in the upper energy state. Therefore a 'π pulse' has the property that the two-level ensemble is fully inverted after excitation. Even more remarkable is the '2π pulse' which causes the Bloch vector to rotate one full circle in the $\hat{\mathbf{v}}-\hat{\mathbf{w}}$ plane to finish up in its initial thermal-equilibrium position. The energy absorbed by the

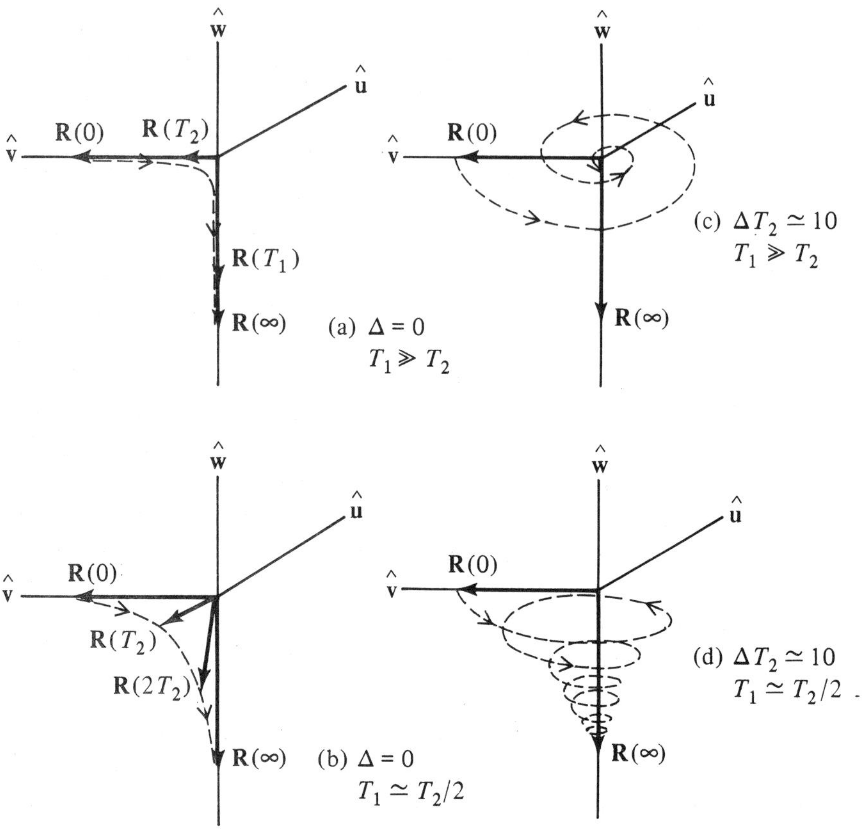

Fig. 6.3 The evolution of the Bloch vector **R**(t) towards thermal equilibrium when the driving field has been turned off at time $t = 0$. For illustration, the initial Bloch vector **R**(0) is taken to lie along the $\hat{\mathbf{v}}$ direction. In the presence of relaxation, the magnitude of **R**(t) is generally not conserved. The vector **R**(t) eventually reaches its equilibrium state **R**(∞).

(a)(b)The rotation frequency is resonant with the transition frequency ($\Delta = 0$).

(a) Collisional effects destroy the phase memory in a short time compared to the excited-state lifetime. The tip of the **R**(t) vector relaxes towards the $\hat{\mathbf{w}}$ axis at the rate $1/T_2$ and then relaxes towards its equilibrium position **R**(∞) at the slower rate $1/T_1$.

(b) This situation applies when population decay occurs on a time scale similar to that for dephasing. The tip of the **R**(t) vector traces a curve in the $\hat{\mathbf{v}}-\hat{\mathbf{w}}$ plane.

(c)(d)These situations occur when the driving fields and frame rotation frequency are off resonance ($\Delta \neq 0$).

(c) For fast dephasing, **R**(t) spirals in towards the $\hat{\mathbf{w}}$ axis before the population inversion begins to decay towards its equilibrium value w_0.

(d) When the population decay rate is significant compared to the dephasing rate, the tip of the **R**(t) vector follows a whirlpool-shaped trajectory.

(Adapted from Levenson and Kano (1988), by permission.)

ensemble during the first half of the 2π pulse, so as to invert the two-level systems fully, is then redelivered back to the optical field during the second half of the pulse until, finally, the ensemble is returned to its equilibrium state. In this way the 2π pulse is able to propagate unattenuated through the absorbing medium. This effect is known as 'self-induced transparency'. These and other coherent pulse-propagation phenomena have been observed experimentally in various media (Brewer, 1977; Levenson and Kano, 1988). The reason that the parameter β characterising the strength of the matter-field interaction is termed the Rabi 'frequency' is now apparent; in the case of an exact-resonance interaction ($\Delta=0$), β is the frequency of the periodic oscillations of the energy-level occupation probabilities (called 'Rabi oscillations').

For a more general resonant time-dependent field $E(t)$ which is applied suddenly at $t=0$, the angle of rotation of the Bloch vector is given by

$$\theta(t) = \int_0^t \mathrm{d}\tau\, \beta(\tau)\,. \tag{6.56}$$

The quantity $\theta(t)$ is termed the 'field area'; it will be recalled that this same quantity appeared in the integral expansion (6.17) for the polarisation in §6.1. If the field incident on the medium takes the form of a single pulse, then the total field area (found by setting $t\to\infty$ in (6.56)) is known as the 'pulse area', as has been noted already in the case of a rectangular pulse.

General analytical solutions of the vector-precession equation (6.54) in the off-resonant case, even in the limit of negligible relaxation, are not simple – details are given by Allen and Eberly (1975). The addition of relaxation terms makes the Bloch equations more realistic, but yet more complicated. Analytical solutions of (6.52) were found by Torrey (1949) for the case of a constant field envelope (the Rabi case), and the more important features and limits of these solutions are discussed by Allen and Eberly. The essential point is that, when relaxation is included, the magnitude of the Bloch vector $\mathbf{R}$ no longer remains constant. In particular, when the applied field is turned off, the components of $\mathbf{R}$ relax towards their corresponding thermal-equilibrium values at different rates: the u and v components relax toward zero at the transverse relaxation rate T_2, and w relaxes towards the equilibrium value w_0 at the longitudinal relaxation rate T_1. Examples of the time evolution of $\mathbf{R}$ for various relative values of relaxation times are illustrated in Fig. 6.3.

A great many coherent phenomena of interest in nonlinear optics can be described and explained in terms of the two-level model and the vector-precession equation (6.54). By means of the defining relations (6.40) and (6.42), the components of the Bloch vector can be transformed back to obtain the matrix elements of the density operator. Then, by means of (3.32), the expectation value for the operator representing any desired dynamical variable can be found. The formalism is very flexible, since the matrix elements and vector components that appear in the formulae may be calculated in any basis and labelled with any number of physical parameters.

6.2.8 Adiabatic-following regime

Important approximate solutions of the Bloch equations, including relaxation, were found by Crisp (1973) for the case that the variations in the applied field $\mathbf{E}(t)$ are slow compared to the rate at which $\mathbf{R}$ precesses around $\boldsymbol{\Omega}$. In this situation $\mathbf{R}$ follows $\boldsymbol{\Omega}$ adiabatically as the field strength and frequency are varied – hence this is termed the 'adiabatic-following' regime. This is illustrated in Fig. 6.4 in the limit of negligible relaxation and taking $w_0 = -1$. Initially the Bloch vector $\mathbf{R}$ is in its thermal-equilibrium position (as shown in Fig. 6.2(a)). An off-resonant field is then gradually applied. Because the field is off resonance (in Fig. 6.4 it is assumed arbitrarily that Δ is negative), initially $\boldsymbol{\Omega}$ also points nearly-vertically downwards in the $-\hat{\mathbf{w}}$ direction. As the optical intensity is increased, $\mathbf{R}$ is gradually lifted from its equilibrium position. A feature of adiabatic following is that, although the vector $\mathbf{R}$ precesses about $\boldsymbol{\Omega}$ as before, the cone angle is so very small that $\mathbf{R}$ can be assumed to remain almost parallel to $\boldsymbol{\Omega}$ – thus the Bloch vector $\mathbf{R}$ follows the motion of $\boldsymbol{\Omega}$. In the adiabatic-following limit illustrated in Fig. 6.4, the cone angle $\alpha \rightarrow 0$ (*i.e.*, $v \rightarrow 0$), and the constant of motion for the Bloch vector becomes $u^2 + w^2 = 1$. By applying simple trigonometry, it is easy to show (for the case of $\Delta < 0$ and $w < 0$, as illustrated in Fig. 6.4) that $u \simeq -\sin\phi = -\beta/|\boldsymbol{\Omega}|$ and $w \simeq -\cos\phi = -|\Delta/\boldsymbol{\Omega}|$. By constructing further diagrams similar to Fig. 6.4, it is quite straightforward to show that in the general case:

$$u \simeq -\operatorname{sgn}(w)\operatorname{sgn}(\Delta)\,\beta/|\boldsymbol{\Omega}| \tag{6.57}$$

and

$$w \simeq \operatorname{sgn}(w)\,|\Delta/\boldsymbol{\Omega}|, \tag{6.58}$$

where $\operatorname{sgn}(x)$ denotes the sign of x. These results are used in the next section.

The adiabatic-following regime of the material response can be related to the concept of the adiabatic field, introduced in §6.1. As

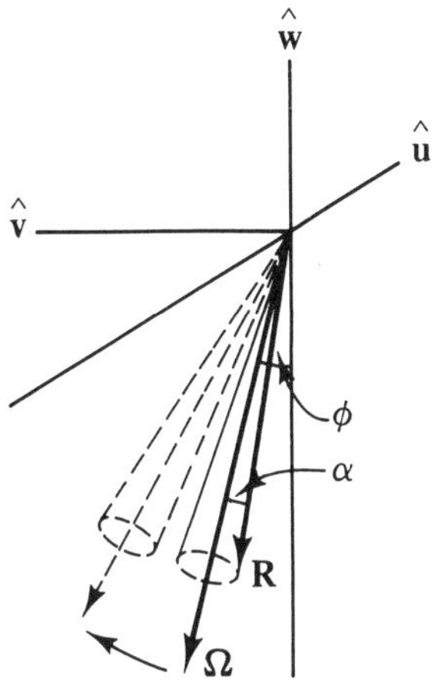

Fig. 6.4 Precession of the Bloch vector **R** in a narrow cone about the pseudo-field vector **Ω** in the adiabatic-following regime. The vector **R** follows the motion of **Ω**.

previously, we denote the characteristic time scale for changes of the field envelope by τ_c. When the off-resonance detuning $|\Delta|$ is large enough, say several inhomogeneous linewidths $(T_2{}^*)^{-1}$, we can assume that the precession frequency Ω is large enough that $\tau_c >> \Omega^{-1} \equiv |\Delta|^{-1}$ as required for adiabatic following. But from (6.10), the adiabatic field limit is described by $\tau_c >> |\Delta - i/T_2{}^*|^{-1} \simeq |\Delta|^{-1}$, and so we see in this case that the two limits are indeed closely related (see also §6.6).

6.3 Intensity-dependent refractive index

As an example of the application of the dynamic two-level model to describe resonant nonlinear-optical phenomena, we now describe a simple yet very important effect; namely, the intensity-dependent refractive index. An optical field incident on the nonlinear medium experiences a refractive index consisting of the linear (low-intensity) refractive index together with an additional contribution whose magnitude depends on the incident light intensity. The resonant interaction of light with matter can be an efficient mechanism giving rise to nonlinear refraction which, in turn, is responsible for numerous important phenomena such as self-focusing and -defocusing, detuning of resonant optical cavities and self-phase modulation, (§7.5). The intensity-dependent refractive index exhibited by several materials is utilised in the all-optical switching and logic devices being developed currently in many research laboratories (Gibbs, 1985). These phenomena are often well described by the behaviour of a two-level system.

6.3.1 The field-corrected susceptibility

An important objective in this section is to establish the links between the susceptibility formalism developed in earlier chapters and the description of dynamic resonant nonlinear-optical processes discussed in the present chapter. Therefore, before applying the dynamic two-level model, we first obtain the expressions for the nonlinear refractive index in terms of the susceptibility formulae; the latter are derived in Chapter 4 by assuming that the nonlinear medium experiences only very small perturbations from its thermal-equilibrium state.

Once again, the incident field is taken be the quasi-monochromatic wave (6.36): $\mathbf{E}(t)=\frac{1}{2}[\mathbf{E}_\omega(t)\exp(-i\omega t)+\mathbf{E}_\omega^*(t)\exp(i\omega t)]$. As indicated in Table 2.1, the lowest-order susceptibility that gives rise to an intensity-dependent refractive-index change is $\boldsymbol{\chi}^{(3)}(-\omega;\omega,-\omega,\omega)$, and the corresponding expression for $\mathbf{P}_\omega(t)$, the amplitude of the polarisation wave, is

$$\begin{aligned}\mathbf{P}_\omega(t) &\simeq \mathbf{P}_\omega^{(1)}(t) + \mathbf{P}_\omega^{(3)}(t)\\ &= \varepsilon_0\left[\boldsymbol{\chi}^{(1)}(-\omega;\omega) + \tfrac{3}{4}\boldsymbol{\chi}^{(3)}(-\omega;\omega,-\omega,\omega)|\mathbf{E}_\omega(t)\mathbf{E}_\omega^*(t)|\right]\mathbf{E}_\omega(t).\end{aligned} \tag{6.59}$$

We assume here that the field envelope $\mathbf{E}_\omega(t)$ is applied adiabatically, and thus (6.59) is similar in form to (2.69). (To simplify the notation, we drop the explicit time-dependence (t) in the following.) The definition of the dielectric tensor $\boldsymbol{\varepsilon}(\omega)$ found in standard texts on electromagnetic theory may be generalised to include the nonlinear contribution to the polarisation: we therefore define $\boldsymbol{\varepsilon}(\omega)$ through the relation

$$\mathbf{P}_\omega = \varepsilon_0[\boldsymbol{\varepsilon}(\omega) - \mathbf{1}]\cdot\mathbf{E}_\omega\,, \tag{6.60a}$$

where $\mathbf{1}$ is the unit dyadic. For simplicity, the tensorial aspects are avoided in the following discussion by defining the effective dielectric constant $\varepsilon(\omega)$ as

$$\begin{aligned}\varepsilon(\omega) &= \mathbf{e}^*\cdot\boldsymbol{\varepsilon}(\omega)\cdot\mathbf{e}\\ &\simeq 1 + \chi^{(1)}(-\omega;\omega) + \tfrac{3}{4}\chi^{(3)}(-\omega;\omega,-\omega,\omega)|E|^2\,,\end{aligned} \tag{6.60b}$$

where $\mathbf{E}_\omega=\mathbf{e}E$ and the scalar susceptibilities are defined by (2.58). (For birefringent media, a small correction to (6.60b) should be made, as indicated by (7.7).) The linear refractive index is given by

$$n_0(\omega) = \sqrt{1 + \mathrm{Re}\,\chi^{(1)}(-\omega;\omega)}\,. \tag{6.61a}$$

It follows, then, that the *change* in refractive index δn induced by the incident field is given by

$$\delta n(\omega) = \sqrt{\mathrm{Re}\,\varepsilon(\omega)} - n_0(\omega)$$
$$\simeq \frac{1}{2}\,\frac{\frac{3}{4}\,\mathrm{Re}\,\chi^{(3)}(-\omega;\omega,-\omega,\omega)|E|^2}{n_0(\omega)}, \tag{6.61b}$$

where the second line follows from the binomial expansion of (6.60) with the assumption that $\delta n << n_0$. The relation (6.61b) is often written in terms of a nonlinear refraction coefficient $n_2(\omega)$, defined by

$$n(\omega) = n_0(\omega) + \delta n(\omega) = n_0(\omega) + n_2(\omega)|E|^2, \tag{6.62}$$

and it follows that

$$n_2(\omega) = \frac{3\,\mathrm{Re}\,\chi^{(3)}(-\omega;\omega,-\omega,\omega)}{8n_0(\omega)}. \tag{6.63}$$

(Common alternative definitions for n_2 and relations between the different systems of units are given in Appendix 5.)

In the adiabatic regime, it is convenient to rewrite the expression (6.59) for the total polarisation in the abbreviated form:

$$\mathbf{P}_\omega(t) = \varepsilon_0\,\overline{\boldsymbol{\chi}}(\omega;\mathbf{E}_\omega)\,\mathbf{E}_\omega(t), \tag{6.64}$$

where $\overline{\boldsymbol{\chi}}(\omega;\mathbf{E}_\omega)$ is a dimensionless susceptibility 'corrected' to include the effect of the field. We can therefore write the expression:

$$\varepsilon(\omega) = 1 + \overline{\chi}(\omega;E) = n(\omega)\,[n(\omega) + i2\kappa(\omega)], \tag{6.65}$$

where $n(\omega)$ is the refractive index, and $\kappa(\omega)$ is related to the intensity absorption coefficient $\alpha(\omega)$ (see §7.1.2). The nonlinear refractive-index change may now be written as

$$\delta n(\omega) = \sqrt{1 + \mathrm{Re}\,\overline{\chi}(\omega;E)} - n_0(\omega). \tag{6.66}$$

In (6.59) only the lowest-order correction (the $n=3$ term) was included. As has been discussed earlier, for a strong incident field or for the case when the field frequency ω approaches a resonance frequency of the medium, (6.59) is no longer an adequate approximation and higher-order nonlinear terms must be included. In that case, the 'field-corrected' susceptibility could be written as a series of terms of increasing order of nonlinearity:

$$\overline{\chi}(\omega;E) = \chi^{(1)}(-\omega;\omega) + \tfrac{3}{4}\chi^{(3)}(-\omega;\omega,-\omega,\omega)|E|^2$$
$$+ \tfrac{5}{8}\chi^{(5)}(-\omega;\omega,-\omega,\omega,-\omega,\omega)|E|^4 + \cdots$$
$$\cdots + \frac{n!\,2^{1-n}}{(\frac{n-1}{2})!\,(\frac{n+1}{2})!}\chi^{(n)}(-\omega;\omega,-\omega,\ldots,-\omega,\omega)|E|^{n-1} + \cdots. \tag{6.67}$$

The numerical factors associated with the fifth- and higher-order terms in (6.67) are calculated using (2.56).

6.3.2 A two-level saturable absorber

We now show that by using the dynamic two-level model for the dominant resonant transition, it is possible to obtain a much more simple and compact expression for the intensity-dependent refractive index; in effect the series (6.67) can be summed exactly. The two-level model also provides insight into the origin of the nonlinearity.

From (6.48), the macroscopic polarisation induced in the ensemble of two-level systems by an incident field of frequency ω is

$$P(t) = -\operatorname{sgn}(w)\operatorname{sgn}(\Delta)\frac{Ne^2|\mathbf{r}_{ab}\cdot\mathbf{e}|^2}{\hbar\sqrt{\Delta^2+\beta^2}}E(t). \tag{6.68}$$

We have assumed that the field is applied adiabatically and also, for the moment, that relaxation is negligible. We have therefore taken the expressions for v and u that apply in the adiabatic-following regime; namely, $v\simeq 0$ and (6.57). When (6.68) is compared with (6.64), we obtain the expression:

$$\begin{aligned}\overline{\chi}(\omega;E) &= -\operatorname{sgn}(w)\operatorname{sgn}(\Delta)\frac{Ne^2|\mathbf{r}_{ab}\cdot\mathbf{e}|^2}{\varepsilon_0\hbar\sqrt{\Delta^2+\beta^2}}\\ &= -\operatorname{sgn}(w)\operatorname{sgn}(\Delta)\frac{Ne^2|\mathbf{r}_{ab}\cdot\mathbf{e}|^2}{\varepsilon_0\sqrt{(\hbar\Delta)^2+[e|\mathbf{r}_{ab}\cdot\mathbf{e}|E(t)]^2}}.\end{aligned} \tag{6.69}$$

We can establish the link between this formula (6.69) for the 'corrected' susceptibility and the earlier expression (6.67) by making a direct comparison in the off-resonance regime. For $w\simeq w_0=-1$ and $\Delta^2>>\beta^2$, (6.69) can be expanded as a power series, to yield

$$\overline{\chi}(\omega;E) = \frac{Ne^2}{\varepsilon_0\hbar\Delta}|\mathbf{r}_{ab}\cdot\mathbf{e}|^2\left[1-\tfrac{1}{2}(\beta/\Delta)^2+\tfrac{3}{8}(\beta/\Delta)^4-\cdots\right]. \tag{6.70}$$

By comparing terms in equal powers of the field $E(t)$ appearing in the two series (6.67) and (6.70), we can make the following identifications:

$$\chi^{(1)}(-\omega;\omega) = \frac{Ne^2}{\varepsilon_0\hbar\Delta}|\mathbf{r}_{ab}\cdot\mathbf{e}|^2 \tag{6.71}$$

$$\chi^{(3)}(-\omega;\omega,-\omega,\omega) = \frac{-2Ne^4}{3\varepsilon_0(\hbar\Delta)^3}|\mathbf{r}_{ab}\cdot\mathbf{e}|^4 \tag{6.72}$$

$$\chi^{(5)}(-\omega;\omega,-\omega,\omega,-\omega,\omega) = \frac{3Ne^6}{5\varepsilon_0(\hbar\Delta)^5}|\mathbf{r}_{ab}\cdot\mathbf{e}|^6 \tag{6.73}$$

....

Equation (6.71) is the familiar expression for the linear susceptibility, and

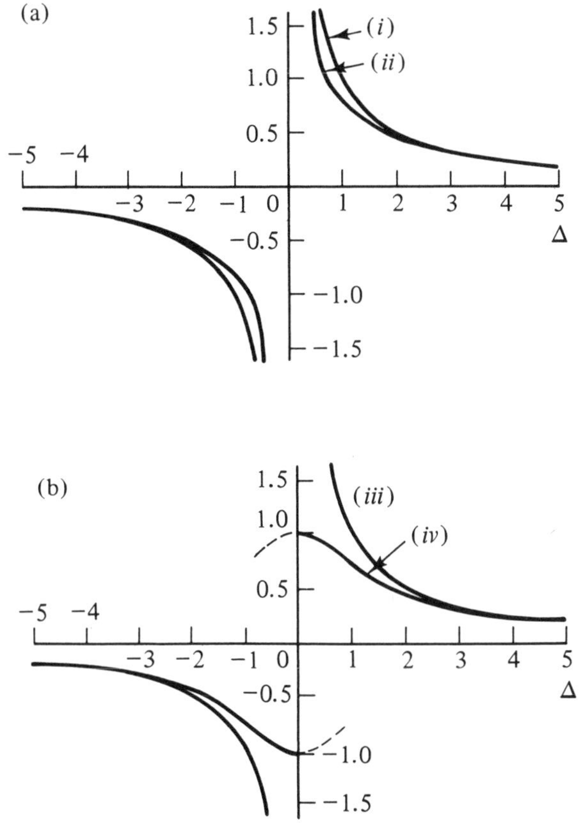

Fig. 6.5 The various expressions for the intensity-dependent refractive index of a two-level system, neglecting relaxation, are compared.

(a) The susceptibilities calculated assuming negligible perturbation of the system from its lower-energy state. Curve (*i*) is a plot of the function Δ^{-1}, which is proportional to the real part of the linear susceptibility $\chi^{(1)}(\omega;\omega)$. Curve (*ii*) is a plot of $\Delta^{-1}(1-k_1\Delta^{-2}+k_2\Delta^{-4})$, which illustrates the 'field-corrected' susceptibility, as given by the first three terms in the series expression (6.67) for $\overline{\chi}(\omega;E)$. (For purposes of illustration, the values $k_1=0.25$ and $k_2=0.05$ are assumed.)

(b) The 'field-corrected' susceptibility derived from the dynamic two-level model. Curves (*iii*) and (*iv*) show plots of the function $\mathrm{sgn}(\Delta)\sqrt{\Delta^2+\beta^2}$, which is proportional to $\overline{\chi}(\omega;E)$ given by (6.69). The two cases shown are (*iii*): vanishing field ($\beta \to 0$), and (*iv*): a field of strength $\beta=1$.

Curves (*i*) and (*iii*) are identical. The solid curves in Fig. 6.5(b) are drawn assuming that the two-level system is predominantly in the lower energy state ($w<0$). The dashed lines show the effect of sweeping the field frequency through resonance, causing a population inversion $w>0$ (adiabatic rapid passage).

is identical to the resonant term in (4.58). Equation (6.72) is the expression that can be derived from the previous formula (4.120) for the third-order susceptibility when only one-photon resonant terms are retained and the two-level system is taken to be in its ground state. By inserting (6.72) into (6.63), we obtain the following useful first-order expression for the nonlinear refraction coefficient n_2:

$$n_2(\omega) = \frac{-Ne^4}{4\varepsilon_0 n_0(\omega)(\hbar\Delta)^3}|\mathbf{r}_{ab}\cdot\mathbf{e}|^4 . \tag{6.74}$$

The succession of higher-order susceptibilities ((6.72), (6.73) and so on) gives successively higher-order field corrections to the linear susceptibility. Thus, by using the two-level model to derive (6.69), we have been able to express the series (6.67) in closed form.

In Fig. 6.5 we compare the various expressions for the nonlinear refractive index, assuming negligible relaxation. The susceptibilities shown in Fig. 6.5(a) are those derived in Chapter 4 assuming a very small perturbation of the nonlinear medium. In Fig. 6.5(a)(*i*), the linear susceptibility $\chi^{(1)}(-\omega;\omega)$ given by (6.71) is shown plotted as a function of detuning Δ from resonance. The addition of higher-order nonlinear terms, illustrated by Fig. 6.5(a)(*ii*), has the effect of introducing the field-dependent corrections. The susceptibilities calculated assuming negligible perturbation are valid for large resonance detunings $|\Delta| >> 0$. Figure 6.5(b) shows the results calculated using the dynamic two-level model. The field-dependent susceptibility $\overline{\chi}(\omega;E)$ contains the resonant denominator term $\sqrt{\Delta^2+\beta^2}$. Therefore, as shown in Fig. 6.5(b), the influence of a finite field ($\beta>0$) is such as to remove the pole at $\Delta=0$, and β is identified as the power-dependent linewidth, an effect known as 'power broadening'. The essential difference between the small-perturbation analysis and the dynamic two-level model, illustrated by Figs. 6.5(a) and (b), is that the latter properly takes into account the finite limit to the excitation that can be induced.

As illustrated by Figs. 6.5(a) and (b) respectively, the effect of the field is to depress the refractive index for $\Delta>0$ and to increase it for $\Delta<0$. In other words (since $\Delta\equiv\Omega_{ba}-\omega$), the nonlinear refractive index is negative for field frequencies below resonance, and positive above resonance. This does, however, assume that the two-level system is initially in the lower energy state ($w_0=-1$); if instead $w_0=1$ then the signs are reversed. We note that the field-dependent susceptibility $\overline{\chi}(\omega;E)$ shown in Fig. 6.5(b) is finite everywhere, but is apparently discontinuous at $\Delta=0$. This discontinuity stems from the assumption made in drawing the figure that $w<0$, and in fact this assumption is invalid if Δ is swept through zero. The reason for this can be seen by referring to the vector model of adiabatic following, depicted in Fig. 6.4; as Δ passes through zero (*i.e.*, changes sign), the pseudo-field vector $\mathbf{\Omega}$ swings through the $\hat{\mathbf{u}}-\hat{\mathbf{v}}$ plane followed adiabatically by the Bloch vector $\mathbf{R}$, so that w changes sign also. Thus the polarisation $P(t)$ given by (6.68), which contains the term $\mathrm{sgn}(w)\,\mathrm{sgn}(\Delta)$, is a smoothly-continuous function as Δ is swept through zero, but simultaneously the two-level population inversion changes sign. We should emphasise the assumption here that the field-amplitude variations are slow (adiabatic following). The inversion that can be achieved simply by sweeping the frequency of the applied optical field through resonance under these conditions is termed 'adiabatic rapid passage'; the effect is well known in magnetic resonance, and has been demonstrated in the optical region (Loy, 1974).

Relaxation processes were neglected in the two-level model used to derive the expression (6.69) for the field-corrected susceptibility. We have seen in the previous section that, in general, the addition of relaxation processes greatly complicates the model, although solutions can be found readily in the steady-state regime (*cf.* §6.2.6). By inserting the solutions (6.53) into (6.48), and comparing with (6.64), the result for the 'field-corrected' susceptibility is

$$\overline{\chi}(\omega;E) = \frac{-Ne^2|\mathbf{r}_{ab}\cdot\mathbf{e}|^2 w_0 T_2(\Delta T_2+i)}{\varepsilon_0\hbar[1+(\Delta T_2)^2+\beta^2T_1T_2]}. \tag{6.75}$$

The power-broadened linewidth $\delta\Omega_{ba}$ due to saturation of the homogeneously-broadened resonant transition is given by

$$\delta\Omega_{ba} = \sqrt{1+\beta^2T_1T_2}\,/T_2. \tag{6.76}$$

The formula (6.75) allows us to make a simple and revealing interpretation of the optical nonlinearity exhibited by the two-level system under steady-state conditions. We recall, from (2.51), that the intensity of the applied field is given by

$$I_\omega = \tfrac{1}{2}\varepsilon_0 c\, n(\omega)E^2, \tag{6.77}$$

and we define a saturation intensity I_{sat} as being the steady-state intensity which reduces the population inversion w to one-half of its thermal-equilibrium value w_0. From (6.53b) it follows directly that when $I_\omega=I_{\mathrm{sat}}$, the Rabi frequency β is given by

$$\beta^2 = \frac{1+(\Delta T_2)^2}{T_1T_2}. \tag{6.78}$$

With the aid of (6.37) and (6.77), we thus find an expression for I_{sat}:

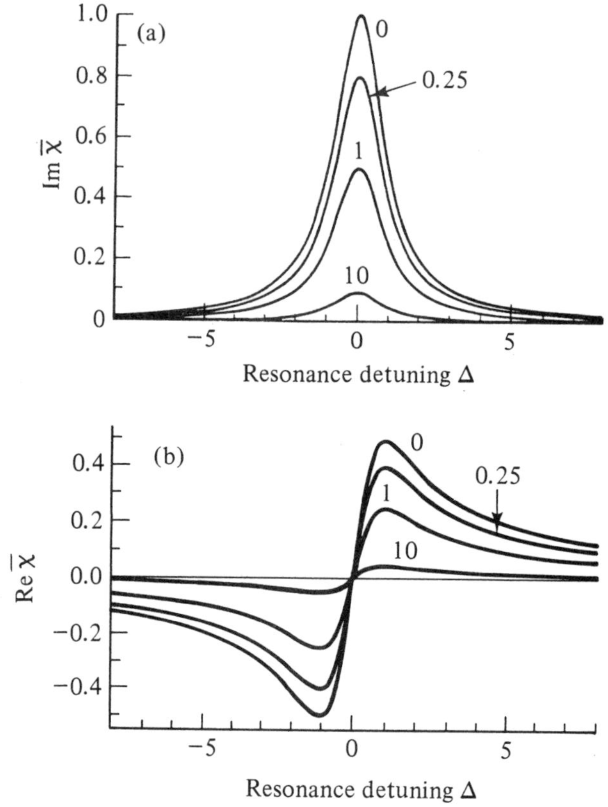

Fig. 6.6 Curves showing the dispersion in the (a) imaginary and (b) real parts of the field-corrected susceptibility $\overline{\chi}(\omega;E)$ for the homogeneously-broadened two-level system in the steady state. The curves are plots of the imaginary and real parts of the function $\{(\Delta-i/T_2)[1+I_\omega/I_{sat}(\Delta)]\}^{-1}$ which is proportional to $\overline{\chi}(\omega;E)$ given by (6.80), for the case $T_2=1$, and for optical intensities $I_\omega/I_{sat}(\Delta)=0$, 0.25, 1 and 10, as labelled. The effect of increasing intensity is therefore to saturate both the absorption and refractive index, giving rise to a nonlinear refraction coefficient which is negative for $\omega<\Omega_{ba}$ $(\Delta>0)$, and positive for $\omega>\Omega_{ba}$.

$$I_{sat}(\Delta) = \left\{\frac{2e^2T_1T_2}{\varepsilon_0 n_0(\omega)c\hbar^2[1+(\Delta T_2)^2]}|\mathbf{r}_{ab}\cdot\mathbf{e}|^2\right\}^{-1}. \tag{6.79}$$

It is now straightforward to verify that the formula (6.75) for the field-corrected susceptibility $\overline{\chi}(\omega;E)$ can be expressed in the simple form:

$$\overline{\chi}(\omega;E) = \frac{\chi^{(1)}(-\omega;\omega)}{1+I_\omega/I_{sat}(\Delta)}, \tag{6.80}$$

and similarly (6.76) can be rewritten as:

$$\delta\Omega_{ba} = \sqrt{1+I_\omega/I_{sat}(\Delta)}/T_2. \tag{6.81}$$

Thus the meaning of the saturation intensity I_{sat} becomes obvious; it is the steady-state intensity which reduces $\overline{\chi}(\omega;E)$ to one-half of its value at low intensity. The real and imaginary parts of $\overline{\chi}(\omega;E)$ are drawn in Fig. 6.6. The intensity dependence contained in the factor $[1+I_\omega/I_{sat}(\Delta)]^{-1}$ in (6.80) is attributable to the reduction of the population difference $N\{\rho_0(a)-\rho_0(b)\}$ in the familiar linear-susceptibility expression. The dynamic two-level model thus clearly shows that the origin of the optical nonlinearity is saturation of the homogeneous transition between energy states E_a and E_b.

It is sometimes useful to express I_{sat} in a more compact form:

$$I_{sat} = \frac{\hbar\omega}{2\sigma_a(\omega)T_1}, \tag{6.82a}$$

where $\sigma_a(\omega)$ is the absorption cross-section defined as (*cf.* §7.1.2)

$$\sigma_a(\omega) = \frac{\omega}{c\,n_0(\omega)N}\,\mathrm{Im}\left[\chi^{(1)}(-\omega;\omega)\Big|_{\rho_0(a)=1}\right]. \tag{6.82b}$$

The expression (6.82a) for I_{sat} is easily verified by writing down the explicit formula for the linear susceptibility $\chi^{(1)}(-\omega;\omega)$ of the two-level system:

$$\chi^{(1)}(-\omega;\omega) = \frac{-w_0Ne^2}{\varepsilon_0\hbar(\Delta-i/T_2)}|\mathbf{r}_{ab}\cdot\mathbf{e}|^2, \tag{6.83}$$

which is obtained from (4.111) in the rotating-wave approximation.

A result similar to (6.80) is given in §6.5.3 to describe saturation effects in another nonlinear-optical process, namely stimulated Raman scattering. As a further illustration, the dynamic two-level model is used in §8.4.3 to describe saturation of the interband absorption in semiconductors.

6.3.3 Optical Stark effect

It is instructive to consider now the energy-level shifts δE_a and δE_b, which were neglected for simplicity in the treatment of the two-level system (§6.2.2). We show that these level shifts–which must always occur to some degree–can be viewed as another manifestation of the refractive index changes.

Let us first consider the linear response of a medium to a quasi-monochromatic optical field: $\mathbf{E}(t)=\frac{1}{2}[\mathbf{E}_\omega(t)\exp(-\omega t)+\mathbf{E}_\omega^*(t)\exp(i\omega t)]$. The medium consists of N two-level atoms per unit volume and, for simplicity, in thermal equilibrium all the atoms are taken to be in the ground state E_a (*i.e.*, $w_0=-1$). We suppose that the field is applied sufficiently slowly that the response of the system is adiabatic (*cf.* §6.1.1).

We also suppose, to begin with, that the field is sufficiently weak (or equivalently, sufficiently far from resonance) that the perturbation of the system from the ground state is small ($w \simeq w_0$). According to (5.6), the real power input $W(t)$ to unit volume of the medium at any point is given by

$$W(t) = <\mathbf{E}(t)\cdot\frac{\mathrm{d}}{\mathrm{d}t}\mathbf{P}(t)> = \tfrac{1}{2}\,\omega\,\mathrm{Im}\,[\mathbf{E}_\omega{}^*(t)\cdot\mathbf{P}_\omega^{(1)}(t)], \qquad (6.84)$$

where the angle brackets denote a cycle average. From (4.104), the linear susceptibility of the system is given by $\chi^{(1)}(-\omega;\omega) = N\alpha_{aa}(-\omega;\omega)/\varepsilon_0$, where $\alpha_{aa}(-\omega;\omega)$ denotes the linear polarisability of the two-level atom in its lower state. (We note that, from the outset in §6.2.1, we assumed for simplicity that the two-level atoms are non-interacting, so that the local-field factors in (4.104) can be set equal to unity.) Inserting $P_\omega^{(1)} = \varepsilon_0\chi^{(1)}(-\omega;\omega)E_\omega$ into (6.84) we obtain:

$$\begin{aligned} W(t) &= \tfrac{1}{2}\,N\omega\,\mathrm{Im}\,\alpha_{aa}(-\omega;\omega)\,|E_\omega(t)|^2 \\ &= N\omega\,\mathrm{Im}\,\alpha_{aa}(-\omega;\omega)\,U_0(t)/\varepsilon_0\,, \end{aligned} \qquad (6.85)$$

where $U_0(t) = \frac{1}{2}\,\varepsilon_0|E_\omega(t)|^2$ is the energy density of the electric field in vacuum. The power flow given by (6.85) is dependent on the *imaginary* part of the ground-state polarisability. This flow of power to the medium is taken up by two-level atoms as they make real transitions from the ground to upper energy levels; *i.e.*, the process of absorption.

The *real* part of the polarisability $\alpha_{aa}(-\omega;\omega)$ determines the refractive index of the medium through the expression (6.61a):

$$\begin{aligned} n_0(\omega) &= \sqrt{1 + \mathrm{Re}\,\chi^{(1)}(-\omega;\omega)} \\ &= \sqrt{1 + N\,\mathrm{Re}\,\alpha_{aa}(-\omega;\omega)/\varepsilon_0} \\ &\simeq 1 + N\,\mathrm{Re}\,\alpha_{aa}(-\omega;\omega)/2\varepsilon_0\,. \end{aligned} \qquad (6.86)$$

From (6.77), we see that the energy density of the quasi-monochromatic field $\mathbf{E}(t)$ in the medium is $U(t) = \frac{1}{2}\,\varepsilon_0 n_0^2|E_\omega(t)|^2$. In other words, by virtue of the refractive index, the energy density of the optical field is greater than its vacuum value $U_0(t)$. This increase in the optical energy density is a continuous function of the adiabatically-applied field. In the total atom-field system in which energy is conserved, there is therefore a reversible flow of energy from the medium to the field – slowly turning off the field restores the atom to its original state. The action of the field is to create an 'adiabatic ground state'; the atom experiences an adiabatic variation of the ground-state energy by an amount $\delta\mathrm{E}_a$. The energy shift $\delta\mathrm{E}_a$ is determined by the overall energy-conservation relation:

$$\tfrac{1}{2}\varepsilon_0\,|E_\omega|^2 = (N\,\delta\mathrm{E}_a + \tfrac{1}{2}\,\varepsilon_0 n^2|E_\omega|^2)/n\,. \qquad (6.87)$$

With the aid of (6.71) and (6.86) we may express the result as:

$$\delta \mathrm{E}_a = -\frac{e^2}{4\hbar\Delta}|\mathbf{r}_{ab}\cdot\mathbf{E}_\omega|^2 . \tag{6.88}$$

This is the usual formula for the optical (a.c.) Stark effect to first order. According to this, the energy shift scales linearly with intensity (or quadratically with the field), and this phenomenon is sometimes also referred to as the 'quadratic optical Stark effect'. The upper level of the two-level atom experiences an equal and opposite shift: $\delta \mathrm{E}_b = -\delta \mathrm{E}_a$.

The analysis of the quadratic optical Stark shift for a two-level atom can be extended to the more general case of a multi-level system. By recognising that the polarisability $\alpha_{jj}(-\omega;\omega)$ is the first-order transition hyperpolarisability given by (4.115), we obtain from (6.86)–(6.87) a more general result for the lowest-order Stark shift of a level with an unshifted energy E_j:

$$\begin{aligned}\delta \mathrm{E}_j &= -\frac{e^2}{4\hbar}\sum_i|\mathbf{r}_{ji}\cdot\mathbf{E}_\omega|^2\left[\frac{1}{\Omega_{ij}-\omega}+\frac{1}{\Omega_{ij}+\omega}\right]\\ &= -\frac{e^2}{2\hbar}\sum_i|\mathbf{r}_{ji}\cdot\mathbf{E}_\omega|^2\,\frac{\Omega_{ij}}{(\Omega_{ij})^2-\omega^2},\end{aligned} \tag{6.89}$$

where $\hbar\Omega_{ij} = \mathrm{E}_i - \mathrm{E}_j$ is the energy difference between the unshifted levels. The shift $\delta \mathrm{E}_j$ can be interpreted as a shift of the energy level j in the sense that if a weak probe field were used to induce transitions between level j and other levels, one would find these probe resonances shifted by $\delta \mathrm{E}_j$. The previous result (6.88) for the two-level atom is recovered from (6.89) by noting that, in the rotating-wave approximation, the second term in brackets in the first line of (6.89) is negligible. When $\omega \to 0$ in (6.89), this reduces to the quadratic d.c. Stark effect when the change $\frac{1}{2}|E_\omega|^2 \to E_{\mathrm{dc}}^2$ is made.

The refractive index seen by the field and the energy-level Stark shift seen by the atom are thus two aspects of the same phenomenon. Having recognised this correspondence, it is a simple task to introduce as necessary higher-order corrections to the formulae for the optical Stark shift. In principle, the series expression (6.70) for $\overline{\chi}(\omega;E)$ enables us to calculate the energy-level shifts to any order of approximation. For increasing intensity, the Stark effect turns over from a quadratic dependence to one that is approximately linear in field strength ($\sim\sqrt{I_\omega}$). In the limit of strong fields or close resonance the power-series expansion is no longer practical (as discussed earlier in this chapter), and we then have recourse to the formalism in terms of a saturated two-level transition; when relaxation processes are included, the appropriate expression is (6.75).

An alternative way of looking at the power-broadening effect discussed in §6.3.2 is in terms of the optical Stark shifts of the resonant energy levels. Close to resonance, power broadening and the level shifts are indistinguishable.

6.3.4 Inhomogeneous broadening

So far in this discussion of the nonlinear refractive index, we have assumed that all the saturable absorbers in the ensemble share the same resonant frequency Ω_{ba}, *i.e.,* it has been assumed that the two-level transition is homogeneously broadened. Whilst this does sometimes occur in nature, more often the absorbing line is broadened inhomogeneously. In an inhomogeneous system the absorbers are distinguishable, such that each one has a unique resonant frequency, and a statistical distribution characterises the spectrum of resonant frequencies in the ensemble. Two principal examples of this type of broadening were mentioned previously in §6.2.5. In solids, the energy levels of impurity ions depend on the immediate surroundings of each ion. Because of random variations in the local strain field, as well as other types of crystal imperfections, the crystal surroundings vary from one ion to the next, resulting in a spread of resonant frequencies. A second example is that of molecules in gases. The molecules are free to move around so that the resonant frequency of an individual molecule, as seen by a stationary observer, is Doppler-shifted. Provided the gas pressure is low enough that there is little broadening due to collisions and other kinds of interactions between the molecules, the observed spread of resonant frequencies reflects the statistical distribution of molecular velocities. For a gas in thermal equilibrium this is given by the Maxwell-Boltzmann distribution. Further examples of inhomogeneous broadening in artificial semiconductor materials are mentioned in Chapter 9; significant broadening of the absorption lines in multiple quantum wells and glasses containing semiconductor microcrystals may occur because of the statistical distribution of well widths and microcrystal sizes.

The key differences between systems with homogeneous and inhomogeneous broadening are seen in the manner of their saturable absorption. The intensity-dependence and spectral-dependence of the saturation are quite different in the two cases, as we now describe.

The description to be used for an inhomogeneous system is quite similar to that for the homogeneous system, outlined in the previous sections. The important difference is that the resonant frequencies of the two-level absorbers, which we denote by Υ, are statistically distributed. Very often, the frequencies Υ are modified by some external influence having randomness (*e.g.*, molecular velocity or local crystal field), so that the probability distribution for Υ is aptly described by a normal (Gaussian) distribution $g(\Upsilon)$, with the mean value Ω_{ba} and standard deviation $\delta\Omega_{ba}^{\mathrm{I}}/\sqrt{2}$:

$$g(\Upsilon) = (\sqrt{\pi}\,\delta\Omega_{ba}^{\mathrm{I}})^{-1} \exp\left[-\left(\frac{\Omega_{ba}-\Upsilon}{\delta\Omega_{ba}^{\mathrm{I}}}\right)^2\right]. \tag{6.90}$$

The distribution is normalised, so that $\int_{-\infty}^{+\infty} \mathrm{d}\Upsilon\, g(\Upsilon) = 1$. The standard deviation and the inverse of the inhomogeneous lifetime T_2* (referred to in §6.2.2) are approximately equal. (The superscript on $\delta\Omega_{ba}^{\mathrm{I}}$ is used here to distinguish the inhomogeneous linewidth from the homogeneous width $\delta\Omega_{ba}$ given by (6.76).) An example of (6.90) is the Doppler-broadened line shape which follows from the Maxwell-Boltzmann velocity distribution; in that case $\delta\Omega_{ba}^{\mathrm{I}} = \Omega_{ba}\overline{v}/c$, where $\overline{v}$ is the root-mean-square thermal velocity given by $\overline{v} = \sqrt{2kT/M}$, M is the molecular mass, T is the temperature, and k is Boltzmann's constant. The Doppler linewidth $\delta\Omega_{\mathrm{D}}$ (full width at half maximum) is expressed in terms of the standard deviation as $\delta\Omega_{\mathrm{D}} = 2\sqrt{\ln 2}\,\delta\Omega_{ba}^{\mathrm{I}} = 2\sqrt{2\ln 2\,kT/M}\,\Omega_{ba}/c$.

The expression for the polarisation $\mathbf{P}(t) = N\,\mathrm{Tr}(\rho e\mathbf{r})$ should now be written as

$$\mathbf{P}(t) = N \int_{-\infty}^{+\infty} \mathrm{d}\Upsilon\, \mathrm{Tr}[\rho(\Upsilon)\, e\mathbf{r}]\, g(\Upsilon)\,, \tag{6.91}$$

where $\rho(\Upsilon)$ denotes the density matrix for the group of saturable absorbers which exhibit a resonant frequency in the interval $\Upsilon, \Upsilon + \mathrm{d}\Upsilon$. In the steady state, the trace in (6.91) can be evaluated by following the steps described in §6.2 that lead to the solutions (6.53). Thus we obtain the following expression for the polarisation-wave amplitude $P(t)$:

$$P(t) = -w_0 N \frac{(e\mathbf{r}_{ba}\cdot\mathbf{e})^2}{\hbar} T_2 \int_{-\infty}^{+\infty} \mathrm{d}\Upsilon \left[\frac{(\Delta_\Upsilon T_2 + i)\, g(\Upsilon)}{1 + (\Delta_\Upsilon T_2)^2 + \beta^2 T_1 T_2}\right] E(t)\,, \tag{6.92}$$

where $\Delta_\Upsilon = \Upsilon - \omega$ is the optical detuning from resonance for those molecules with resonant frequency Υ. In the case when $g(\Upsilon)$ is the normal distribution (6.90), the integral in (6.92) has the following value (Abramowitz and Stegun, 1965):

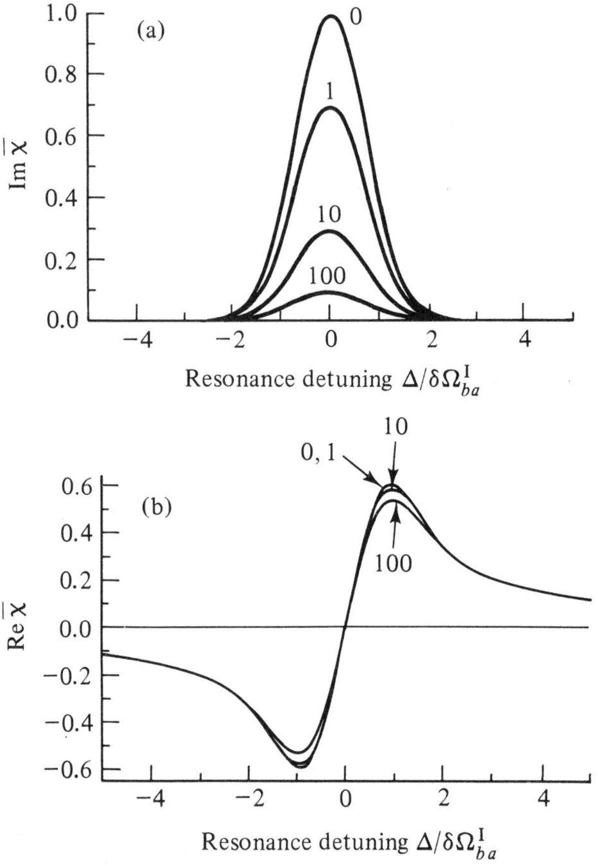

Fig. 6.7 Absorption and refractive-index curves for a Gaussian inhomogeneously-broadened transition, represented by the (a) imaginary and (b) real parts, respectively, of the susceptibility $\overline{\chi}$ given by (6.95). The curves were calculated assuming the ratio of inhomogeneous to homogeneous linewidth $\delta\Omega^{\mathrm{I}}_{ba}/T_2^{-1}=10$, and optical intensities $I_\omega/I_{\mathrm{sat}}(0)$ = 0, 1, 10 and 100, as labelled. The susceptibility is shown normalised with respect to the term outside the large brackets on the right-hand side of (6.95). The intensity dependence of the refractive index (*i.e.*, the *change* in Re$\overline{\chi}$ with increasing intensity) is rather small in comparison with that shown in Fig. 6.6 for a homogeneous transition.

$$\int_{-\infty}^{+\infty} \mathrm{d}\Upsilon \left[\frac{(\Delta_\Upsilon T_2+i)g(\Upsilon)}{1+(\Delta_\Upsilon T_2)^2+\beta^2 T_1 T_2}\right] = \frac{\sqrt{\pi}}{T_2\,\delta\Omega^{\mathrm{I}}_{ba}}\left\{\frac{i}{\sqrt{1+\beta^2T_1T_2}}\mathrm{Re}[W(x+iy)] + \mathrm{Im}[W(x+iy)]\right\} \tag{6.93}$$

where $x=\Delta/\delta\Omega^{\mathrm{I}}_{ba}$ and $y=\sqrt{1+\beta^2T_1T_2}\,/T_2\,\delta\Omega^{\mathrm{I}}_{ba}=\delta\Omega_{ba}/\delta\Omega^{\mathrm{I}}_{ba}$. As in the previous case of a homogeneous transition, $\Delta=\Omega_{ba}-\omega$ is the detuning from the absorption line centre. The function $W(z)$ is closely related to the error function, and is defined by

$$W(z) = \exp(-z^2)\left[1 + \frac{2i}{\sqrt{\pi}}\int_0^z ds\, \exp(s^2)\right]. \tag{6.94}$$

By inserting (6.93) into (6.92), and comparing the result with the definition (6.64) for the 'field-corrected' susceptibility $\overline{\chi}(\omega;E)$, we obtain the expression

$$\overline{\chi}(\omega;E) = -w_0 N \frac{(e\mathbf{r}_{ba}\cdot\mathbf{e})^2\sqrt{\pi}}{\varepsilon_0 \hbar\, \delta\Omega_{ba}^{\mathrm{I}}} \times\left\{\frac{i}{\sqrt{1+\beta^2 T_1 T_2}}\mathrm{Re}[W(x+iy)] + \mathrm{Im}[W(x+iy)]\right\}. \tag{6.95}$$

The parameters x and y which appear in these expressions relate the magnitudes of the important frequency factors. Thus, in words, x is the ratio of the optical detuning from the line-centre to the width of the inhomogeneously-broadened line, whilst y is the ratio of the power-broadened homogeneous linewidth, given by (6.76), to the inhomogeneous linewidth.

The expression (6.95) for the susceptibility $\overline{\chi}(\omega;E)$ has a particularly simple form in the regime where the inhomogeneous linewidth is very much broader than the homogeneous width, *i.e.* $y\rightarrow 0$. In that limit the function $W(z)$ defined by (6.94) takes the form

$$\mathrm{Re}[W(x+iy)] \rightarrow \exp(-x^2) = \sqrt{\pi}\,\delta\Omega_{ba}^{\mathrm{I}}\, g(\omega) \tag{6.96}$$

and

$$\mathrm{Im}[W(x+iy)] \rightarrow \frac{2\exp(-x^2)}{\sqrt{\pi}}\int_0^x ds\, \exp(s^2) = 2\,\delta\Omega_{ba}^{\mathrm{I}}\, g(\omega)\int_0^x ds\, \exp(s^2). \tag{6.97}$$

Substituting these expressions in (6.95), we obtain the following expression for the susceptibility which applies in the limit $y\rightarrow 0$:

$$\overline{\chi}(\omega;E) = -w_0 N \frac{(e\mathbf{r}_{ba}\cdot\mathbf{e})^2\sqrt{\pi}}{\varepsilon_0\hbar}\left[\frac{i\sqrt{\pi}}{\sqrt{1+\beta^2 T_1 T_2}} + 2\int_0^x ds\, \exp(s^2)\right] g(\omega). \tag{6.98}$$

The imaginary part can be written in the form

$$\mathrm{Im}\,\overline{\chi}(\omega;E) = -w_0 N \frac{(e\mathbf{r}_{ba}\cdot\mathbf{e})^2\pi}{\varepsilon_0\hbar\sqrt{1+I_\omega/I_{\mathrm{sat}}(0)}}\, g(\omega), \tag{6.99}$$

where $I_{\mathrm{sat}}(0)$ is given by (6.79).

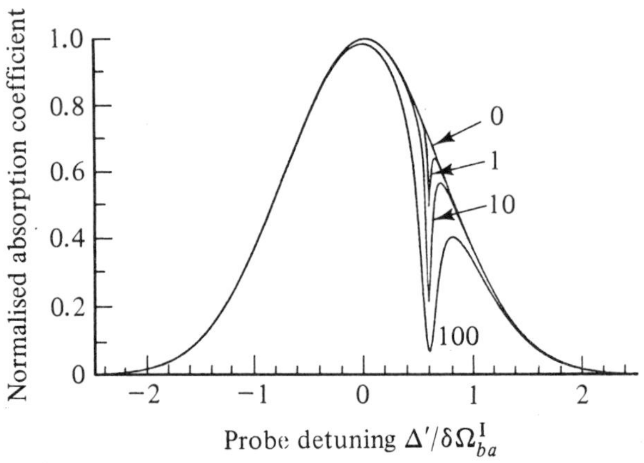

Fig. 6.8 The absorption spectrum of an inhomogeneously-broadened line measured by a weak probe beam reveals a hole 'burnt' at the frequency of the pump source (intensity I_ω). Here $\Delta' = \Omega_{ba} - \omega'$ where ω' is the probe frequency, and the pump frequency is fixed at the detuning $\Delta/\delta\Omega^{\mathrm{I}}_{ba} = 0.6$. As in Fig. 6.7, the curves were calculated assuming the ratio of inhomogeneous to homogeneous linewidth $\delta\Omega^{\mathrm{I}}_{ba}/T_2^{-1} = 10$, and $I_\omega/I_{\mathrm{sat}}(0)$ = 0, 1, 10 and 100, as labelled. The area of the hole, which represents the number of molecules excited resonantly by the pump, grows as I_ω is increased. The depth of the hole is calculated from (6.95), and its spectral width is approximately equal to $\delta\Omega_{ba}$ given by (6.76).

Thus the saturation behaviour for an absorbing transition which is severely broadened inhomogeneously differs in some important respects from that of a homogeneous transition. Firstly, in the inhomogeneous system, unlike a homogeneous one, the saturation intensity does not depend on the detuning Δ, *i.e.*, the saturation intensity $I_{\mathrm{sat}}(0)$ is independent of the position within the spectral line. Secondly, for an inhomogeneously-broadened system, the saturation of absorption has a characteristic $[1+I_\omega/I_{\mathrm{sat}}]^{-1/2}$ dependence, as compared to the $[1+I_\omega/I_{\mathrm{sat}}]^{-1}$ dependence in the homogeneous case, *cf.* (6.80). A simple explanation for this is that, with increasing intensity, the power-broadened homogeneous linewidth $\delta\Omega_{ba}$ increases according to (6.76), and thus a growing number of molecules are able to interact with the incident field. If the form of (6.76) is multiplied by that of the homogeneous saturation (6.80), the result is the inverse square-root dependence shown in (6.99). Thus, in principle, an experimental measurement of the power-law dependence on intensity can be used to distinguish between systems which are homogeneously- and inhomogeneously-broadened; in practice, however, due allowance must be made for the variation of intensity with propagation distance in the medium so that the ratio $I_\omega/I_{\mathrm{sat}}$ is also a function of distance (Frantz and Nodvik, 1963), and this may partially obscure the power-law dependence.

The third point to notice from (6.98), which is perhaps less widely appreciated, is that in the limit where the inhomogeneous linewidth is very much broader than the homogeneous linewidth $\delta\Omega_{ba}$ (*i.e.*, $y \to 0$), the real part of $\overline{\chi}(\omega;E)$ is independent of β – in other words, the intensity-dependence of the refractive index is negligible. However, if the optical intensity I_ω is increased sufficiently (increasing β, and thus also increasing $\delta\Omega_{ba}$ according to (6.76)), there comes a point at which the assumption $y \simeq 0$ is no longer valid. The intensity-dependence of the refractive index must then be found by evaluating the real part of (6.95). Examples are shown in Fig. 6.7. These demonstrate the need to avoid inhomogeneous broadening if a large resonant intensity-dependence of the refractive index is desired.

The effect we have just described is known as 'spectral hole-burning'. To illustrate this more clearly, suppose that our source of intensity I_ω (the pump) has its frequency fixed at a value such that $\Delta/\delta\Omega^{\mathrm{I}}_{ba} = 0.6$. The resulting absorption spectrum measured by a weak probe beam as it is tuned through the inhomogeneously-broadened line is depicted in Fig. 6.8. The pump excites selectively the group of molecules with resonant frequencies Υ in the vicinity of $(\Omega_{ba} - \Upsilon)/\delta\Omega^{\mathrm{I}}_{ba} \simeq 0.6$, so causing the partial saturation of absorption which is observed as a dip or 'hole' in the absorption spectrum. This observation provides a method for resolving the homogeneous linewidth from within an inhomogeneously-broadened transition, and is used in high-resolution saturation spectroscopy (Levenson and Kano, 1988). Spectral hole-burning is not observed in a homogeneously-broadened absorption line because the whole line profile is saturated homogeneously (Fig. 6.6).

6.4 Multiple driving fields

The intensity-dependent refractive index, discussed in the last section, is a particularly simple nonlinear-optical process in the sense that only one optical wave is involved. Often in nonlinear optics we are concerned with processes that involve two or more waves, and in this section and the next one we consider examples of such processes when resonances are involved. The examples given illustrate ways in which the basic two-level model can be adapted to these more complex situations.

An important example which illustrates the principles is resonant degenerate four-wave mixing. This process involves the resonant interaction of four waves, each having the same frequency ω, with the two-level system. In general the waves are distinguishable by their

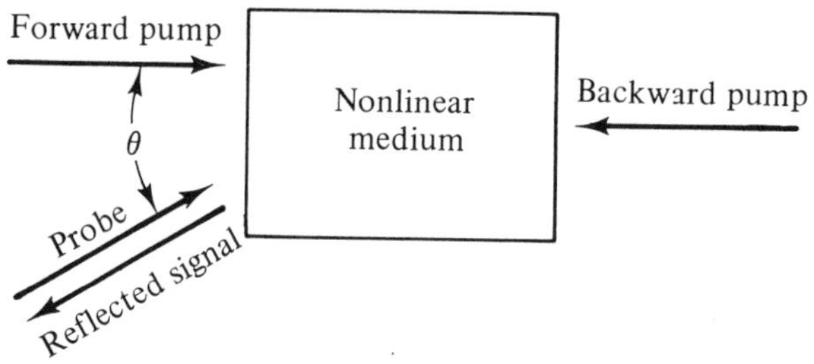

Fig. 6.9 Geometry for degenerate four-wave mixing

different propagation and polarisation directions. Degenerate four-wave mixing has important applications in nonlinear laser spectroscopy and optical phase conjugation (§7.4). Many different geometrical configurations are possible; the most common one is illustrated in Fig. 6.9. A description of the wave-propagation aspects is deferred until Chapter 7; here we mainly focus attention on the matter-field interaction, leading to an expression for the induced polarisation.

In the limit of weak fields whose frequencies are removed from resonance, we can use the susceptibility formalism developed in earlier chapters. The lowest-order term that describes degenerate four-wave mixing is the susceptibility $\chi^{(3)}(-\omega;\omega,-\omega,\omega)$, which also describes nonlinear refraction. The difference between these two effects, as we have just noted, is that degenerate four-wave mixing involves four waves which share the same frequency ω but which are nevertheless distinguishable.

We now consider the resonant case. Let us suppose that the total field $\mathbf{E}(t)$ has four quasi-monochromatic components, denoted by $\mathbf{E}_\mathrm{f}$, $\mathbf{E}_\mathrm{b}$, $\mathbf{E}_\mathrm{p}$ and $\mathbf{E}_\mathrm{r}$; these are the forward, backward, probe and reflected waves, respectively, depicted in Fig. 6.9. The total field in the nonlinear medium is therefore written:

$$\mathbf{E}(t) = \tfrac{1}{2}[(\mathbf{E}_\mathrm{f} + \mathbf{E}_\mathrm{b} + \mathbf{E}_\mathrm{p} + \mathbf{E}_\mathrm{r})\exp(-i\omega t) + \text{c.c.}], \tag{6.100}$$

where the symbol c.c. denotes the complex conjugate. As in (6.34), the perturbation Hamiltonian is given by

$$H_\mathrm{I}(t) = \begin{bmatrix} \delta\mathrm{E}_a & -e\mathbf{r}_{ab}\cdot\mathbf{E}(t) \\ -e\mathbf{r}_{ba}\cdot\mathbf{E}(t) & \delta\mathrm{E}_b \end{bmatrix}. \tag{6.101}$$

This Hamiltonian must then be inserted into the equation of motion for the density operator (6.26) in order to evaluate the density-matrix elements and hence the polarisation. The solution of the four-field problem is difficult in the general case when the fields can take arbitrary amplitudes. Very commonly, however, the situation of most interest involves strong pump fields $\mathbf{E}_f$ and $\mathbf{E}_b$, whilst the intensity of the probe and reflected fields, $\mathbf{E}_p$ and $\mathbf{E}_r$, remain small compared to the saturation intensity. In this case the pump fields are mainly responsible for driving the resonant interaction, possibly giving rise to significant changes of population in the coupled energy levels, whilst the probe and reflected fields have a relatively insignificant effect on the level populations. In this limit, the problem can be greatly simplified by using the methods of §6.2 to calculate the response of the system to the strong pump waves alone, and then the effect of the weaker probe and reflected waves can be calculated as a first-order perturbation.

We now give a more detailed description of this approach, which was developed by Abrams *et al* (1983). We consider the steady-state response of the simplest possible system – a nondegenerate two-level atomic ensemble which is homogeneously broadened – and draw on the results derived in §6.2. The first stage is to derive the equations of motion for the density operator – the master equations. The strong and weak fields are separated by writing (6.100) as

$$\mathbf{E}(t) = \tfrac{1}{2}[(\mathbf{E}_0+\mathbf{E}_1)\exp(-i\omega t) + \text{c.c.}], \qquad (6.102)$$

where $\mathbf{E}_0=\mathbf{E}_f+\mathbf{E}_b$ and $\mathbf{E}_1=\mathbf{E}_p+\mathbf{E}_r$. The various fields $\mathbf{E}_{\omega_j}=\mathbf{e}_j E_j$ may take arbitrary polarisations $\mathbf{e}_j$ and the corresponding envelope functions E_i are complex (this differs from the simplified treatment in §6.2 in which the fields are assumed to have linear polarisations and real envelope functions). Similarly, the expression (6.101) for the perturbation Hamiltonian is separated into two parts:

$$H_{\mathrm{I}}(t) = H_{\mathrm{I}}^{(0)}(t) + H_{\mathrm{I}}^{(1)}(t), \qquad (6.103)$$

where

$$H_{\mathrm{I}}^{(0)}(t) = \begin{bmatrix} \delta\mathrm{E}_a & -\tfrac{1}{2}e\mathbf{r}_{ab}\cdot\{\mathbf{E}_0\exp(-i\omega t)+\text{c.c.}\} \\ -\tfrac{1}{2}e\mathbf{r}_{ba}\cdot\{\mathbf{E}_0\exp(-i\omega t)+\text{c.c.}\} & \delta\mathrm{E}_b \end{bmatrix} \qquad (6.104)$$

and

$$H_{\mathrm{I}}^{(1)}(t) = \begin{bmatrix} 0 & -\tfrac{1}{2}e\mathbf{r}_{ab}\cdot\{\mathbf{E}_1\exp(-i\omega t)+\text{c.c.}\} \\ -\tfrac{1}{2}e\mathbf{r}_{ba}\cdot\{\mathbf{E}_1\exp(-i\omega t)+\text{c.c.}\} & 0 \end{bmatrix}. \qquad (6.105)$$

Lastly, the density operator ρ is written as

$$\rho = \rho^{(0)} + \rho^{(1)}, \qquad (6.106)$$

where the term $\rho^{(0)}$ is found from the zeroth-order equation of motion:

$$i\hbar\,\mathrm{d}\rho^{(0)}/\mathrm{d}t = [\{H_0 + H_{\mathrm{I}}^{(0)} + H_{\mathrm{R}}\}, \rho^{(0)}], \tag{6.107}$$

and similarly the first-order term $\rho^{(1)}$ is found from

$$i\hbar\,\mathrm{d}\rho^{(1)}/\mathrm{d}t = [\{H_0 + H_{\mathrm{I}}^{(0)} + H_{\mathrm{R}}\}, \rho^{(1)}] + [H_{\mathrm{I}}^{(1)}, \rho^{(0)}]. \tag{6.108}$$

By making the following substitutions (equivalent to (6.40)):

$$\begin{aligned}\rho_{ab}^{(0)} &= \rho_{ab}^{(0)\Omega}\exp[i(\Omega_{ba}-\Delta)t]\\ \rho_{ab}^{(1)} &= \rho_{ab}^{(1)\Omega}\exp[i(\Omega_{ba}-\Delta)t],\end{aligned} \tag{6.109}$$

where $\Delta = \Omega_{ba} - \omega$, and applying the rotating-wave approximation, we obtain a set of master equations – similar to (6.41):

$$\mathrm{d}\rho_{aa}^{(0)}/\mathrm{d}t = \tfrac{1}{2}i\hbar^{-1}e(\rho_{ba}^{(0)\Omega}\mathbf{r}_{ab}\cdot\mathbf{E}_0 - \rho_{ab}^{(0)\Omega}\mathbf{r}_{ba}\cdot\mathbf{E}_0^*) - i\hbar^{-1}[H_{\mathrm{R}}, \rho^{(0)}]_{aa} \tag{6.110a}$$

$$\begin{aligned}\mathrm{d}\rho_{ab}^{(0)\Omega}/\mathrm{d}t = i\Delta\rho_{ab}^{(0)\Omega} &+ \tfrac{1}{2}i\hbar^{-1}e(\rho_{bb}^{(0)} - \rho_{aa}^{(0)})\mathbf{r}_{ab}\cdot\mathbf{E}_0(t)\\ &- i\hbar^{-1}[H_{\mathrm{R}}, \rho^{(0)}]_{ab}\exp(-i\omega t)\end{aligned} \tag{6.110b}$$

$$\mathrm{d}\rho_{bb}^{(0)}/\mathrm{d}t = -\tfrac{1}{2}i\hbar^{-1}e(\rho_{ba}^{(0)\Omega}\mathbf{r}_{ab}\cdot\mathbf{E}_0 - \rho_{ab}^{(0)\Omega}\mathbf{r}_{ba}\cdot\mathbf{E}_0^*) - i\hbar^{-1}[H_{\mathrm{R}}, \rho^{(0)}]_{bb} \tag{6.110c}$$

$$\begin{aligned}\mathrm{d}\rho_{aa}^{(1)}/\mathrm{d}t = {}&\tfrac{1}{2}i\hbar^{-1}e(\rho_{ba}^{(0)\Omega}\mathbf{r}_{ab}\cdot\mathbf{E}_1 - \rho_{ab}^{(0)\Omega}\mathbf{r}_{ba}\cdot\mathbf{E}_1^*)\\ &+ \tfrac{1}{2}i\hbar^{-1}e(\rho_{ba}^{(1)\Omega}\mathbf{r}_{ab}\cdot\mathbf{E}_0 - \rho_{ab}^{(1)\Omega}\mathbf{r}_{ba}\cdot\mathbf{E}_0^*) - i\hbar^{-1}[H_{\mathrm{R}}, \rho^{(1)}]_{aa}\end{aligned} \tag{6.110d}$$

$$\begin{aligned}\mathrm{d}\rho_{ab}^{(1)\Omega}/\mathrm{d}t = {}&i\Delta\rho_{ab}^{(1)\Omega} + \tfrac{1}{2}i\hbar^{-1}e(\rho_{bb}^{(0)} - \rho_{aa}^{(0)})\mathbf{r}_{ab}\cdot\mathbf{E}_1(t)\\ &+ \tfrac{1}{2}i\hbar^{-1}e(\rho_{bb}^{(1)} - \rho_{aa}^{(1)})\mathbf{r}_{ab}\cdot\mathbf{E}_0(t) - i\hbar^{-1}[H_{\mathrm{R}}, \rho^{(1)}]_{ab}\exp(-i\omega t)\end{aligned} \tag{6.110e}$$

$$\begin{aligned}\mathrm{d}\rho_{bb}^{(1)}/\mathrm{d}t = {}&-\tfrac{1}{2}i\hbar^{-1}e(\rho_{ba}^{(0)\Omega}\mathbf{r}_{ab}\cdot\mathbf{E}_1 - \rho_{ab}^{(0)\Omega}\mathbf{r}_{ba}\cdot\mathbf{E}_1^*)\\ &-\tfrac{1}{2}i\hbar^{-1}e(\rho_{ba}^{(1)\Omega}\mathbf{r}_{ab}\cdot\mathbf{E}_0 - \rho_{ab}^{(1)\Omega}\mathbf{r}_{ba}\cdot\mathbf{E}_0^*) - i\hbar^{-1}[H_{\mathrm{R}}, \rho^{(1)}]_{bb}.\end{aligned} \tag{6.110f}$$

To these should be added the equations $\rho_{ba}^{(0)\Omega} = (\rho_{ab}^{(0)\Omega})^*$ and $\rho_{ba}^{(1)\Omega} = (\rho_{ab}^{(1)\Omega})^*$. Here, as in §6.2.2, we have neglected the Stark shifts $\delta\mathrm{E}_a$ and $\delta\mathrm{E}_b$ for simplicity.

This set of coupled equations certainly looks daunting. However, the next stage in the analysis is quite straightforward; the steps described in §6.2 are followed to find solutions for the zeroth-order equations (the ones in $\rho^{(0)}$) *i.e.*, taking into account the pump fields alone. In the steady state, exact solutions are found by setting $\mathrm{d}\rho_{ab}^{(0)\Omega}/\mathrm{d}t \to 0$ in (6.110) to yield

$$\rho_{ab}^{(0)\Omega} = \frac{iT_2\beta(\rho_{bb}^{(0)} - \rho_{aa}^{(0)})}{2(1 - i\Delta T_2)}, \tag{6.111}$$

where $\beta = e\mathbf{r}_{ab}\cdot\mathbf{E}_0/\hbar$ is the Rabi flopping frequency corresponding to the strong pump field $\mathbf{E}_0$. Similarly, setting $\mathrm{d}(\rho_{bb}^{(0)} - \rho_{aa}^{(0)})/\mathrm{d}t \to 0$ gives

$$\rho_{bb}^{(0)} - \rho_{aa}^{(0)} = \frac{w_0[1+(\Delta T_2)^2]}{1+(\Delta T_2)^2+|\beta|^2 T_1 T_2}. \qquad (6.112)$$

As before, w_0 denotes the population inversion under thermal-equilibrium conditions. The expression (6.112) has an obvious similarity to (6.53b). In particular we see that the zeroth-order population difference is saturated in the limit of large β.

Having now derived the system response to the strong pump field $\mathbf{E}_0$, the next stage is to obtain the first-order fluctuations $\rho^{(1)}$ described by the master equations (6.110d)–(6.110f), again assuming steady-state conditions. It is this perturbation in the density operator ρ which gives rise to the component of the nonlinear polarisation that is of interest. Setting $\mathrm{d}\rho_{ab}^{(1)\Omega}/\mathrm{d}t \to 0$ in the master equations leads to

$$\rho_{ab}^{(1)\Omega} = \frac{i T_2 e \mathbf{r}_{ab}\cdot[(\rho_{bb}^{(0)} - \rho_{aa}^{(0)})\mathbf{E}_1 + (\rho_{bb}^{(1)} - \rho_{aa}^{(1)})\mathbf{E}_0]}{2\hbar(1-i\Delta T_2)}, \qquad (6.113)$$

and setting $\mathrm{d}(\rho_{bb}^{(1)} - \rho_{aa}^{(1)})/\mathrm{d}t \to 0$ leads to

$$\rho_{bb}^{(1)} - \rho_{aa}^{(1)} = \frac{-i T_1 e \mathbf{r}_{ab}}{\hbar} \cdot [\rho_{ba}^{(0)\Omega}\mathbf{E}_1 - \rho_{ab}^{(0)\Omega}\mathbf{E}_1{}^* + \rho_{ba}^{(1)\Omega}\mathbf{E}_0 - \rho_{ab}^{(1)\Omega}\mathbf{E}_0{}^*]. \qquad (6.114)$$

Using the first-order expressions (6.111) and (6.112), the first-order population difference (6.114) may be rewritten as

$$\rho_{bb}^{(1)} - \rho_{aa}^{(1)} = \frac{-w_0 T_1 T_2 [1+(\Delta T_2)^2] e^2}{[1+(\Delta T_2)^2 + |\beta|^2 T_1 T_2]^2 \hbar^2} [\mathbf{r}_{ab}\cdot\mathbf{E}_0{}^* \, \mathbf{r}_{ab}\cdot\mathbf{E}_1 + \mathbf{r}_{ab}\cdot\mathbf{E}_0 \, \mathbf{r}_{ab}\cdot\mathbf{E}_1{}^*]. \qquad (6.115)$$

This represents the first-order population difference induced by the simultaneous action of the strong pump field $\mathbf{E}_0$ and the weak field $\mathbf{E}_1$.

By inserting (6.111), (6.112) and (6.114) into (6.113), we obtain (after some straightforward but laborious algebra) an expression for the off-diagonal matrix element:

$$\rho_{ab}^{(1)\Omega} = \frac{i w_0 [1+(\Delta T_2)^2](1+i\Delta T_2) T_2 e}{2[1+(\Delta T_2)^2 + |\beta|^2 T_1 T_2]^2 \hbar} \left[\mathbf{r}_{ab}\cdot\mathbf{E}_1 - \frac{\beta^2 T_1 T_2}{1+(\Delta T_2)^2} \mathbf{r}_{ab}\cdot\mathbf{E}_1{}^*\right]. \qquad (6.116)$$

This expression can now be used in the final stage of the calculation, which is to evaluate the macroscopic polarisation $\mathbf{P}(t)$; it is given by

$$\mathbf{P}(t) = N\,\mathrm{Tr}(e\mathbf{r}\rho) = N\left[\mathrm{Tr}(e\mathbf{r}\rho^{(0)}) + \mathrm{Tr}(e\mathbf{r}\rho^{(1)})\right]. \qquad (6.117)$$

The term in $\rho^{(0)}$ is the polarisation induced by the strong pump field $\mathbf{E}_0$ alone, whilst the term in $\rho^{(1)}$ is the perturbation due to the simultaneous action of the strong and weak fields. Following from (6.117), we can write the amplitude of the polarisation wave as

$$\mathbf{P} = \varepsilon_0 \overline{\boldsymbol{\chi}}(\omega;\mathbf{E}_0)\,\mathbf{E}_0 + \delta\mathbf{P}\,, \tag{6.118}$$

where $\overline{\boldsymbol{\chi}}(\omega;\mathbf{E}_0)$ is the field-dependent susceptibility discussed in §6.3.1, and $\delta\mathbf{P}$ denotes the perturbation. By expanding the traces in (6.117), we obtain the expression

$$\delta\mathbf{P} = 2Ne\mathbf{r}_{ba}\,\rho_{ab}^{(1)\Omega}. \tag{6.119}$$

Finally, by substituting for $\rho_{ab}^{(1)}$ from (6.116), we obtain the result:

$$\delta P = \frac{w_0 N(i-\Delta T_2)T_2 e^2\,\mathbf{r}_{ba}\cdot\mathbf{e}_1}{[1+(\Delta T_2)^2+|\beta|^2 T_1T_2]^2\hbar} \times\left\{[1+(\Delta T_2)^2]\,\mathbf{r}_{ab}\cdot\mathbf{E}_1 - \beta^2 T_1T_2\,\mathbf{r}_{ab}\cdot\mathbf{E}_1^*\right\}. \tag{6.120}$$

Here, as previously, $\mathbf{e}_1$ is a unit vector in the direction of polarisation of the field $\mathbf{E}_1$. The polarisation component δP is the complete response of the medium for degenerate four-wave mixing in the presence of arbitrarily strong pump fields but weak signal and probe. The first term in braces on the right-hand side of (6.120) corresponds to the saturated absorption and refractive index experienced by the weak fields but induced by the strong fields, whereas the second term in braces is the polarisation produced by the simultaneous action of the strong and weak fields. It is the second term which is responsible for the generation of the phase-conjugate reflected wave.

Some physical understanding of the mechanism can be gained by noting, from (6.115), that the first-order population difference $\rho_{bb}^{(1)} - \rho_{aa}^{(1)}$ depends on products of the strong and weak fields. When the spatial properties of the various incident fields are taken into account (*cf.* §7.4), it is found that they interfere with each other to form fringes – in other words, the total intensity is modulated in space. Thus, through the interaction with the two-level saturable absorber, the refractive index and absorption are also modulated – a property commonly referred to as 'spatial hole-burning' – giving rise to the formation of 'gratings'. These gratings can be considered to be equivalent to thick holographic phase gratings. Indeed, the process of generating a phase-conjugate wave in resonant four-wave mixing can be viewed as Bragg reflection from the gratings induced in the saturable absorber. We note that the second term in braces on the right-hand side of (6.120) has a real and imaginary part, showing that both the saturated absorption and refractive index

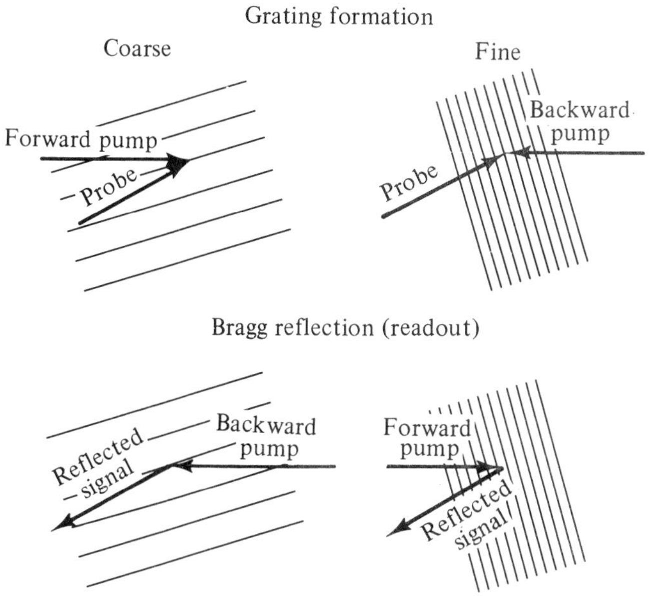

Fig. 6.10 Degenerate four-wave mixing viewed as a process of grating formation and readout. The grating pitch is $(\lambda/2)\sin(\theta/2)$ and $(\lambda/2)\cos(\theta/2)$ for the coarse and fine gratings, respectively, where θ is the angle of intersection of the beams shown in Fig. 6.9. (After Lind *et al*, 1982.)

contribute. We notice also that, for strong pump fields (*i.e.*, large β), the first term in braces is reduced to zero (similar to (6.75)), whilst the second term merely acquires a constant maximum value. This is also physically understandable in terms of the spatial population gratings; when $\beta \rightarrow \infty$ the gratings acquire the maximum possible amplitude modulation (equal to 100% modulation, in the absence of relaxation). Many experiments have been performed which confirm this interpretation in terms of Bragg gratings (Fisher, 1983; Eichler *et al*, 1986).

Figure 6.10 shows the two important grating patterns that occur when the input fields are plane waves. The probe and forward pump waves, $\mathbf{E}_p$ and $\mathbf{E}_f$, form a grating of relatively coarse period which has the correct spacing and orientation to Bragg-scatter the backward pump $\mathbf{E}_b$ into the reflected field $\mathbf{E}_r$. The input waves $\mathbf{E}_p$ and $\mathbf{E}_b$, on the other hand, form a fine grating with the period and alignment necessary for scattering $\mathbf{E}_f$ into $\mathbf{E}_r$. A third grating, due to the interference of $\mathbf{E}_f$ and $\mathbf{E}_b$ alone, does not contribute to the four-wave process because it does not have the appropriate alignment for coherent scattering of $\mathbf{E}_p$ or $\mathbf{E}_r$. Further control can be exercised by using orthogonal polarisation directions for some of the input waves (since orthogonally-polarised waves cannot interfere). For example, by arranging for $\mathbf{E}_b$ to be polarised orthogonally to $\mathbf{E}_f$ and $\mathbf{E}_p$, the fine grating in Fig. 6.10 is suppressed; the reflected wave $\mathbf{E}_r$ is then entirely due to scattering of $\mathbf{E}_b$ by the coarse grating (Jain and Lind, 1983).

A similar situation occurs in *nearly*-degenerate four-wave mixing. Like the degenerate case discussed above, this process can be viewed in terms of induced spatial population gratings. The difference, however, is that the gratings no longer have a fixed position, but move through the nonlinear medium. The frequency difference between the incident and diffracted waves is then equivalent to a Doppler shift.

6.5 Multi-photon resonances

In the two-level model of a nonlinear medium, the energy levels E_a and E_b (depicted in Fig. 6.1) are coupled by an electric-dipole transition and all other levels are neglected. This model is therefore appropriate to the case when the incident optical field is in close *single*-photon resonance with the transition Ω_{ba}. However, as discussed in §4.5, nonlinear-optical processes may be enhanced by tuning the frequencies of the incident waves into *multi*-photon resonance with a pair of energy levels of the medium. An example of this, for third-harmonic generation, is depicted in Fig. 4.1. Therefore, it would be very useful to extend the two-level model to provide a description of these multi-photon-resonant processes. A general and powerful procedure for doing this, due to Friedmann and Wilson-Gordon (1978), is described below, and is illustrated by the example of stimulated Raman scattering. The advantage of so extending the two-level model is that many of the most useful results derived for single-photon resonant processes may be carried forward into the multi-quantum regime.

One aim of the discussion here is to show the links that exist between the classical and quantum descriptions. Therefore, we first review the classical model of stimulated scattering.

6.5.1 Classical model

The classical description of vibrational Raman scattering is based on the phenomenological model of Placzek (1934). The dipole moment $e\mathbf{r}$ induced in a molecule by a local field $\mathbf{E}$ is given by $e\mathbf{r}=\boldsymbol{\alpha}\cdot\mathbf{E}$, where the molecular polarisability $\boldsymbol{\alpha}$ (a second-rank tensor) is regarded as a function of the nuclear conformation $\mathbf{X}$. In a diatomic molecule, such as H_2, the magnitude X is simply the displacement of the interatomic separation from its equilibrium value. The basis of the Placzek model is that, for small displacements, the optical polarisability is a linear function of $\mathbf{X}$:

$$\boldsymbol{\alpha}(\mathbf{X}) = \boldsymbol{\alpha}_0 + (\partial\boldsymbol{\alpha}/\partial\mathbf{X})_0 \cdot \mathbf{X}, \tag{6.121}$$

where the subscript $_0$ denotes evaluation at the equilibrium conformation $\mathbf{X}=0$. The differential polarisability $\partial\boldsymbol{\alpha}/\partial\mathbf{X}$ is a third-rank tensor. If the molecule is asymmetric, then it may possess a permanent dipole moment $e\mathbf{r}^0$, which may also be expressed as a linear function of $\mathbf{X}$:

$$e\mathbf{r}^0(\mathbf{X}) = e\mathbf{r}_0^0 + (\partial e\mathbf{r}^0/\partial\mathbf{X})_0 \cdot \mathbf{X}. \tag{6.122}$$

According to the theory of electrostatics, the potential energy of the molecule in the local field $\mathbf{E}$ is represented by the interaction Hamiltonian:

$$\begin{aligned} H_{\mathrm{I}}(t) &= -e\mathbf{r}\cdot\mathbf{E} \\ &= -\left[e\mathbf{r}_0^0 + (\partial e\mathbf{r}^0/\partial\mathbf{X})_0 \cdot \mathbf{X} + \boldsymbol{\alpha}_0 \cdot \mathbf{E} + (\partial\boldsymbol{\alpha}/\partial\mathbf{X})_0 : \mathbf{X}\,\mathbf{E}\right] \cdot \mathbf{E}. \end{aligned} \tag{6.123}$$

For simplicity, we take X to represent the displacement of a single vibrational coordinate from the equilibrium position; this allows us to drop the vector and tensor notation. Let us consider the various terms in $H_{\mathrm{I}}(t)$. The first and third terms in the square brackets in (6.123) are independent of X, and do not therefore allow any exchange of energy between the field and the internal energy of the molecule. The first term, involving the permanent static moment er_0^0 is the energy of molecular alignment in the applied field. The second term in square brackets gives rise to absorption when the optical frequency approaches the resonant vibrational frequency Ω_{v}. The dipole moment induced by the local field E leads to Rayleigh scattering through the third term $\alpha_0 E$. It is the last term in the square brackets in (6.123) that gives rise to Raman scattering. If the local field contains a wave at ω_{P} with the amplitude E_{P}, Raman scattering occurs through the term $(\partial\alpha/\partial X)_0 X E_{\mathrm{P}}$ because X vibrates at its resonant frequency Ω_{v}, and thus XE_{P} oscillates at the upper and lower sideband frequencies $\omega_{\mathrm{P}} - \Omega_{\mathrm{v}}$ and $\omega_{\mathrm{P}} + \Omega_{\mathrm{v}}$; these are the Stokes and anti-Stokes frequencies, respectively. Here we focus attention on the Stokes process. Picking out the appropriate term from (6.123), the classical Raman-interaction Hamiltonian is

$$H_{\mathrm{I}}^{\mathrm{R}}(t) = -\tfrac{1}{4}\,(\partial\alpha/\partial X)_0 X_\Omega^* E_{\mathrm{S}}^* E_{\mathrm{P}} + \text{c.c.} \tag{6.124}$$

We are assuming here that the local field $E(t)$ consists of two quasi-monochromatic waves:

$$E(t) = \tfrac{1}{2}\left[E_{\mathrm{P}}\exp(-i\omega_{\mathrm{P}}t) + E_{\mathrm{S}}\exp(-i\omega_{\mathrm{S}}t) + \text{c.c.}\right], \tag{6.125}$$

where $\omega_{\mathrm{P}} > \omega_{\mathrm{S}}$, and the pump field ω_{P} and Stokes field ω_{S} are in close

Raman resonance with the vibrational frequency, *i.e.*, $\omega_P - \omega_S \simeq \Omega_v$. The vibrational coordinate X has also been written in terms of its slowly-varying envelope X_Ω as

$$X = X_\Omega \exp[-i(\Omega_v - \Delta)t] + X_\Omega^* \exp[i(\Omega_v - \Delta)t], \tag{6.126}$$

where $\Delta = \Omega_v - \omega_P + \omega_S$. Conventionally, X is chosen to be dimensionless and normalised such that in the spontaneous scattering regime $X_\Omega = 1$. The influence of the local fields E_P and E_S on the molecule, *via* the Raman interaction, is such that a force $-\partial H_I^R/\partial X$ acts on the coordinate X. Since, for small displacements, the motion of X is approximately harmonic, we have

$$\frac{\partial^2 X}{\partial t^2} + 2\Gamma\frac{\partial X}{\partial t} + \Omega_v^2 X = -\frac{2\Omega_v}{\hbar}\frac{\partial H_I^R(t)}{\partial X}, \tag{6.127}$$

where Γ is a damping factor, and $\hbar/2\Omega_v$ is the reduced mass associated with the vibration (*cf.* (1.1)). By substituting $H_I^R(t)$ from (6.124) into (6.127), and rewriting in terms of the vibration amplitude X_Ω^*, we obtain

$$\begin{aligned}\frac{\partial X_\Omega^*}{\partial t} - i(\Delta + i\Gamma)X_\Omega^* &= (i/\hbar)\frac{\partial H_I^R(t)}{\partial X_\Omega} \\ &= -(i/4\hbar)(\partial\alpha/\partial X)_0^* E_S E_P^* .\end{aligned} \tag{6.128}$$

Here we have assumed that the resonance detuning is small, so that $\Omega_v >> |\Delta|$, and also that the vibration amplitude is a slowly-varying quantity ($\partial^2 X_\Omega/\partial t^2 << \Omega_v^2 X_\Omega$). From (6.123), the macroscopic polarisation at the Stokes frequency resulting from the Raman interaction is

$$P_S = N(\partial\alpha/\partial X)_0 X_\Omega^* E_P, \tag{6.129}$$

where N is the molecular number density, and local-field corrections are neglected here (as in the remainder of this section). Then, for example, in the steady-state limit of stimulated Raman scattering ($\partial X_\Omega^*/\partial t \simeq 0$), we find from (6.128) and (6.129) that

$$\begin{aligned}P_S &= \frac{N}{4\hbar}(\partial\alpha/\partial X)_0 E_P\left(\frac{(\partial\alpha/\partial X)_0^* E_S E_P^*}{\Delta + i\Gamma}\right) \\ &= \frac{N|(\partial\alpha/\partial X)_0|^2 |E_P|^2 E_S}{4\hbar(\Delta + i\Gamma)} .\end{aligned} \tag{6.130}$$

In §7.3 it is shown that this polarisation at the Stokes frequency leads to the exponential growth of the Stokes field as it propagates through the pumped medium.

The connection with the susceptibility formalism is made by comparing (6.129) with the earlier expression (2.65) for the polarisation in terms of the wave response function. In the adiabatic limit, we find the link:

$$\chi^{(3)}(-\omega_S;\omega_P,-\omega_P,\omega_S) = \frac{N\,|(\partial\alpha/\partial X)_0|^2}{6\hbar\varepsilon_0(\Delta+i\Gamma)}. \tag{6.131}$$

This is identical to the previous expression (4.118), with the correspondence:

$$(\partial\alpha/\partial X)_0 = \alpha_{ba}(-\omega_S;\omega_P). \tag{6.132}$$

By thus relating the differential polarisability $\partial\alpha/\partial X$, the first-order transition hyperpolarisability $\alpha_{ba}(-\omega_S;\omega_P)$ and the third-order susceptibility, we have established the connections between the classical and quantum-mechanical descriptions. Also, the 'material excitation' referred to in the pictorial description of two-photon-resonant third-order processes in §4.5.2 can now be identified as equivalent to the classical excitation $X_\Omega{}^*$. The meaning of this excitation in a more general context, in quantum-mechanical terms, emerges towards the end of the section.

Although the above discussion of the classical model is in terms of a Raman resonance, the same considerations apply to other types of multi-photon-resonant processes. For example, for each of the two-photon-resonant processes depicted in Fig. 4.3, there are expressions corresponding to (6.128)–(6.132), in which the optical frequencies involved correspond to the pairs of frequencies listed in Table 4.1.

In §4.5 it is shown how the multi-photon-resonant susceptibilities factorise into products of transition hyperpolarisabilities. It can now be understood a little more clearly how this arises by considering the implications of (6.128) and (6.129). The polarisation at the Stokes frequency arises from the modulation of the polarisability of the medium due to some internal motion X. The polarisation is proportional to the differential polarisability (or transition hyperpolarisability) multiplied by the excitation, and this is itself proportional to the differential polarisability. It is also apparent from (6.129) that it is the coherent excitation $X_\Omega{}^*$ alone which modulates the pump wave, and hence creates the Stokes polarisation – quite independent of the *way* in which $X_\Omega{}^*$ was created. We could, for instance, set up $X_\Omega{}^*$ by one pair of waves (denoted as pair (ii) in Table 4.1), and this excitation $X_\Omega{}^*$ may then modulate a third wave to create a polarisation, which in turn is coupled to the fourth wave through the wave equation. Because the generation of $X_\Omega{}^*$ is a process distinct from the subsequent coupling back to the field, it is not necessary that these two stages occur simultaneously. In the transient limit of four-wave mixing, for example, the coherent excitation $X_\Omega{}^*$ set up by one pair of waves will last for a time on the order of Γ^{-1} after the waves are suddenly switched off. As implied in §2.4.3, this excitation may be probed at a later time by a second pair of waves, and so information on the relaxation time can be found.

There is thus an obvious analogy between the excitation set up by a multi-photon resonance, and the single-photon resonant excitation described by the two-level model. This analogy is explored in more detail in §6.5.3.

Finally, the classical model of stimulated Raman scattering is usually presented, as above, in terms of the vibration of the nuclei in a molecule. This has the advantage that the physical mechanism can be easily visualised. However, there are many alternatives. Depending on the mechanism, the Stokes frequency shift may range between radio frequencies and values almost as large as the pump frequency itself. Different names are given to the Raman process depending on the type of material excitation involved. In vibrational-rotational Raman scattering, the Stokes shifts associated with intramolecular (nuclear) motions fall in the infrared and microwave spectral regions. Stokes shifts in the visible and ultraviolet regions are associated with electronic transitions, and the process is then known as electronic Raman scattering. At the other extreme, very small shifts (less than about 10 GHz) are usually related to molecular or bulk-lattice motion. An example of the latter is stimulated Brillouin scattering, which involves the excitation of an acoustic, or density, wave in the medium by the process of electrostriction; processes of this type are described conveniently in terms of a parametric coupling between the light and material-excitation waves.

6.5.2 Effective two-level Hamiltonian

In the electric-dipole approximation, the energy of the interaction between an applied field $\mathbf{E}(t)$ and the medium is represented by the interaction Hamiltonian:

$$H_{\mathrm{I}}(t) = -e\mathbf{r}\cdot\mathbf{E}(t)\,, \tag{6.133}$$

where $e\mathbf{r}$ denotes the dipole-moment operator. The discussion of the nonlinear susceptibilities in Chapter 4 shows that, in this approximation, a multi-photon resonance with Ω_{ba} *requires* the presence of intermediate energy levels coupled to E_a and E_b via sequences of electric-dipole transitions. Therefore, for multi-photon processes we can no longer consider the two resonant energy levels in isolation. However, the situation of greatest interest occurs when the incident waves are in close multi-photon resonance with the transition frequency Ω_{ba}, but there are no other significant resonances, either single- or multi-photon ones. It was pointed out by Heitler (1954) that, in this situation, new operators can be defined which allow the problem to be cast in the form of an effective two-level system. This has the advantage that all the results derived previously for single-photon resonance in a two-level system, including those discussed in §6.2, may be carried forward into the multi-photon regime. Friedmann and Wilson-Gordon (1978) devised a powerful and general technique which allows this to be done. We now summarise the general method, and give an example of its application in the next subsection.

The basic idea is to use projection operators to extract from the Hamiltonian for the *total* matter-field interaction an *effective* Hamiltonian for the resonant two-level system. The quantum-mechanical apparatus needed to follow the method is not difficult, and has been set up in Chapter 3.

The starting point for the calculation is the Schrödinger equation for the system in the interaction picture (*cf.* §3.7.3):

$$\begin{aligned} i\hbar\, \mathrm{d}\psi'(t)/\mathrm{d}t &= \exp(iH_0 t/\hbar)\, H_\mathrm{I}(t) \exp(-iH_0 t/\hbar)\, \psi'(t) \\ &= H_\mathrm{I}'(t)\, \psi'(t)\,, \end{aligned} \tag{6.134}$$

where $\psi'(t)$ is obtained by premultiplying the wave function $\psi(t)$ in the Schrödinger picture by the inverse time-development operator, thus:

$$\psi'(t) = \exp(iH_0 t/\hbar)\, \psi(t)\,. \tag{6.135}$$

The equation of motion (6.134) can be verified easily by differentiating (6.135), with the aid of (3.3) and (3.76), and by recalling that the Hamiltonian H_0 for the unperturbed molecular system is the same in both the Schrödinger and interaction pictures. The second step in (6.134) follows from the defining relation (3.76) for the perturbation Hamiltonian $H_\mathrm{I}'(t)$ in the interaction picture. As before, the perturbation Hamiltonian $H_\mathrm{I}(t)$ describes the *total* matter-field interaction, and is given by (6.133). A major difference from the previous work in §6.2, however, is that we must take into account the electric-dipole-allowed nonresonant transitions between the states labelled a or b and all of the other molecular states (previously the nonresonant states were neglected altogether).

The key step in the Friedmann and Wilson-Gordon approach is to isolate the dynamics of the m-photon-resonant two-level portion of the system from the remainder. This is done by defining two new projection operators (*cf.* §3.4), one of which points to the resonant two-level system, and the other points to the remaining states:

$$R = P(u_a) + P(u_b) \tag{6.136}$$

and

$$S = \sum_c P(u_c)\,; \quad c \neq a,b\,, \tag{6.137}$$

where $\{u_j\}$ are the energy eigenstates, $P(\psi)$ is the usual projection operator defined in (3.25), and $R+S$ is the unit operator. The equation of motion (6.133) can then be rewritten in the form

$$i\hbar\frac{\mathrm{d}}{\mathrm{d}t}[R\psi'(t)] = [RH_I'(t)R + RH_I'(t)S]\psi'(t) \tag{6.138}$$

$$i\hbar\frac{\mathrm{d}}{\mathrm{d}t}[S\psi'(t)] = [SH_I'(t)S + SH_I'(t)R]\psi'(t). \tag{6.139}$$

These equations may be solved by formal integration.

We now assume that the applied field consists of a superposition of quasi-monochromatic waves given by (6.1). The situation of interest occurs when one particular frequency ω_Ω is in close resonance with the transition frequency Ω_{ba}, where ω_Ω is taken from the set $\{\omega''\}$ of linear combinations of the incident frequencies and their multiples. Redefining the detuning factor Δ in the multi-photon-resonant case as $\Delta=\Omega_{ba}-\omega_\Omega$, the requirement that the resonance involving ω_Ω should dominate the system can be expressed formally as

$$|\Delta| \ll |\Omega_{jk}-\omega''| \quad \text{for all } j \text{ and } k, \text{ with } \omega''\neq\omega_\Omega. \tag{6.140}$$

We also assume that the adiabatic condition

$$\left|(\Omega_{jk}-\omega')^{-1}E_{\omega'}(t)^{-1}\frac{\mathrm{d}}{\mathrm{d}t}E_{\omega'}(t)\right| \ll 1 \tag{6.141}$$

is satisfied for all incident optical frequencies ω' and intermediate resonances. Although it is assumed that the incident field is applied adiabatically with respect to the intermediate single-photon transitions, it is not necessarily adiabatic with respect to the m-photon-resonant transition Ω_{ba}.

Under these conditions, Friedmann and Wilson-Gordon (1978) found, by formally integrating and manipulating the second of the two equations of motion (6.139), that the first equation (6.138) can be expressed in the form

$$i\hbar\frac{\mathrm{d}}{\mathrm{d}t}[R\psi'(t)] = \tilde{H}_I'(t)\,[R\psi'(t)] \tag{6.142}$$

where

$$\tilde{H}_I'(t) = \exp(-iH_0t/\hbar)\Big\{RH_I'(t)R - i\hbar^{-1}RH_I'(t)S$$
$$\times\int_{-\infty}^{t}\mathrm{d}\tau\,\mathfrak{I}\exp\left[-i\hbar^{-1}\int_{\tau}^{t}\mathrm{d}\tau'\,SH_I'(\tau')S\right]SH_I'(\tau)R\Big\}\exp(iH_0t/\hbar). \tag{6.143}$$

The exponential preceded by the symbol $\mathfrak{I}$ in (6.143) is defined by the chronologically-ordered expansion:

$$\mathfrak{J}\exp\left[-i\hbar^{-1}\int_{\tau}^{t}\mathrm{d}\tau'\, S\,H_{\mathrm{I}}'(\tau')S\right] =$$

$$1 + (-i\hbar^{-1})\int_{\tau}^{t}\mathrm{d}\tau'\, S\,H_{\mathrm{I}}'(\tau')S$$

$$+ (-i\hbar^{-1})^2\int_{\tau}^{t}\mathrm{d}\tau'\int_{\tau}^{\tau'}\mathrm{d}\tau''\, S\,H_{\mathrm{I}}'(\tau')SH_{\mathrm{I}}'(\tau'')S$$

$$+ \cdots . \tag{6.144}$$

It is assumed that the field $\mathbf{E}(t)$ vanishes in the distant past, so the integral with respect to τ in (6.143) is finite. By substituting for R from (6.136), it can be seen that (6.142) is identical to the Schrödinger equation (6.134) in the case that $\psi(t)$ is the wave function (6.25) for a two-level atom. Thus we have been able to express the multi-photon resonance in terms of a much more simple two-level system, in which the operator $\tilde{H}_{\mathrm{I}}(t)$ is identified as an effective two-level perturbation Hamiltonian. (The tilde is used here to distinguish it from $H_{\mathrm{I}}(t)$, the corresponding operator for the total matter-field system. The prime in (6.143) is used to indicate the operator in the interaction picture; *cf.* (6.134).) As before, the off-diagonal matrix elements of $\tilde{H}_{\mathrm{I}}(t)$ induce transitions between the coupled levels a and b, and the diagonal elements correspond to optical Stark shifts: $\delta \mathrm{E}_a = \tilde{H}_{\mathrm{I}}(t)_{aa}$ and $\delta \mathrm{E}_b = \tilde{H}_{\mathrm{I}}(t)_{bb}$.

When evaluating the effective matrix elements for a *specific* nonlinear-optical process, it is sufficient only to evaluate $\tilde{H}_{\mathrm{I}}(t)$ at the appropriate lowest non-vanishing order in the fields. In other words, when the applied field is taken to be in near m-photon resonance with Ω_{ba}, it is sufficient that the ordered exponential in (6.143) is substituted by the term in $(H_{\mathrm{I}}')^{m-2}$ from its expansion given by (6.142). This approximation is justified by the assumption that the m-photon-resonant process is the dominant mechanism which induces transitions between levels a and b. Also, it is necessary only to retain the frequency components in $\tilde{H}_{\mathrm{I}}(t)_{ab}$ which correspond to $\omega_{\Omega} \simeq \Omega_{ba}$. These points are made clearer by considering a specific example.

6.5.3 Stimulated Raman scattering – the links between the different approaches

To illustrate the general procedure described above, we now consider the two-photon-resonant third-order processes depicted in Fig. 4.3. As a specific example, we derive the effective two-level Hamiltonian describing stimulated Raman scattering, and show how this can be incorporated into the two-level model. In the course of doing this, we define the associated Rabi frequency which characterises the strength of the two-photon-resonant interaction, and also establish the links between this approach, the susceptibility, and the classical theory of Raman scattering.

We therefore take the optical field to comprise the superposition of a pump field at frequency ω_P and a Stokes field at ω_S :

$$\mathbf{E}(t) = \tfrac{1}{2}\Big[\mathbf{E}_{\omega_P}(t)\exp(-i\omega_P t) + \mathbf{E}_{\omega_S}(t)\exp(-i\omega_S t) + \text{c.c.}\Big], \quad (6.145)$$

where the pump frequency ω_P and Stokes frequency ω_S are in close Raman-resonance with the transition frequency Ω_{ba}. In this case the detuning factor is defined as

$$\Delta = \Omega_{ba} - (\omega_P - \omega_S). \quad (6.146)$$

The effective two-level Hamiltonian describing this class of process is bilinear in the incident field amplitudes, and therefore we need only retain the leading term (which is 1) in the exponential expansion (6.144). We recall from (3.7) that the matrix elements are defined by

$$\Big[\widetilde{H}_I(t)\Big]_{ij} = \int d\varsigma\, u_i^* \widetilde{H}_I(t) u_j\,, \quad (6.147)$$

where the labels i and j each takes values a and b. The matrix elements so defined are evaluated in a straightforward way by inserting the expressions (6.135)–(6.137) into (6.143). In so doing, we make repeated use of the energy eigenvalue equation (3.41) and the orthonormality condition (3.4). The significant components are found by extracting the terms which oscillate as $\cos[(\Omega_{ba}-\Delta)t]$. Finally, in the rotating-wave approximation, only the terms in $\exp[i(\Omega_{ba}-\Delta)t]$ are retained. The result for the off-diagonal element is

$$\Big[\widetilde{H}_I(t)\Big]_{ab}^{\text{RWA}} = \frac{-e^2}{4\hbar}\sum_c \left[\frac{\mathbf{r}_{ac}\cdot\mathbf{E}_{\omega_P}{}^* \,\mathbf{r}_{cb}\cdot\mathbf{E}_{\omega_S}}{\Omega_{ca}-\omega_P} + \frac{\mathbf{r}_{ac}\cdot\mathbf{E}_{\omega_S}\,\mathbf{r}_{cb}\cdot\mathbf{E}_{\omega_P}{}^*}{\Omega_{ca}+\omega_S}\right] \times \exp[i(\Omega-\Delta)t], \quad (6.148)$$

where the rotating-wave approximation implies $\Omega_{ba}+\omega_S \gg \Delta$ and $\Omega_{ca}-\omega_P \gg \Delta$. This result can be written in the form:

$$\Big[\widetilde{H}_I(t)\Big]_{ab}^{\text{RWA}} = -\tfrac{1}{4}\alpha_R{}^* E_P{}^* E_S \exp[i(\Omega-\Delta)t], \quad (6.149)$$

in which α_R is termed the Raman polarisability. By comparing (6.148)–(6.149) with (4.115), we see that α_R is equivalent to $\alpha_{ba}(-\omega_S;\omega_P)$, the first-order transition hyperpolarisability. We reached the same conclusion by a different route in §6.5.1.

To complete the analogy with the single-photon-resonant two-level model, we must define a Rabi frequency for the multi-photon regime. In the rotating-wave approximation, the expression (6.37) for the single-photon-resonant Rabi frequency can be rewritten in the form:

$$\beta(t) = -2\,[H_{\mathrm{I}}(t)]_{ab}^{\mathrm{RWA}} \exp[-i(\Omega_{ba}-\Delta)t]/\hbar\,, \tag{6.150}$$

where the resonance detuning is defined as $\Delta = \Omega_{ba} - \omega$. By analogy with (6.150), the multi-photon Rabi frequency in the rotating-wave approximation may be defined as

$$\beta(t) = -2\,[\widetilde{H}_{\mathrm{I}}(t)]_{ab}^{\mathrm{RWA}} \exp[-i(\Omega_{ba}-\Delta)t]/\hbar\,, \tag{6.151}$$

where, as in the previous subsection, the multi-photon resonance detuning takes the more general definition $\Delta = \Omega_{ba} - \omega_{\Omega}$. In the particular case of stimulated Raman scattering, Δ is given by (6.146), and the expression (6.151) for the Rabi frequency becomes

$$\beta(t) = \frac{\alpha_{\mathrm{R}}^{*} E_{\mathrm{P}}^{*} E_{\mathrm{S}}}{2\hbar}. \tag{6.152}$$

We have thus defined both an effective two-level Hamiltonian and a Rabi frequency for the multi-quantum resonant regime. The two-level model in this regime is described, as before, by the master equations of motion (6.41) or, in the T_1, T_2 approximation, by (6.52).

For stimulated Raman scattering, the equation of motion for the off-diagonal density matrix element ρ_{ab}^{Ω} in the rotating-wave approximation is

$$\mathrm{d}\rho_{ab}^{\Omega}/\mathrm{d}t - i(\Delta + i/T_2)\rho_{ab}^{\Omega} = i\beta(t)w/2 = (i/4\hbar)\alpha_{\mathrm{R}}^{*} E_{\mathrm{S}} E_{\mathrm{P}}^{*} w\,. \tag{6.153}$$

In the limit of small perturbation, when the molecules are assumed to be predominantly in the ground state ($w = -1$), we find that (6.153) is identical to the classical equation (6.128), in which there is a correspondence between ρ_{ab}^{Ω} and the classical coordinate X_{Ω}^{*}. In fact, ρ_{ba}^{Ω} can be identified as the well-defined generalised excitation that is appropriate to *any* system. For example, it represents the polarisability of electrons in an atom (arising from their being in a coherent superposition of the states u_a and u_b, *cf.* §6.2.1) or, as we have just seen, the vibrational coordinate X_{Ω}^{*} for a molecule. In the Born-Oppenheimer approximation, a clear distinction can be made between the vibrational and electronic contributions, as described in §4.5.4.

The various results derived previously for the two-level model in the single-photon-resonant regime may thus be carried forward into the multi-photon regime. To take an example, we consider stimulated Raman scattering in the steady-state regime. Similar to the field-corrected susceptibility $\overline{\chi}(\omega;E)$ which describes the nonlinear refractive index (in §6.3.1), we can define a field-dependent Raman susceptibility denoted by $\overline{\chi}_R(\omega_S;E_P,E_S)$. In the steady state, the field-dependent susceptibility is given by

$$\overline{\chi}_R(\omega_S;E_P,E_S) = \frac{\chi^{(3)}(-\omega_S;\omega_P,-\omega_P,\omega_S)}{1+I_P I_S/I_{sat}^2}, \tag{6.154}$$

where I_P and I_S are the pump and Stokes intensities, respectively, and the saturation is characterised by

$$I_P I_S/I_{sat}^2 = T_1 T_2 |\alpha_R E_S^* E_P/2\hbar|^2. \tag{6.155}$$

This expression for the saturation parameter I_{sat}^2 is obtained by substituting for the Rabi frequency β from (6.152) in the general relation (6.78). The saturation parameter may also be written in the form:

$$I_{sat}^2 = 3\hbar\omega_S/2(g_R/N)T_1, \tag{6.156}$$

where g_R is the Raman gain coefficient in the limit of negligible perturbation of the Raman medium from its ground state (we note the analogy with (6.82a)). In §7.3, the gain coefficient is derived in terms of the Raman susceptibility:

$$g_R = -\frac{3\omega_S}{\varepsilon_0 c^2 n_S n_P} \operatorname{Im}\chi^{(3)}(-\omega_S;\omega_P,-\omega_P,\omega_S)\Big|_{\rho_0(a)=1}. \tag{6.157}$$

The final connection between the two-level model and the susceptibility formalism is made by noting, with the aid of (4.118), that

$$\operatorname{Im}\chi^{(3)}(-\omega_S;\omega_P,-\omega_P,\omega_S)\Big|_{\rho_0(a)=1} = \frac{-N}{6\hbar\varepsilon_0} < |\alpha_R|^2 > \frac{(T_2)^{-1}}{\Delta^2+(T_2)^{-2}}$$

$$\simeq -N T_2 < |\alpha_R|^2 > /6\hbar\varepsilon_0 \quad \text{when } |\Delta| \ll (T_2)^{-1}. \tag{6.158}$$

Comparison between these results and the previous expressions (6.79)–(6.83) for a single-photon-resonance demonstrates the close analogy between the single- and multi-photon-resonant situations. A generalised Rabi frequency for multi-photon-resonant interactions is defined by Hanna *et al* (1979). The vector model (§6.2.7) may also be extended into the realm of multi-photon resonant nonlinear optics. Grischkowsky *et al* (1975) consider the extension into the two-photon regime.

Finally, we mention that the two-level model of resonant multi-photon interactions is no longer valid or appropriate when there are more than two discrete energy levels simultaneously in resonance. Such situations do occur in nonlinear optics. An example is the third-harmonic generation process, depicted in Fig. 4.1, in the case when two or more of the frequencies ω, 2ω and 3ω are simultaneously in close resonance with molecular transitions. The situation is then greatly complicated by the likely interdependence of relaxation processes involving intermediate levels, and correlations between the different transitions. Brewer and Hahn (1975) give complete solutions for the three-level case (an example of which is the multi-resonant third-harmonic generation just referred to), whilst Bloembergen *et al* (1978) provide a general formula for the third-order nonlinear susceptibility of a four-level system. The latter approach takes proper account of subtle coherence effects encountered when the interacting levels are degenerate (as reviewed by Omont (1977)). Further examples are given by Levenson and Kano (1988).

6.6 Summary of mechanisms and time regimes in nonlinear optics

In this and earlier chapters we have seen that nonlinear-optical phenomena can be considered in two broad categories. In the first category are those phenomena in which the optical frequencies are far from resonances of the medium (known variously as 'nonresonant', 'nondissipative' or 'passive' nonlinearities), and in that case the exchange of energy between the field and the medium is involved in *virtual* excitation (see §§4.3.3, 5.1.1, 6.3.3). Examples of processes in this category are second- and third-harmonic generation, optical Kerr effect, and the quadratic optical Stark effect. In the second category are those phenomena which are dissipative ('active' or 'dynamic' nonlinearities); in this case the flow of energy from and to the optical field is taken up in *real* energy transitions of the medium, by processes of absorption and emission. Examples of nonlinear-optical phenomena in this category are stimulated Raman scattering, two-photon absorption, and the nonlinear refractive index arising from saturated absorption (see §§4.5 and 6.3).

In this chapter we have mainly considered those dissipative processes which involve optical resonance with a transition between bound electronic or vibrational states of the medium. In general, the use of resonances enhances the magnitude of nonlinear effects and can also lead to many new phenomena, such as coherent-transient effects.

Other types of dissipative nonlinear-optical mechanism may involve free-carrier states; one example is absorption by free carriers in semiconductors (§8.3.2). Other examples are 'charge-transport-assisted' optical nonlinearities – such as the photorefractive effect and the 'self-electrooptic' effect; these occur when free carriers which are generated by the incident optical field subsequently migrate and separate, resulting in the setting up of an internal space-charge field in the material. In the case of the photorefractive effect, the charge separation occurs when free

Table 6.1: Characteristic time scales

Optical field	τ_P	Pulse duration
	τ_c	Coherence or fluctuation time (see text)
Medium	T_1	Excited state lifetime
	T_2	Dephasing time
Matter-field interaction	Δ^{-1}	Inverse detuning from resonance
	β^{-1}	Inverse Rabi frequency

carriers generated in particular regions of material – defined by an optical-interference pattern of the incident light – become trapped there at defect sites. On the other hand, devices based on the self-electrooptic effect (SEEDs) use an applied d.c. electric field to cause the separation of the optically-generated free carriers (Miller *et al*, 1988a). In either case, the optically-induced space-charge field causes changes in the absorption and refractive index of the medium by means of its electrooptic properties.

An important practical distinction between these various nonlinear mechanisms is the widely differing speed of response that is obtained. The response time is a vital limiting factor for some important applications, such as optical switching and signal processing. In the case of passive (nonresonant) interactions which involve virtual electronic transitions, the speed of response of the medium can be extremely fast: response times on the order of one Bohr orbital, or ~1 fs (10^{-15} s). In that case, for light pulses of duration longer than ~10 fs, the induced polarisation $\mathbf{P}(t)$ follows the temporal variation of the field.

When resonances are used to increase the nonlinearity, real transitions occur and the important parameters are then the equivalent relaxation times T_1 and T_2. In gaseous media, these typically fall within the nanosecond or microsecond range. In condensed matter, because of the greater likelihood of collisions and other many-body interactions, relaxation times in the picosecond and sub-picosecond range are typical. Alternatively, if the displacement of free electrons is involved – as in the photorefractive effect – then the response (determined by carrier drift)

Table 6.2: Important regimes in nonlinear optics

Limit	Regime
PASSIVE NONLINEARITY	
$\Delta^{-1} << T_2$ $\beta^{-1} >> T_1, T_2$	Nonresonant, weak interaction
ACTIVE NONLINEARITY	
$\Delta^{-1} \gtrsim T_2$	Resonant interaction
$\beta^{-1} \sim T_1, T_2$	Strong interaction
$\tau_c >> T_1, T_2$	Steady state
$\tau_c, T_1 >> T_2$	Rate equation
$\beta^{-2} < T_1 T_2$	Saturation
$\tau_c >> 1/\sqrt{\Delta^2 + T_2^{-2}}$	Adiabatic field
$\tau_c >> 1/\sqrt{\beta^2 + \Delta^2}$	Adiabatic following
$\tau_c < T_1, 1/\sqrt{\Delta^2 + T_2^{-2}}$	Coherent transient

can be relatively slow – microseconds or longer. In some liquids, the reorientation of molecules by the incident optical field can give rise to a significant nonlinear refractive index; in some cases (*e.g.*, CS_2) the response time can be as short as a few picoseconds, but times in the nanosecond range are more typical for larger molecules.

Table 6.1 lists the important characteristic time scales for resonant interactions. The meaning of τ_c depends on the circumstances. It represents in some sense the characteristic time scale for fluctuations of the field envelope. For single pulses with significant temporal structure, it is usually appropriate to take $\tau_c \simeq 1/\delta\omega$ where $\delta\omega$ is the optical bandwidth – in other words, τ_c is identified with the coherence time. For 'smooth' single pulses, then $\tau_c \simeq \tau_P$. Alternatively, in situations involving multiple-pulse excitation (*e.g.*, pump-and-probe), τ_c might represent the time delay between pulses.

Table 6.2 summarises the important regimes of nonresonant and resonant interactions, which are more fully explained elsewhere in the text (see Subject index).

Coherent-transient effects have been studied extensively, mainly in dilute media such as gases (Slusher, 1974; Shoemaker, 1978). Far fewer experiments have been done in the solid state, however, owing to the much shorter time scales involved. The recent availability of sources of ultrashort optical pulses (in the range of 100 fs or less) has made possible the observation of coherent phenomena associated with the excitonic resonances in semiconductors, using pump-probe techniques (Joffre *et al,* 1988), four-wave mixing and photon echoes (Schultheis *et al,* 1985, 1986).

7

Wave propagation and processes in nonlinear media

In earlier chapters we described how optical waves interact with matter through which they pass by inducing a polarisation. If the light is sufficiently intense, the induced polarisation exhibits a nonlinear dependence on the field strength; this is the realm of nonlinear optics, in which the optical properties of the medium depend on the intensity and other characteristics of the light waves, and different waves may interact with each other as well as with the medium. It is precisely these properties which make nonlinear optics interesting and useful.

To examine the propagation behaviour we need to consider the coupling between the light waves and the induced polarisation, and we can do this by introducing the polarisation as a source term in Maxwell's equations. In the first section of this chapter we derive the linear and nonlinear wave equations in a variety of forms; each of these provides a convenient starting point for analysing practical problems in different regimes, as illustrated in later sections. In order to show the physical ideas most clearly we simplify the treatment by considering only plane waves in detail, although the important aspects of a more comprehensive treatment involving transversely-varying fields are also described.

7.1 Coupled wave equations

7.1.1 General form

Our starting point is Maxwell's equations for the macroscopic variables, and in the general case these take the form:

$$\begin{aligned} \nabla\times\mathbf{E}(t) &= -\frac{\partial}{\partial t}\mathbf{B}(t) \\ \nabla\times\mathbf{H}(t) &= \varepsilon_0\frac{\partial}{\partial t}\mathbf{E}(t) + \mathbf{J}(t)\,. \end{aligned} \tag{7.1}$$

Here $\mathbf{J}(t)$ denotes the total volume-current density which, in general, is a sum of two parts: the polarisation current $\partial\mathbf{P}(t)/\partial t$, and the conduction current $\mathbf{J}_c(t)$. In optics we are usually concerned with the induced polarisation in which $\mathbf{J}_c(t)$ plays no part and we therefore neglect it; $\mathbf{J}_c(t)=0$ rigorously in dielectric media which contain only bound charges. We also neglect any static polarisation in $\mathbf{P}(t)$. In optics we are almost always concerned with nonmagnetic media (*cf.* Appendix 9) for which we can write: $\mathbf{B}(t)=\mu_0\mathbf{H}(t)$, where $\mathbf{B}(t)$ is the magnetic-induction vector, $\mathbf{H}(t)$ is the magnetic field and μ_0 is the free-space permeability.

We can eliminate $\mathbf{H}(t)$ from (7.1) by operating on the first of these equations with $\nabla\times$, operating on the second with $\mu_0\,\partial/\partial t$, and subtracting. Thus we obtain a driven vector wave equation for the electric field:

$$\nabla\times\nabla\times\mathbf{E}(t) = -\frac{1}{c^2}\frac{\partial^2}{\partial t^2}\mathbf{E}(t) - \mu_0\frac{\partial^2}{\partial t^2}\mathbf{P}(t). \tag{7.2}$$

Depending on the particular problem, we may choose to work with (7.2) in the time domain or instead transform to the frequency domain. By expressing $\mathbf{E}(t)$ and $\mathbf{P}(t)$ in (7.2) in terms of their Fourier transforms (given by (2.25), and (2.44) with (2.47), respectively), we obtain a vector wave equation for the frequency component ω:

$$\nabla\times\nabla\times\mathbf{E}(\omega) = \frac{\omega^2}{c^2}\mathbf{E}(\omega) + \omega^2\mu_0\,\mathbf{P}(\omega). \tag{7.3}$$

7.1.2 Wave propagation in the linear regime

To establish the notation and derive results used elsewhere, we begin by considering the solutions of the wave equation in the linear regime. An expression for the polarisation is obtained from (2.46) for the case $n=1$: $\mathbf{P}(\omega)=\varepsilon_0\boldsymbol{\chi}^{(1)}(-\omega;\omega)\cdot\mathbf{E}(\omega)$. When this is inserted in (7.3), we obtain:

$$\nabla\times\nabla\times\mathbf{E}(\omega) = \frac{\omega^2}{c^2}\boldsymbol{\varepsilon}(\omega)\cdot\mathbf{E}(\omega), \tag{7.4}$$

where the dielectric tensor is defined as

$$\boldsymbol{\varepsilon}(\omega) = \mathbf{1} + \boldsymbol{\chi}^{(1)}(-\omega;\omega). \tag{7.5}$$

The latter definition is equivalent to (6.60a) when the nonlinear contribution to the dielectric tensor is neglected. A possible set of solutions of (7.4) are the running waves

$$\mathbf{E}(\omega) = \hat{\mathbf{E}}(\omega)\exp(i\mathbf{k}\cdot\mathbf{r}), \tag{7.6}$$

where the wave vector $\mathbf{k}=\{\omega[n(\omega)+i\kappa(\omega)]/c\}\,\mathbf{s}$, $\mathbf{s}$ is a real unit vector in the direction of propagation ($\mathbf{s}\cdot\mathbf{s}=1$), and $n(\omega)$ and $\kappa(\omega)$ are the real and imaginary refractive indices, respectively (the phase velocity is c/n). It is convenient to write $\hat{\mathbf{E}}(\omega)=\mathbf{e}\hat{E}(\omega)$, where $\mathbf{e}$ is the unit vector in the polarisation direction of the $\mathbf{E}$ vector (with $\mathbf{e}\cdot\mathbf{e}^*=1$; see §2.3.4). In general, the envelope $\hat{E}(\omega)$ is a function of all three space coordinates, and there are two independent directions of $\mathbf{e}$ for each direction $\mathbf{s}$. When (7.6) is substituted into (7.4), we obtain Fresnel's equation which connects the direction of propagation $\mathbf{s}$ with the refractive index in that direction:

$$[(\mathbf{s}\cdot\mathbf{e})\mathbf{s} - \mathbf{e}][n(\omega) + i\kappa(\omega)]^2 + [\boldsymbol{\varepsilon}'(\omega) + i\boldsymbol{\varepsilon}''(\omega)]\cdot\mathbf{e} = 0. \qquad (7.7)$$

In deriving this equation we have used the vector identity: $\mathbf{s}\times(\mathbf{s}\times\mathbf{e}) = (\mathbf{s}\cdot\mathbf{e})\mathbf{s} - (\mathbf{s}\cdot\mathbf{s})\mathbf{e}$, and the dielectric tensor is separated into its real and imaginary parts: $\boldsymbol{\varepsilon}(\omega) = \boldsymbol{\varepsilon}'(\omega) + i\boldsymbol{\varepsilon}''(\omega)$. In a propagating lossless medium, κ and $\boldsymbol{\varepsilon}''$ are zero. Our main concern is with propagating media in which losses are small, and for the moment we neglect κ and $\boldsymbol{\varepsilon}''$ altogether; we shall take loss into account shortly.

The three components of (7.7) provide us with three homogeneous linear equations for the components of $\mathbf{e}$. If they are to be consistent with one other, the determinant of their coefficients must vanish. Thus we obtain a determinantal equation for the square of the refractive index. In writing out the components of (7.7), the choice of coordinate axes is at our disposal. Since (as shown in §5.2) $\boldsymbol{\varepsilon}'(\omega)$ is a symmetrical tensor, there exists a set of coordinate axes **Oxyz** with respect to which $\boldsymbol{\varepsilon}'(\omega)$ is diagonal; these are the principal axes of the dielectric tensor. We may then denote the corresponding principal refractive indices in terms of the dielectric constants as follows: $n_x = \sqrt{\varepsilon'_{xx}}$, $n_y = \sqrt{\varepsilon'_{yy}}$ and $n_z = \sqrt{\varepsilon'_{zz}}$. When the determinant of coefficients in (7.7) is equated to zero, we find:

$$s_x^2(n^2 - n_y^2)(n^2 - n_z^2) + s_y^2(n^2 - n_x^2)(n^2 - n_z^2) + s_z^2(n^2 - n_x^2)(n^2 - n_y^2) = 0. \qquad (7.8)$$

This is a quadratic in n^2, giving four roots, two each of the same magnitude but opposite sign. That is, in general, there are two phase velocities for propagation in a given direction $\mathbf{s}$, and the medium is said to be birefringent. There are only two (orthogonal) polarisations that can propagate, called the ordinary and extraordinary waves, and these experience different refractive indices: the solutions of (7.8). The ordinary wave propagates isotropically – its phase velocity is independent of the wave direction.

The simplest situation arises in a medium for which $n_x = n_y = n_z = n$, *i.e.*, an isotropic medium or crystal belonging to the cubic system (see Table A3.1, Appendix 3). In this case, we find from (7.7) that $\mathbf{e}$ is orthogonal to $\mathbf{s}$ and is otherwise unrestricted. The energy flux, given by the Poynting vector $\mathbf{N} = \mathbf{E}\times\mathbf{H}$, shares the same direction as the wave vector $\mathbf{s}$.

In a uniaxial birefringent medium, $n_x = n_y\,(\equiv n_{\rm O})$ is the ordinary index, and $n_z\,(\neq n_{\rm O})$ is denoted $n_{\rm E}$. The orientation of the **x** and **y** principal axes is arbitrary. Let us choose them so that the wave-vector direction **s** lies in the **Oyz** plane; then $s_x = 0$, $s_y = \sin\theta$ and $s_z = \cos\theta$, where θ is the angle between **s** and the principal axis **z**. The ordinary wave comprises the component of the field which is polarised orthogonally to **z**, and its refractive index is n_{O} regardless of the direction of propagation. The refractive index n_{e} for the extraordinary wave is

$$n_{\mathrm{e}}(\theta) = \left[\frac{\sin^2\theta}{n_{\mathrm{E}}^2} + \frac{\cos^2\theta}{n_{\mathrm{O}}^2}\right]^{-1/2} . \tag{7.9}$$

In a positively-birefringent uniaxial medium, $n_{\mathrm{O}} \leq n_{\mathrm{e}} \leq n_{\mathrm{E}}$; if the birefringence is negative, $n_{\mathrm{E}} \leq n_{\mathrm{e}} \leq n_{\mathrm{O}}$. Only if $\theta = 0$ are the refractive indices for the ordinary and extraordinary waves the same. The wave-vector direction **s** and the direction of power flow for the extraordinary wave, determined by by the Poynting vector **N**, differ by an angle ρ, called the 'walk-off angle', which is given by:

$$\tan\rho = \tfrac{1}{2}[n_{\mathrm{e}}(\theta)]^2 \left[\frac{1}{(n_{\mathrm{E}})^2} - \frac{1}{(n_{\mathrm{O}})^2}\right] \sin 2\theta . \tag{7.10}$$

For typical birefringent crystals, $\rho \sim 2°$ at $\theta = 45°$.

Biaxial media, for which $n_x \neq n_y \neq n_z$, are those with orthorhombic symmetry or lower; uniaxial materials include crystals with tetragonal, trigonal or hexagonal symmetry, and naturally-isotropic materials which are strained in one direction or subject to a strong electric field.

Finally, we consider the effect of the imaginary part of the dielectric tensor, $\boldsymbol{\varepsilon}''(\omega) = \mathrm{Im}\,\boldsymbol{\chi}^{(1)}(-\omega;\omega)$, which we neglected earlier. As described in §6.3.3, this is responsible for the absorption of optical power in the medium. For all media useful in optics, $n \gg \kappa$; this is equivalent to stating that the absorption over a distance of one optical wavelength is small. The effect of $\boldsymbol{\varepsilon}''(\omega)$ can therefore be calculated as a perturbation in the above analysis, with $(n + i\kappa)^2$ in (7.7) replaced by $n(n + i2\kappa)$, where n is the real refractive index calculated previously. Neglecting diffraction, the absorption causes the optical intensity I to decrease with distance l in the medium according to Beer's law: $\partial I/\partial l = -\alpha I$, where $\alpha = 2\kappa\omega/c$ is the absorption coefficient.

7.1.3 Nonlinear regime

We now turn to the solutions of the wave equation (7.2) or (7.3) in the regime where nonlinear processes occur. It is customary to separate the polarisation $\mathbf{P}(\omega)$ into its linear and nonlinear parts by writing, with the aid of (2.33) and (2.47):

$$\mathbf{P}(\omega) = \varepsilon_0 \boldsymbol{\chi}^{(1)}(-\omega;\omega) \cdot \mathbf{E}(\omega) + \mathbf{P}^{\mathrm{NL}}(\omega) , \tag{7.11}$$

where $\mathbf{P}^{\mathrm{NL}}(\omega)=\sum_{n=2}^{\infty}\mathbf{P}^{(n)}(\omega)$ and the $\mathbf{P}^{(n)}(\omega)$ are given by (2.40) together with (2.45). With this substitution, (7.3) becomes

$$\nabla\times\nabla\times\mathbf{E}(\omega) = \frac{\omega^2}{c^2}\,\boldsymbol{\varepsilon}(\omega)\cdot\mathbf{E}(\omega) + \omega^2\mu_0\mathbf{P}^{\mathrm{NL}}(\omega)\,, \tag{7.12}$$

where here $\boldsymbol{\varepsilon}(\omega)$ denotes the linear dielectric tensor given by (7.5).

Alternatively, if we choose to work in the time domain, it is useful to write the Fourier transform (2.25) of $\mathbf{E}(t)$ in the form:

$$\mathbf{E}(t) = \int_{-\infty}^{+\infty} \mathrm{d}\omega'\;\mathbf{E}(\omega+\omega')\exp[-i(\omega+\omega')t]\,. \tag{7.13}$$

Then, on separating the linear and nonlinear parts of the polarisation $\mathbf{P}(t)$ in (7.2), and taking the transform of $\mathbf{P}^{(1)}(t)$ as indicated in (7.13), we obtain

$$\begin{aligned}\nabla\times\nabla\times\mathbf{E}(t) &= \mu_0\int_{-\infty}^{+\infty}\mathrm{d}\omega'\,(\omega+\omega')^2\,\boldsymbol{\varepsilon}(\omega+\omega')\cdot\mathbf{E}(\omega+\omega')\exp[-i\,(\omega+\omega')t]\\ &\quad - \mu_0\frac{\partial^2}{\partial t^2}\mathbf{P}^{\mathrm{NL}}(t)\,,\end{aligned} \tag{7.14}$$

with $\mathbf{P}^{\mathrm{NL}}(t)=\sum_{n=2}^{\infty}\mathbf{P}^{(n)}(t)$.

As in §7.1.2, we can investigate the travelling-wave solutions of the wave equation by taking the electric field in the form of a superposition of waves (7.6). Inserting this into the wave equation (7.12), for example, we obtain

$$\nabla\times\nabla\times\left[\hat{\mathbf{E}}(\omega)\exp(i\mathbf{k}\cdot\mathbf{r})\right] = \frac{\omega^2}{c^2}\boldsymbol{\varepsilon}(\omega)\cdot\hat{\mathbf{E}}(\omega)\exp(i\mathbf{k}\cdot\mathbf{r}) + \omega^2\mu_0\mathbf{P}^{\mathrm{NL}}(\omega). \tag{7.15}$$

The envelope function $\hat{\mathbf{E}}(\omega)$ is complex; it incorporates both amplitude and phase information about the wave and, in general, it is a function of all three space coordinates. To simplify the spatial analysis in the following discussion, we shall make two assumptions: first, that the components $\hat{\mathbf{E}}(\omega)$ in (7.15) are infinite plane waves that propagate collinearly in some arbitrary direction denoted by $\mathbf{z}$ so that $\hat{\mathbf{E}}(\omega)$ is a function of z only and $\mathbf{k}\cdot\mathbf{r}=\pm kz$ (+ for forward-travelling waves in the $\mathbf{z}$ direction and − for backward waves); and second, that the field vector and wave vector are orthogonal ($\mathbf{e}\cdot\mathbf{s}=0$). As shown in §7.1.2, the latter condition is strictly valid only in the case of isotropic media and those having cubic symmetry; the effect of double refraction in birefringent media can be incorporated when necessary (an example occurs in §7.2.1), but we neglect this additional complication here. This simplification allows us to replace $\nabla\times\nabla\times\mathbf{E}$ by $-\partial^2\mathbf{E}/\partial z^2$, where $\mathbf{E}$ denotes either $\mathbf{E}(t)$ or its Fourier transform $\mathbf{E}(\omega)$.

7.1.4 Slowly-varying envelope approximation

The wave envelope $\hat{\mathbf{E}}(\omega)$ varies with distance through the medium as a result of both linear and nonlinear processes. If the variations of $\hat{\mathbf{E}}(\omega)$ (both in amplitude and phase) are sufficiently slow with distance z, we can assume

$$\left| \frac{\partial^2}{\partial z^2} \hat{\mathbf{E}}(\omega) \right| << \left| k \frac{\partial}{\partial z} \hat{\mathbf{E}}(\omega) \right| . \tag{7.16a}$$

This is known as the slowly-varying envelope approximation. It implies that we may neglect the second derivative of $\hat{\mathbf{E}}(\omega)$ with respect to z in (7.15), so that for a forward-travelling wave we have

$$\frac{\partial}{\partial z} \hat{\mathbf{E}}(\omega) = \frac{i\omega^2 \mu_0}{2k} \mathbf{P}^{\mathrm{NL}}(\omega) \exp(-ikz) . \tag{7.17}$$

The second-order differential wave equation (7.15) is thus reduced to a simple first-order equation. In optics the approximation (7.16a) is almost always a good one. Many authors justify it by stating that the variations of $\hat{\mathbf{E}}(\omega)$ can be considered small over distances of the order of a wavelength. However, as pointed out by Shen (1984b), the real significance of (7.16a) is that it is equivalent to neglecting the component of the field generated by $\mathbf{P}^{\mathrm{NL}}(\omega)$ which propagates in the $-\mathbf{z}$ direction. (Of course, we are at liberty to consider the propagation of a backward-travelling wave $\mathbf{E}(\omega) = \hat{\mathbf{E}}(\omega)\exp(-ikz)$ by making the changes $z \rightarrow -z$, $\partial z \rightarrow -\partial z$ in (7.17); in that case, the slowly-varying envelope approximation is equivalent to neglecting a forward-travelling field component in the new differential equation.) An equivalent way of expressing the slowly-varying envelope approximation in the time domain is to neglect the second-order time derivative of the envelope:

$$\left| \frac{\partial^2}{\partial t^2} \hat{\mathbf{E}}(\omega) \right| << \left| \omega \frac{\partial}{\partial t} \hat{\mathbf{E}}(\omega) \right| . \tag{7.16b}$$

Equation (7.17) is just one of an infinite set of equations for the Fourier components of the field, $\mathbf{E}(\omega)$. The equations within this set are coupled together through the source term $\mathbf{P}^{\mathrm{NL}}(\omega)$ which, as may be seen in (2.40), contains integrals over all the frequency components of the field. Fortunately, in most situations of practical interest the coupling between a small subset of equations only need be considered. When a particular process dominates in the interaction (perhaps due to phase matching or resonance enhancement), the coupled equations describing the few fields involved may be treated in isolation. If, in addition, the nonlinear interaction is so weak that the envelopes of the incident waves are not significantly changed (*i.e.*, 'pump depletion' is negligible), the equations of the form (7.17) which describe the variation in the incident fields may be neglected, whilst the corresponding equations for the fields generated within the medium are effectively decoupled. This is illustrated by examples in the following sections.

7.1.5 Monochromatic waves

Equation (7.17) is the starting point for most plane-wave calculations of the growth of waves as they propagate through a nonlinear medium. Its most straightforward application occurs when the incident field consists of a superposition of collinear monochromatic waves:

$$\mathbf{E}(\omega) = \frac{1}{2}\sum_{\omega_j \geq 0}\left[\hat{\mathbf{E}}_{\omega_j}\exp(ik_jz)\,\delta(\omega-\omega_j) + \hat{\mathbf{E}}^*_{\omega_j}\exp(-ik_jz)\,\delta(\omega+\omega_j)\right], \tag{7.18}$$

where the wave-vector magnitude $k_j = \omega_j\,[n(\omega_j) + i\kappa(\omega_j)]/c$, and k_j is also known as the propagation constant for the wave at the frequency ω_j. Very often the imaginary part of the refractive index, $\kappa(\omega_j)$, is small enough to be negligible, and we assume this to be the case here. When we select a particular component at a frequency ω_σ, (7.17) becomes:

$$\frac{\partial}{\partial z}\hat{\mathbf{E}}_{\omega_\sigma} = \frac{i\omega_\sigma^2\mu_0}{2k_\sigma}\mathbf{P}^{\mathrm{NL}}_{\omega_\sigma}\exp[-ik_\sigma z], \tag{7.19}$$

where $\mathbf{P}^{\mathrm{NL}}_{\omega_\sigma} = \sum_{n=2}^{\infty}\mathbf{P}^{(n)}_{\omega_\sigma}$ and the $\mathbf{P}^{(n)}_{\omega_\sigma}$ are given by (2.55). Equation (7.19) must be satisfied by all the monochromatic waves present in the nonlinear medium, whether applied or generated. In many practical situations, because of resonance enhancement, phase matching, spectral selectivity or some other discriminating feature, attention can be focused on one particular order of nonlinearity (order n, say). Then, with the aid of (2.55), we may write:

$$\mathbf{P}^{\mathrm{NL}}_{\omega_\sigma} = \varepsilon_0 K(-\omega_\sigma;\omega_1,\ldots,\omega_n)\,\boldsymbol{\chi}^{(n)}(-\omega_\sigma;\omega_1,\ldots,\omega_n)\,|\,\hat{\mathbf{E}}_{\omega_1}\cdots\hat{\mathbf{E}}_{\omega_n}\exp(ik_{\mathrm{P}}z) \tag{7.20}$$

where $\omega_\sigma = \omega_1 + \cdots + \omega_n$ and $k_{\mathrm{P}} = k_1 + \cdots + k_n$. (In forming the latter sum, the rule is that $-k_j$ is associated with a negative argument $-\omega_j$; this follows from the fact that the field $\mathbf{E}(t)$ is real and consequently $\hat{\mathbf{E}}^*_{\omega_j} = \hat{\mathbf{E}}_{-\omega_j}$ in (7.18).) When (7.20) is substituted in (7.19) and the result is expressed in scalar form, we have:

$$\frac{\partial}{\partial z}\hat{E}_\sigma = \frac{i\omega_\sigma^2}{2k_{\mathrm{P}}c^2}K(-\omega_\sigma;\omega_1,\ldots,\omega_n) \times\chi^{(n)}(-\omega_\sigma;\omega_1,\ldots,\omega_n)\,\hat{E}_1\cdots\hat{E}_n\exp(i\Delta kz), \tag{7.21}$$

where $\hat{\mathbf{E}}_{\omega_j} = \hat{E}_j \mathbf{e}_j$, and $\chi^{(n)}$ is defined by (2.58). We recall that the intensity of each monochromatic wave is given by (2.51): $I_{\omega_j} = \frac{1}{2}\varepsilon_0 c n_j |\hat{E}_j|^2$. In (7.21), $\Delta k = k_P - k_\sigma$ is termed the phase mismatch; its significance for 'parametric' nonlinear-optical processes is discussed in §7.2. It is easy to include simultaneously several orders of nonlinearity in (7.21) if necessary; the notation is cumbersome, however, and we therefore neglect such cases. For second-order processes ($n = 2$), the versions of (7.21) in terms of the **d**-tensor (5.23) or d_{eff} (5.28) may be preferred.

7.1.6 Time-dependent processes

We now consider the application of the wave equation to time-dependent problems. Equation (7.17) provides a complete integro-differential formalism for the description of such processes. The terms in the series expansion of $\mathbf{P}^{\text{NL}}(\omega)$ are given by (2.40) and (2.45) with complete generality. This is sometimes the best approach to analysing ultrafast phenomena that involve a wide spectrum of frequencies. However, it is often better to formulate time-dependent problems by writing the field as a superposition of quasi-monochromatic travelling waves (*cf.* §2.4):

$$\mathbf{E}(t) = \frac{1}{2}\sum_{\omega_j \geq 0}\left\{\hat{\mathbf{E}}_{\omega_j}(t)\exp[-i(\omega_j t - k_j z)] + \hat{\mathbf{E}}^*_{\omega_j}(t)\exp[i(\omega_j t - k_j z)]\right\}. \tag{7.22}$$

Here, as in (2.60), the $\hat{\mathbf{E}}_{\omega_j}(t)$ are complex time-dependent envelope functions, and ω_j are nominal centre frequencies (sometimes called 'carrier frequencies' in borrowed radio terminology). This formalism has the advantage that the evolution of a particular wave with centre frequency ω_j can be more easily examined. It does, however, require some approximation to be made before an appropriate time-dependent differential wave equation can be written down. The reason for this is that, because the envelope function $\hat{\mathbf{E}}_{\omega_j}(t)$ is time-dependent, the wave occupies some spectral width in the vicinity of the nominal centre frequency ω_j. Therefore the magnitude of the wave vector k_j can no longer be considered constant; we must take into account the spectral dispersion of the dielectric tensor $\boldsymbol{\varepsilon}(\omega)$ within the bandwidth of the wave. This can be done in a simple way by expanding $\boldsymbol{\varepsilon}(\omega)$ in a Taylor series about the centre frequency:

$$\boldsymbol{\varepsilon}(\omega_j + \delta\omega) = \boldsymbol{\varepsilon}(\omega_j) + \delta\omega\frac{\mathrm{d}}{\mathrm{d}\omega}\boldsymbol{\varepsilon}(\omega)\bigg|_{\omega_j} + \frac{1}{2}(\delta\omega)^2\frac{\mathrm{d}^2}{\mathrm{d}\omega^2}\boldsymbol{\varepsilon}(\omega)\bigg|_{\omega_j} + \cdots. \tag{7.23}$$

When (7.22) and (7.23) are substituted in (7.14), we obtain the following equation for the wave $\hat{\mathbf{E}}_{\omega_\sigma}$ in the slowly-varying envelope approximation (7.16a,b) and infinite-plane-wave limit:

$$i\left(\frac{\partial}{\partial z}+\left.\frac{\mathrm{d}k}{\mathrm{d}\omega}\right|_{\omega_\sigma}\frac{\partial}{\partial t}\right)\hat{\mathbf{E}}_{\omega_\sigma}(t)-\frac{1}{2}\left.\frac{\mathrm{d}^2k}{\mathrm{d}\omega^2}\right|_{\omega_\sigma}\frac{\partial^2}{\partial t^2}\hat{\mathbf{E}}_{\omega_\sigma}(t)+\cdots$$
$$=-\frac{\omega_\sigma^2\mu_0}{2k_\sigma}\mathbf{P}^{\mathrm{NL}}_{\omega_\sigma}(t)\exp[-ik_\sigma z]. \tag{7.24}$$

In writing (7.24) we have retained only the first and second differentials in the Taylor series (7.23). It is clear that the velocity of propagation of the wave envelope is $(\mathrm{d}k/\mathrm{d}\omega)^{-1}$; this, of course, is the group velocity v_{g}. The second differential, $\mathrm{d}^2k/\mathrm{d}\omega^2$, is a measure of the group-velocity dispersion. Only if $\hat{\mathbf{E}}_{\omega_j}(t)$ varies slowly (both in amplitude and phase) so that its spectral width is narrow, or if the dielectric tensor is only weakly dispersive, may we neglect the second-order dispersion $\mathrm{d}^2k/\mathrm{d}\omega^2$ in (7.24). On the other hand, to describe pulse propagation in dispersive media in general we must retain the second-order dispersion, and for ultrashort pulses or those with a wide frequency spectrum it may sometimes be necessary to also include higher-order terms.

The nonlinear polarisation in (7.24) can be constructed from $\mathbf{P}^{\mathrm{NL}}_{\omega_\sigma}(t)=\sum_{r=2}^{\infty}\mathbf{P}^{(r)}_{\omega_\sigma}(t)$ where $\mathbf{P}^{(r)}_{\omega_\sigma}(t)$ is expressed in terms of the material response functions by (2.65). For the same reasons mentioned in §7.1.5 for the case of monochromatic waves, it is often sufficient to consider only one order of nonlinearity (n, say). Then, in the adiabatic limit, $\mathbf{P}^{\mathrm{NL}}_{\omega_\sigma}(t)$ is given by (2.69), and (7.24) can be written as

$$i\left(\frac{\partial}{\partial z}+\left.\frac{\mathrm{d}k}{\mathrm{d}\omega}\right|_{\omega_\sigma}\frac{\partial}{\partial t}\right)\hat{\mathbf{E}}_{\omega_\sigma}(t)-\frac{1}{2}\left.\frac{\mathrm{d}^2k}{\mathrm{d}\omega^2}\right|_{\omega_\sigma}\frac{\partial^2}{\partial t^2}\hat{\mathbf{E}}_{\omega_\sigma}(t)+\cdots$$
$$=-\frac{\omega_\sigma^2}{2k_\sigma c^2}K(-\omega_\sigma;\omega_1,\ldots,\omega_n)\,\boldsymbol{\chi}^{(n)}(-\omega_\sigma;\omega_1,\ldots,\omega_n)$$
$$|\,\hat{\mathbf{E}}_{\omega_1}(t)\cdots\hat{\mathbf{E}}_{\omega_n}(t)\exp(i\,\Delta k\,z) \tag{7.25}$$

where, as before, $\Delta k=k_{\mathrm{P}}-k_\sigma$ and $k_{\mathrm{P}}=k_1+\cdots+k_n$. This is the analogue of the monochromatic wave equation (7.21) in the adiabatic limit. A particular case with $n=3$ has important solitary wave solutions, which are described in §7.5.

The wave equation in the form (7.24) is often used to describe propagation effects in resonant nonlinear interactions. Rather than use the form (7.25) in terms of the susceptibilities, it is then more appropriate to calculate the nonlinear polarisation $\mathbf{P}^{\mathrm{NL}}_{\omega_\sigma}(t)$ directly from the equation of motion for the density operator; for example, by using the methods described in Chapter 6. For a resonant two-level system, the equations (6.52) for the time evolution of the density operator, coupled with (7.24), are known as the Maxwell-Bloch equations.

If it should happen that the nonlinear interaction induces a large refractive-index change (see §6.3), then the treatment of the wave equation by separating the polarisation into its linear and nonlinear parts may not be the best approach; it may be preferable to work from (7.2) directly, incorporating the slowly-varying envelope approximation.

7.2 Parametric processes

7.2.1 Harmonic generation

The simplest nonlinear process that illustrates the application of the wave equation in nonlinear optics is second-harmonic generation with a monochromatic input wave of frequency ω. Equation (7.21) provides us with the two coupled equations for the second-harmonic and fundamental waves:

$$\begin{aligned} \frac{\partial}{\partial z}\hat{E}_{2\omega} &= \frac{i\,2\omega}{2n_{2\omega}c}\left[\tfrac{1}{2}\,\chi^{(2)}(-2\omega;\omega,\omega)\,\hat{E}_{\omega}^{2}\right]\exp(i\,\Delta k\,z) \\ \frac{\partial}{\partial z}\hat{E}_{\omega} &= \frac{i\,\omega}{2n_{\omega}c}\left[\chi^{(2)}(-2\omega;\omega,\omega)\,\hat{E}_{\omega}^{*}\hat{E}_{2\omega}\right]\exp(-i\,\Delta k\,z), \end{aligned} \tag{7.26}$$

where $\Delta k = 2k_{\omega} - k_{2\omega} = 2\omega(n_{\omega} - n_{2\omega})/c$, the K values in (7.21) have been evaluated using (2.56), and we have invoked Kleinman symmetry (§5.1.1). We have also introduced the shorthand notation n_j to denote the refractive index for the wave with frequency ω_j. In the absence of absorption, the flow of power between the waves is governed by the Manley-Rowe relation (5.15), which may be written as $(1/2\omega)\partial I_{2\omega}/\partial z = -\frac{1}{2}(1/\omega)\partial I_{\omega}/\partial z$, where $I_{2\omega}$, I_{ω} are the intensities of the second-harmonic and fundamental waves, respectively.

The parameter Δk is termed the phase mismatch. We now explain its physical meaning and significance. As described in Chapter 1, the effect of a coherent optical field on the medium is to induce an assembly of electric dipoles which oscillate coherently. Such an assembly of oscillating dipoles comprises a macroscopic polarisation. In the case of second-harmonic generation, we are concerned with the component of the polarisation which oscillates at 2ω. The *phase* of this polarisation component is not constant throughout the medium; instead, fronts of constant phase move through the medium in the z direction with a phase velocity ω/k_{ω} (as given by the exponential term in (7.20) with $\omega_{\sigma} = 2\omega$, $k_{\mathrm{P}} = 2k_{\omega}$). This gives rise to the idea of a polarisation 'wave' (although

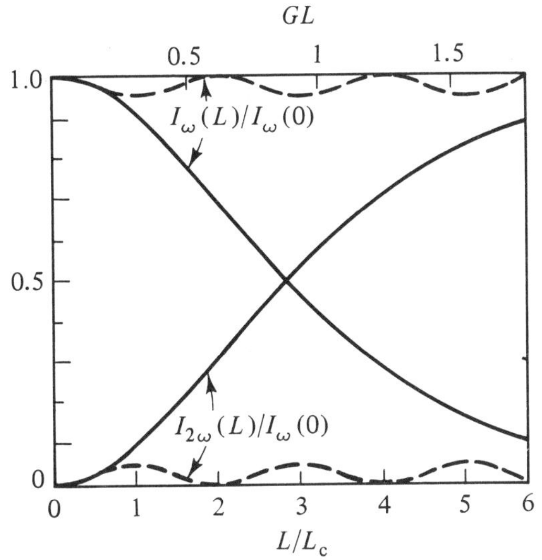

Fig. 7.1 Normalised plots of second-harmonic intensity $I_{2\omega}$ and fundamental intensity I_ω versus length of medium L for perfect phase matching (continuous curves and upper horizontal scale) and for a finite phase mismatch (dashed curves and lower horizontal scale).

the polarisation does not actually move in space). The second-harmonic light wave, on the other hand, travels with a phase velocity $2\omega/k_{2\omega}$. The phase-matching condition $\Delta k=2k_\omega-k_{2\omega}=0$ occurs when these two phase velocities are equal. The polarisation wave and light wave at 2ω are then in phase synchronism; this maximises the transfer of energy to the second-harmonic light wave. If, however, $\Delta k\neq 0$ then the polarisation wave and light wave at 2ω do not remain in phase synchronism; they become out of phase by 180° after a distance $L_c=|\pi/\Delta k|$, known as the coherence length. (An alternative interpretation of phase matching in terms of photon-momentum conservation is given at the end of §7.2.2.)

Because of dispersion, $\Delta k\neq 0$ in general, and therefore $I_{2\omega}$ is an oscillatory function of z; $I_{2\omega}$ increases over the first distance L_c, then decreases to zero over the next L_c while the intensity is transferred back to I_ω, and so on. This is depicted by the dashed curves in Fig. 7.1. In such a case, the conversion efficiency $I_{2\omega}/I_\omega$ remains small and in the coupled-wave analysis we can take I_ω to be constant; we thus neglect pump depletion. The solution of (7.26) with $\hat{E}_\omega(z)=\hat{E}_\omega(0)$ and the initial condition $\hat{E}_{2\omega}(0)=0$ is a matter of simple integration; after a distance L, $I_{2\omega}$ is given by

$$I_{2\omega}(L) = \frac{(2\omega)^2}{8\epsilon_0 c^3}\frac{|\chi^{(2)}(-2\omega;\omega,\omega)|^2}{n_\omega^2\, n_{2\omega}} I_\omega^2 L^2 \mathrm{sinc}^2(\Delta k L/2)\,, \qquad (7.27)$$

where $\mathrm{sinc}(x) = \sin(x)/x$, so that $\mathrm{sinc}(0) = 1$. The argument of this function in (7.27) may be written as $\pi L/2L_c$, showing that $I_{2\omega}$ reaches maxima when L equals odd multiples of the coherence length L_c; for even multiples the intensity is zero (see Fig. 7.1). For typical nonbirefringent crystals (*e.g.*, GaAs), $L_c \sim 10-100\ \mu$m.

In the case when the process is phase-matched ($\Delta k = 0$, $L_c \to \infty$), $I_{2\omega}(L)$ is maximised and is proportional to L^2. With sufficient input intensity and length L, the conversion efficiency may become large (as depicted in Fig. 7.1). One can no longer neglect pump depletion, and instead the coupled equations (7.26) must be solved exactly. For exact phase matching, the results are (Armstrong *et al*, 1962):

$$I_{2\omega}(L) = I_\omega(0)\tanh^2(GL)$$
$$I_\omega(L) = I_\omega(0)\,\mathrm{sech}^2(GL)\,, \qquad (7.28)$$

where $G^2 = (2\omega^2/\varepsilon_0 c^3)(d_{\mathrm{eff}}^2/n^3)I_\omega(0)$, we have taken $n_{2\omega} = n_\omega = n$, and d_{eff} is given by (5.28). In the uniform plane-wave approximation, total conversion is predicted by (7.28). In practice it is limited by the non-uniform intensity distribution in the laser beam, although efficiencies exceeding 50% can be achieved. The factor d_{eff}^2/n^3 is commonly used as a materials figure-of-merit for second-order processes.

The most convenient method for achieving phase matching in crystals is to use the birefringence to offset dispersion. (Other techniques such as inverting the crystal about its symmetry axis every coherence length (Armstrong *et al*, 1962; Byer, 1989) or employing waveguide dispersion are also used.) For second-harmonic generation the phase-matching condition reduces to $n_\omega = n_{2\omega}$, which can be satisfied if one frequency is an ordinary wave and the other extraordinary, provided the crystal birefringence is large enough. The case of phase-matched second-harmonic generation in the organic crystal urea (Halbout and Tang (1987).) is depicted for example in Fig. 7.2. Urea is a positively-birefringent uniaxial material ($n_{\mathrm{E}} > n_{\mathrm{O}}$). The birefringence ($n_{\mathrm{E}} - n_{\mathrm{O}} \simeq 0.1$) is sufficiently large that phase matching can be obtained when the input radiation is launched at an angle θ_{m} to the optic axis so that the condition $n_{\mathrm{e}}(\theta_{\mathrm{m}}, \omega) = n_{\mathrm{O}}(2\omega)$ is satisfied, where $n_{\mathrm{e}}(\theta, \omega)$ is given by (7.9). (For a negatively-birefringent material the rôles of ordinary and extraordinary waves are reversed.) This is an illustration of type I phase matching. Another possibility is to use type II phase matching in which the fundamental field comprises an ordinary and extraordinary wave; for positively-birefringent crystals the required phase-matching condition is $\frac{1}{2}[n_{\mathrm{O}}(\omega) + n_{\mathrm{e}}(\theta_{\mathrm{m}}, \omega)] = n_{\mathrm{O}}(2\omega)$. For a more detailed discussion of phase-matching techniques in crystals, see Byer (1977). Phase matching in

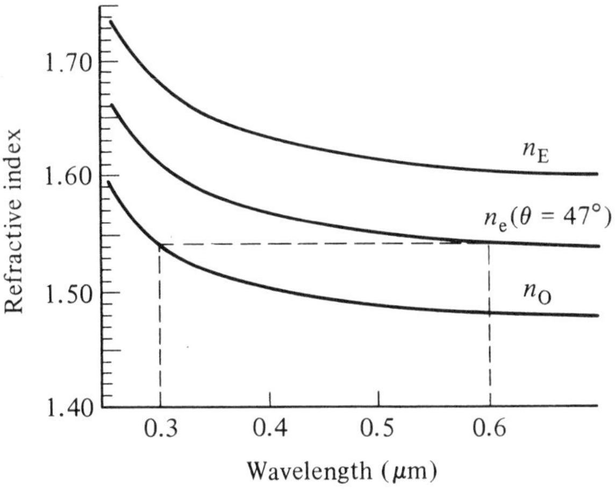

Fig. 7.2 Type I angle phase matching for frequency doubling a 600 nm-wavelength dye laser into the ultraviolet region using a urea crystal. The crystal is oriented for $\theta_m = 47°$. (Calculated from data of Halbout and Tang, 1987.)

biaxial crystals is treated in detail by Hobden (1967).

A vital consideration when choosing the crystal orientation is to ensure that the corresponding value of d_{eff} is satisfactory (see Appendix 4); for some ill-chosen orientations d_{eff} is zero. Two further aspects of angle phase matching should be considered: the acceptance angle, and double refraction (walk-off). If the angle θ to the optic axis deviates from its optimum value θ_m, Δk is no longer zero and the harmonic-generation efficiency decreases. The angular sensitivity is measured by an acceptance angle $\Delta\theta$ defined by the condition $|\Delta k\, L| = 2\pi$ at which the sinc^2 function in (7.27) reaches its first zero. For type I phase matching (positive birefringence), this may be written as

$$\begin{aligned}(2\omega/c)\left|n_e(\theta_m + \Delta\theta, \omega) - n_O(2\omega)\right| L \\ = (2\omega/c)\left|\Delta\theta \frac{d}{d\theta} n_e(\theta, \omega)\Big|_{\theta_m} + (\Delta\theta)^2 \frac{d^2}{d\theta^2} n_e(\theta, \omega)\Big|_{\theta_m} + \cdots \right| L \\ = 2\pi. \qquad (7.29)\end{aligned}$$

If $\theta_m \neq 90°$ (known as critical phase matching), the second- and higher-order differential terms in the Taylor expansion in (7.29) are negligible, and we can thus express the external acceptance angle ($\Delta\theta_{ext} = n\,\Delta\theta$) in terms of the walk-off angle ρ, given by (7.10), as $\Delta\theta_{ext} = \pi c/\omega L\rho$. Phase matching with $\theta_m \neq 90°$ is very sensitive to crystal orientation and also beam divergence. For example, with typical values of $\rho = 2°$, $\lambda = 1\,\mu$m and $L = 5$ mm, we obtain $\Delta\theta_{ext} = 0.16°$. However, if 'noncritical' phase matching ($\theta_m = 90°$) can be obtained, $\rho = 0$ and the acceptance angle $\Delta\theta$ is increased. To calculate $\Delta\theta$ is this case we can no longer neglect the second-order differential in (7.29); typically $\Delta\theta_{ext}$ is increased to $\sim 1-2°$ (Byer, 1977). Noncritical phase matching is therefore preferred; it has the important advantage of allowing a good conversion efficiency even when the input beam contains a spread of angles, as for example, when the input radiation is a focused beam. Phase matching at $\theta_m = 90°$ can be obtained in some materials by controlling the crystal temperature to adjust the birefringence.

The previous coupled-wave analysis assumes that the interacting fields are infinite plane waves. In practice, however, the input beams have finite transverse dimensions. A more comprehensive analysis of the full wave equation (7.2) including diffraction and other transverse effects is then required, and this introduces several complicating factors. If critical phase matching is used ($\theta_m \neq 90°$), double refraction will cause the direction of power flow for the extraordinary beam to differ from θ_m by a small angle ρ given by (7.10). Thus the fundamental and harmonic beams suffer 'walk-off', limiting the interaction length for beams of a given aperture. For an input beam focused to a spot of radius w_0, walk-off is characterised by an aperture length $L_a = \sqrt{\pi} w_0/\rho$, and becomes important for crystal lengths $L > L_a$. For example, values of $w_0 = 50\,\mu m$ and $\rho = 2°$ give $L_a = 2$ mm. Boyd and Kleinman (1968) show that for $L \ll L_a$, $I_{2\omega} \propto L^2$ as predicted by (7.27); for $L \simeq L_a$, $I_{2\omega} \propto LL_a$; and for $L \gg L_a$, $I_{2\omega} \propto L_a^2$, independent of L. Noncritical phase matching ($\theta_m = 90°$) therefore has the added advantage that walk-off does not occur ($L_a \to \infty$). In general the conversion efficiency is increased by focusing the input beam to a small spot inside the crystal to obtain a high intensity. Boyd and Kleinman consider the important practical case when the input field is a focused Gaussian beam characterised by a confocal parameter b (see Appendix 10). As the focal spot is reduced in size, the intensity increases but the interaction length for the nonlinear process is reduced when $b < L$. Boyd and Kleinman show that, provided $L > L_a$, optimum conversion is obtained when $L/b = 2.84$ with $\Delta k = 0$. Practical limitations are the available crystal length L and the maximum intensity that the crystal can withstand without damage.

The generation of third and higher harmonics of visible and ultraviolet lasers in gases and vapours provides a method of generating vacuum ultraviolet radiation (Reintjes, 1984). When compared with second-order processes in solids, third- and higher-order processes in gases suffer the disadvantage of much lower molecular number density; despite this, useful efficiencies can be obtained by using resonance enhancement to increase the susceptibility (described in §4.5; see Fig.

4.2), longer interaction lengths, and the higher intensities that are possible without material damage or breakdown. However, the use of resonance enhancement also leads to saturation and competing processes, and for this reason the conversion efficiencies in gases are generally smaller than for second-order processes in crystals (Hanna *et al*, 1979; Reintjes, 1984).

The plane-wave analysis of third-harmonic generation closely follows that for second-harmonic generation. Starting from the coupled equations for the fundamental and harmonic waves, derived from (7.21), simple integration leads to an expression similar to (7.27):

$$I_{3\omega}(L) = \frac{(3\omega)^2}{16\varepsilon_0^2 c^4}\,\frac{|\chi^{(3)}(-3\omega;\omega,\omega,\omega)|^2}{n_\omega^3 n_{3\omega}} I_\omega^3 L^2 \mathrm{sinc}^2(\Delta k\, L/2) \quad (7.30)$$

with $\Delta k = 3k_\omega - k_{3\omega}$. (Some authors define Δk with the opposite sign.) To obtain a high conversion efficiency, the input beam is usually focused to a small spot in the medium. However, in the focal region of a Gaussian beam the optical phase is no longer a linear function of distance (see Appendix 10). This problem was analysed by Ward and New (1969) and Bjorklund (1975). They found that the generated third-harmonic power $\mathbf{P}_{3\omega}$ at the output of the medium (calculated by integrating the intensity over the cross-section of the beam) is given by

$$\mathbf{P}_{3\omega} = \frac{3\omega^4}{16\pi^2\varepsilon_0^2 c^6}\,\frac{|\chi^{(3)}(-3\omega;\omega,\omega,\omega)|^2}{n_\omega n_{3\omega}} \mathbf{P}_\omega^3 |F|^2, \quad (7.31)$$

where $\mathbf{P}_\omega$ is the fundamental power at the input and F is a dimensionless integral. For weak focusing ($L \ll b$, where b is the Gaussian-beam confocal parameter), $|F|^2$ reduces to $(4L^2/b^2)\,\mathrm{sinc}^2[(\Delta k - 4/b)L/2]$. The sinc^2 term is similar to that in the plane-wave limit (7.30), and its maximum value of unity occurs when $\Delta k\, L/2 = 2L/b \simeq 0$. However, in the tight-focus limit ($b \ll L$), F takes the values $F = 0$ if $\Delta k \leq 0$; $= \pi b\, \Delta k \exp(-b\,\Delta k/2)$ if $\Delta k > 0$. This behaviour is sketched in Fig. 7.3. Therefore, to obtain a third-harmonic output with a tightly-focused input beam, it is essential that $\Delta k > 0$, and $|F|^2$ has a maximum value when $b\,\Delta k = 2$. For third-harmonic generation and other parametric processes in gases and liquids, Δk may be adjusted in three ways: by adding a buffer gas or liquid to form a mixture with the required dispersion; by choosing the optical frequencies to exploit 'anomalous' dispersion in the vicinity of absorbing transitions; and by adjusting the pressure in the case of a gas. For a review of experimental techniques, and also a more detailed discussion of the complicated question of optimisation and avoidance of saturation and competing processes, see Hanna *et al* (1979) and Reintjes (1984).

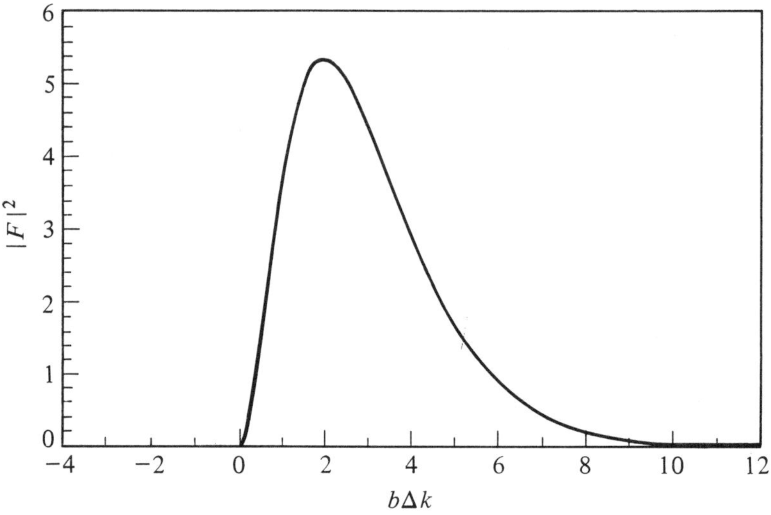

Fig. 7.3 Variation of the phase-matching integral $|F|^2$ with $b\,\Delta k$ for third-harmonic generation in the tight-focus limit.

7.2.2 Frequency mixing and parametric amplification

The treatment of second-harmonic generation ($\omega+\omega\to 2\omega$) can be generalised to three-wave processes, such as sum-frequency ($\omega_1+\omega_2\to\omega_3$) and difference-frequency mixing ($\omega_3-\omega_1\to\omega_2$, $\omega_3-\omega_2\to\omega_1$). For all these processes the phase-matching condition is $\Delta k = k_3 - k_2 - k_1 = 0$. Second-harmonic generation can be regarded as sum-mixing with degenerate input frequencies ($\omega_1 = \omega_2$). However, difference-mixing has fundamentally different properties, as we show shortly.

In difference-frequency mixing (with $\omega_3 > \omega_2 > \omega_1$) it is customary to refer to ω_3 as the pump frequency, ω_2 as the signal, and ω_1 the idler. Let us consider the case (outlined briefly in Chapter 1) when the optical field in the nonlinear medium consists of a strong pump wave ω_3 and a weak signal ω_2. The nonlinear response (7.20) causes the pump and signal fields to beat together to produce a polarisation component at the idler frequency $\mathbf{P}^{NL}_{\omega_1}$. If the phase-matching condition is satisfied, an idler wave can grow according to the coupled wave equation (7.21). The total field then consists of three waves, and the newly-generated idler ω_1 may now beat with the pump ω_3 to produce a polarisation $\mathbf{P}^{NL}_{\omega_2}$ which causes the signal wave to *grow* in amplitude. In this way power is transferred from the pump to the signal and idler, in a manner governed by the Manley-Rowe relation (5.16). The fundamental difference between sum- and difference-mixing processes is now apparent. In the former case power is transferred from the input waves to the generated wave, whereas difference mixing is essentially an *amplification* process. For the parametric amplifier it is not possible to describe the growth of the signal or idler waves alone; we must consider the coupling between the two. The solutions of the coupled-wave equations in the small-signal regime (where pump depletion is neglected by setting I_{ω_3}=constant) show that the signal (ω_2) and idler (ω_3) experience exponential gain (Byer, 1977):

$$
\begin{aligned}
I_{\omega_2}(L) &= I_{\omega_2}(0)\cosh^2(GL) \\
I_{\omega_1}(L) &= \frac{\omega_1}{\omega_2} I_{\omega_2}(0)\sinh^2(GL)\,,
\end{aligned}
\tag{7.32}
$$

where $G^2=(9\omega_1\omega_2 d_{\text{eff}}^2 I_{\omega_3})/(8\varepsilon_0 c^3 n_1 n_2 n_3)$, and Kleinman symmetry is invoked. In fact, it is not even necessary for the signal and idler to be provided as input waves since sources of optical noise are always present (both black-body and zero-point (quantum) noise). Provided the parametric process is phase-matched, the interaction length is long enough and the pump is sufficiently intense, the noise at the signal and idler frequencies can be amplified to a level at which it is readily detectable; this is termed parametric fluorescence. For a fixed pump frequency, the pair of signal and idler frequencies which experiences maximum gain is determined by the phase-matching condition, and may be adjusted by rotating the crystal or in some cases by adjusting its temperature. By placing the nonlinear crystal between aligned mirrors which are good reflectors for ω_1 or ω_2 (or both), a parametric oscillator is obtained. This operates rather like a laser; if the parametric gain exceeds the losses, the field generated by parametric fluorescence can build up during several round trips in the cavity and thus reach a significant proportion of the pump intensity (Byer, 1977).

The noise field which initiates parametric fluorescence and oscillation can be considered to be equivalent to a wave applied at the input to the nonlinear medium. It is useful to estimate the equivalent noise intensity I_{N} so that, for example, the parametric fluorescence efficiency or the rise-time of oscillation can be predicted. This requires a quantum treatment of the optical field (Loudon, 1983; Kleinman, 1968). Except in the far-infrared region, black-body radiation can be neglected at room temperature. It is found that the significant component of the zero-point quantum noise is equivalent to one photon per mode of the generated field. (Disregarding a factor of $\frac{1}{2}$ which sometimes comes in, this simple result is valid for nonlinear processes of any order, and for both parametric and nonparametric effects). In the common situation when the incident waves are collinear Gaussian beams focused inside the medium with a confocal parameter $b \leq L$ (Appendix 10), a useful estimate is $I_{\text{N}}=\hbar\omega_j^2\delta\omega/8\pi^3 cb$, where $\delta\omega$ is the spectral width of the wave with frequency ω_j (Hanna *et al*, 1979). For example, $\hbar\omega_j = 2\times10^{-19}$ J ($\lambda \simeq 1\,\mu$m), $\delta\omega = 10^{10}$ s^{-1}, $b = 5$ mm, gives $I_N = 1\,\mu$W cm^{-2}. Because of the exponential nature of the amplification process, a rough estimate of I_N is usually sufficient for practical purposes.

Similar coupled-wave analyses can be made for third-order (four-wave) sum- and difference-mixing processes. The phase-optimisation integral $|F|^2$ for the sum-frequency process $\omega_1 + \omega_2 + \omega_3 \rightarrow \omega_4$ is identical to that described in §7.2.1 for third-harmonic generation; with tight focusing ($b \ll L$), the optimum condition is $b\Delta k = 2$. Bjorklund (1975) shows that for the process $\omega_1 - \omega_2 - \omega_3 \rightarrow \omega_4$, the maximum integral $|F|^2$ is obtained with $b\Delta k = -2$, while the process $\omega_1 + \omega_2 - \omega_3 \rightarrow \omega_4$ is optimised for $b\Delta k = 0$. The latter process is unique in that, whichever the sign of Δk, $|F|^2$ can be maximised by focusing tightly ($b \rightarrow 0$). Here, as previously, we have defined $\Delta k = k_P - k_4$, where k_P is the sum of the magnitudes of the wave vectors for the input fields (with $-k_j$ associated with a negative frequency $-\omega_j$).

Throughout this section we have assumed, for simplicity, that the waves are collinear. In the more general case of noncollinear waves, the phase-matching condition must be written as a vector relationship: $\Delta\mathbf{k} = \mathbf{k}_P - \mathbf{k}_\sigma = 0$, where $\mathbf{k}_j$ denotes the wave vector of the optical wave with frequency ω_j, and $\mathbf{k}_P = \mathbf{k}_1 + \mathbf{k}_2 + \cdots + \mathbf{k}_n$. An example of such a vector phase-matching condition is depicted in Fig. 7.4.

In our classical treatment of the optical waves in a parametric interaction, we have described the phase matching in terms of the phase velocities of the interacting waves. An alternative interpretation is provided by the quantum theory of light; the photons which make up an electromagnetic wave of wave vector $\mathbf{k}$ each carry a momentum $\hbar\mathbf{k}$. Therefore, the phase-matching condition $\Delta\mathbf{k} = 0$ can be interpreted as meaning that the total photon momentum is conserved during the interaction. This is easily verified; for example, in the case of second-harmonic generation (in which two photons at the fundamental frequency are annihilated and simultaneously one second-harmonic photon is created), the phase-matching condition can be written in the form: $2\hbar\mathbf{k}_\omega = \hbar\mathbf{k}_{2\omega}$.

7.3 Processes independent of phase matching

In the preceding pages we observed that the phase mismatch $\Delta\mathbf{k}$ is crucial in determining the behaviour of many of the travelling-wave interactions that can occur in a nonlinear medium. However, an intrinsic property of some other interactions is that $\Delta\mathbf{k}$ is identically zero, regardless of dispersion or other parameters of the medium. Examples of such processes are the self-induced nonlinear refractive index, stimulated Raman scattering, optical Kerr effect, and multi-photon absorption. The first two of these processes are considered in detail in Chapters 4 and 6, and therefore the discussion here is confined to a description of their wave properties.

We first consider stimulated Raman scattering. From (2.59) and Table 2.1 we may write down the expression for the polarisation induced at the Stokes frequency ω_S by a monochromatic pump field E_P (frequency ω_P):

$$P_S^{(3)} = (3\varepsilon_0 \chi_R/2)\,|E_P|^2 E_S\,, \tag{7.33}$$

where χ_R denotes the Raman susceptibility $\chi^{(3)}(-\omega_S\,;\omega_P\,,-\omega_P\,,\omega_S)$ defined by (2.58). The stimulated Raman polarisation at ω_S is induced by the simultaneous action of the pump and Stokes fields, and the term 'stimulated' arises from the fact that the polarisation depends on the electric field at the same frequency. With the aid of (7.21), the Stokes wave equation can be written down directly:

$$\frac{\partial}{\partial z}\hat{E}_S = \frac{i3\omega_S^2}{4k_S c^2}\chi_R\,|\hat{E}_P|^2\hat{E}_S \exp(i\,\Delta k\,z)\,, \tag{7.34}$$

and we find that the phase mismatch $\Delta k = (k_P - k_P + k_S) - k_S$ vanishes identically (here k_P denotes the pump wave vector). Therefore an intrinsic property of the process is that the relation between the phases of the Stokes polarisation and Stokes field depends only on the argument of the complex scalar quantity χ_R. In the same way that we have described for the parametric processes in §7.2, a simple solution of (7.34) is obtained in the small-signal limit in which depletion of the pump wave is neglected: the Stokes field after a distance L is

$$E_S(L) = E_S(0)\exp\left[\frac{i3\omega_S\chi_R}{2\varepsilon_0 c^2 n_S n_P} I_P(0)\,L\right], \tag{7.35}$$

where I_P is the pump intensity. It is useful to separate χ_R into its real and imaginary parts: $\chi_R = \chi_R' + i\chi_R''$. Then (7.35) may be written in terms of the Stokes intensity: $I_S(L) = I_S(0)\exp(G_R L) = I_S(0)\exp[g_R I_P(0) L]$, where the Raman gain coefficient G_R is given by

$$G_R = \frac{-3\omega_S}{\varepsilon_0 c^2 n_S n_P}\chi_R'' I_P(0)\,, \tag{7.36}$$

and g_R is the Raman gain per unit pump intensity, which appears in (6.156). The expressions for g_R in terms of the Raman polarisability

α_R and transition hyperpolarisability α_{ba} are given in §6.5.3. Whilst the imaginary susceptibility χ_R'' gives rise to Raman gain, the real part χ_R' causes a refractive-index change for the Stokes wave that is induced by the pump:

$$\delta n_S = 3\chi_R' I_P(0)/2\varepsilon_0 c n_S n_P \,. \qquad (7.37)$$

This is a form of the optical Kerr effect, also known as *cross*-phase modulation. In this way, spatial effects (wavefront distortion, focusing or defocusing) and temporal variations (phase- and frequency-modulation) can be induced on a signal beam by an intense control beam. (We note, as an aside, that cross-phase-modulation effects can also take place between two waves which are at the *same* frequency ω but which are distinguishable in some other way – for example by having orthogonal polarisations; as noted in Table 2.1, cross-phase modulation can then occur through the susceptibility $\boldsymbol{\chi}^{(3)}(-\omega;\omega,-\omega,\omega)$.)

In the small-signal limit, (7.35) shows that the Stokes input $I_S(0)$ experiences exponential gain; this is a characteristic feature of a 'stimulated' process. As in the case of parametric amplification (§7.2.2), it is not necessary for an input signal to be applied deliberately; the process can be initiated by spontaneous Raman scattering, which can be represented by $I_S(0) \simeq I_N$ where I_N is the intensity equivalent to one photon per mode of the generated field (see §7.2.2). The high gain that is available in typical experimental conditions amplifies this noise to a readily detectable level, or indeed to a level at which the pump is depleted significantly. In the small-signal regime (*i.e.,* before pump depletion occurs), the Stokes intensity shows a very rapid dependence on pump intensity by virtue of the exponential gain (7.35). The process is therefore described as exhibiting a threshold; *i.e.,* until the pump reaches this value there is little Stokes intensity. In a practical case the definition of threshold intensity is arbitrary, because it depends on the detection sensitivity. However, a simple definition that is particularly useful in theoretical work is to take the threshold pump intensity as that required to give a small-signal Stokes gain that would amplify the noise intensity I_N to a level that *implies* total conversion of the pump, *i.e.,* to the level

$$I_S(L) = (\omega_S/\omega_P) I_P(0) \,. \qquad (7.38)$$

The latter expression follows from the photon-number conservation relation (5.17) in the limit $I_P(0) >> I_N$. With (7.36), this gives an implicit relation for the pump intensity $I_P(0)$ at threshold:

$$I_P(0) = (g_R L)^{-1} \ln[\omega_S I_P(0)/\omega_S I_N] \,. \qquad (7.39)$$

In the large-signal regime, the coupled wave equations for the pump and Stokes field must be solved together, with the result:

$$I_S(L) = I_S(0)\exp(G_R L)/\{1+[\omega_P I_S(0)/\omega_S I_P(0)]\exp(G_R L)\}\,, \tag{7.40}$$

where we have assumed $I_S(0) \ll I_P(0)$. This expression takes proper account of pump depletion; in the limit of large gain ($G_R L \to \infty$), we recover the previous relation (7.38). The threshold pump intensity given by (7.39) is rigorously that at which $I_S(L)$ decreases, because of pump depletion, to 50% of the value predicted by the small-signal formula (7.35). Another important limiting mechanism is the saturation of the Raman transition; this is accounted for by taking the expression (6.154) for the intensity-dependent Raman susceptibility.

The gain coefficient for stimulated electronic Raman scattering in simple atomic systems can be calculated using the available data for electric-dipole matrix elements and transition frequencies. However, for molecular stimulated Raman scattering involving vibrational and rotational transitions, it is more useful to express g_R in terms of the differential Raman scattering cross-section $d\sigma_R/d\Omega$, because the latter quantity can be determined from experimental spontaneous scattering data. The ratio of the number of Stokes photons scattered spontaneously into a solid angle $d\Omega$ to the incident pump-photon flux is (Kaiser and Maier, 1972):

$$d\sigma_R/d\Omega = [\omega_S^3\omega_P/(4\pi\varepsilon_0 c^2)^2] < |\alpha_R|^2 > \tag{7.41}$$

and, with the aid of (6.158) and (7.36), the Raman gain can be expressed as

$$\begin{aligned} g_R &= \frac{8\pi^2 c^2}{\hbar n_S n_P \omega_S^2 \omega_P} \frac{N T_2^{-1}}{\Delta^2 + T_2^{-2}} \frac{d\sigma_R}{d\Omega} \\ &\simeq \frac{8\pi^2 c^2}{\hbar n_S n_P \omega_S^2 \omega_P} N T_2 \frac{d\sigma_R}{d\Omega} \quad \text{at resonance } (|\Delta T_2| \ll 1). \end{aligned} \tag{7.42}$$

The stimulated Raman gain is inversely proportional to the spontaneous linewidth T_2^{-1}. Often, the uncertainty of this value is the main source of error in the prediction of Raman gain, even if the matrix elements and frequencies in the expression (4.117) for χ_R, or the differential cross-section $d\sigma_R/d\Omega$ in (7.42), are known with accuracy.

As mentioned at the close of §5.1.1, the classification of nonlinear processes as either 'parametric' or 'nonparametric' is often blurred. For example, in solids the Raman transition corresponds to a nonlocalised excitation (*i.e.*, phonon). An alternative treatment of stimulated Raman scattering in that case is to treat the phonon as a wave, similar to the electric fields, leading to a phase-matching condition that involves the phonon wave vector; the process may then be thought of as being 'parametric' (Bloembergen and Shen, 1964; White, 1987). Alternatively, we can consider the electric fields alone, as here, in which case the process is 'nonparametric'; the phase-matching condition no longer appears explicitly, but instead an equivalent condition is embodied in the susceptibility as a selection rule for the excitation of the Raman transition. The selection rule arises from the electric-dipole approximation and spatial symmetry properties (§5.3). Stimulated Brillouin scattering (in which the pump and Stokes waves are coupled *via* a density wave created by electrostriction) is usually treated as a three-wave difference-mixing process, similar to that described in §7.2.2 (Kaiser and Maier, 1972).

As a further example, it is a straightforward exercise to derive the expression (6.63) for the nonlinear refraction coefficient n_2 by solving the wave equation (7.21) in the case of a single monochromatic field ω. When ω is far removed from resonances, $\chi^{(3)}$ is real and the system is purely reactive (implying a parametric process). Nevertheless, the phase-matching condition $\Delta k = 0$ is satisfied automatically in such a case.

Finally, the coupled-wave analysis is rather more complicated for multi-photon-resonant processes, such as those depicted in Fig. 4.4. It is found that the coupling between the waves is governed by terms which depend on Δk as well as those that do not, and in general the various terms are inseparable. This complex situation is considered in detail by Hanna *et al* (1979).

7.4 Degenerate four-wave mixing and phase conjugation

We now consider in more detail an interesting example of nonlinear wave coupling. Degenerate four-wave mixing is an important process in nonlinear laser spectroscopy, and provides a means of obtaining phase conjugation or wavefront 'time reversal'. The process involves the coupling of four waves which have the same frequency ω. The waves are distinguishable by their different wave vectors and perhaps also different **E**-polarisation directions. We consider the forward- and backward-mixing arrangements shown in Fig. 7.4. In both cases it is assumed that the waves intersect at a small angle inside the nonlinear medium. In the forward-mixing case, there are just two input beams: the pump and probe. These give rise to a nonlinear polarisation $P_4^{NL} = [3\varepsilon_0\chi^{(3)}(-\omega;\omega,-\omega,\omega)/4]|\hat{E}_1|^2\hat{E}_3{}^* \exp(i\,\Delta k z)$, where $\hat{E}_j$ denotes the envelope of the field with wave vector $\mathbf{k}_j$. This polarisation radiates a signal field at ω in the phase-matched direction $\mathbf{k}_4 = 2\mathbf{k}_1 - \mathbf{k}_3$. Effective

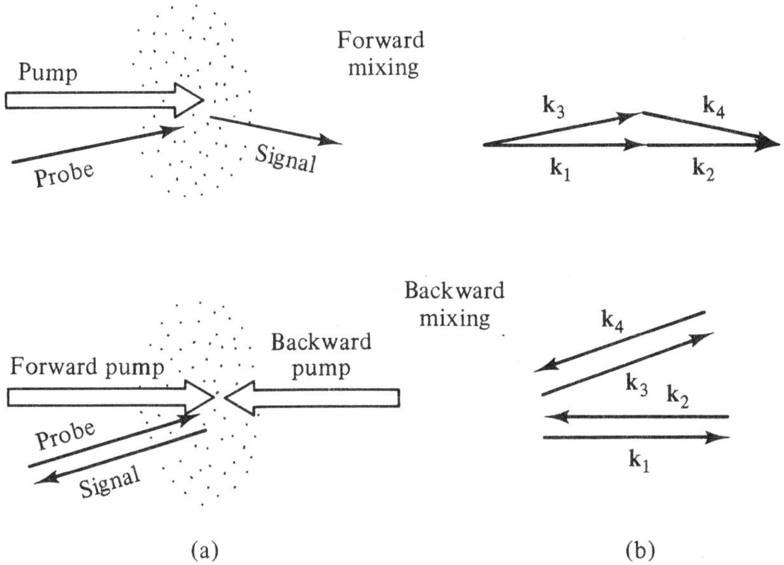

Fig. 7.4 (a) Schematic arrangements and (b) phase-matching vector diagrams for forward- and backward degenerate four-wave mixing. The input waves (pump and probe) intersect in the nonlinear medium at a small angle. The pump waves (with wave vectors $\mathbf{k}_1$ and $\mathbf{k}_2$) are much more intense than either the probe wave ($\mathbf{k}_3$) or the generated signal ($\mathbf{k}_4$). In the case of backward-mixing, the emerging signal retraces exactly the direction of the probe.

coupling between the waves over the length of the medium can occur only if the refractive-index dispersion is such that $\mathbf{k}_4$ is nearly collinear with the input waves. We now focus on the situation of greatest interest when the pump wave is much more intense than the probe and signal. The evolution of the pump wave is then decoupled from the other two, and an analytical solution of the set of coupled wave equations can be found (Maruani, 1980). If we consider the pump wave alone, it is a straightforward task to integrate the wave equation (7.21) to obtain

$$\hat{E}_1(z) = \hat{E}_1(0)\exp[i\gamma I_1(0)z_{\text{eff}} - \alpha z/2], \tag{7.43}$$

where I_1 is the intensity, $\gamma=(3\omega/2n^2\varepsilon_0 c^2)\chi^{(3)}(-\omega;\omega,-\omega,\omega)$, α is the linear absorption coefficient, and $z_{\text{eff}}=2[1-\exp(-\alpha z/2)]/\alpha$. The effective interaction length z_{eff} will be significantly shorter than z if $\alpha z \gtrsim 1$. The real part of γ represents the intensity-dependent refractive-index change, whilst the imaginary part represents the nonlinear loss due to two-photon absorption. When (7.43) is inserted into the coupled equations for the probe and signal waves, we obtain the following expressions for the output intensities of the probe and signal:

$$I_4(L) = \frac{I_3(0)\,|\gamma|^2 I_1^2(0)\,L_{\text{eff}}^2}{2[1+\text{Im}(\gamma)I_1(0)L_{\text{eff}}]^3}\exp(-\alpha L)$$
$$I_3(L) = \frac{I_1(0)}{[1+\text{Im}(\gamma)I_1(0)L_{\text{eff}}]^2}\exp(-\alpha L) + I_4(L). \quad (7.44)$$

These expressions display several effects that occur simultaneously. As well as the linear absorption term $\exp(-\alpha L)$, both waves experience optical nonlinearity induced by the pump: the term that includes $\text{Im}\gamma$ gives rise to nonlinear absorption, whilst the term in $|\gamma|^2$ provides nonlinear gain. Power is transferred from the pump to both the probe and signal waves. We notice that the first term in the expression for $I_3(L)$ is that which is obtained by neglecting the nonlinear gain and retaining only the losses. A physical interpretation of the overall process can be given in terms of grating formation in the nonlinear medium, as illustrated in Fig. 6.10. The pump and probe waves interfere giving rise both to amplitude gratings (resulting from $\text{Im}\,\gamma$) and refractive-index or phase gratings (due to $\text{Re}\,\gamma$), from which the waves scatter coherently.

With this physical picture in mind, we now turn to the backward degenerate four-wave mixing process also depicted in Fig. 7.4. In this case there are three input beams: two collinear and counter-propagating pump beams, and a weaker probe beam which intersects in the medium at a small angle. For practical convenience, it is customary to use a single pump beam which is reflected back by a mirror placed beyond the nonlinear medium. Once again, the pump and probe beams induce a nonlinear polarisation that generates the signal field. The signal wave vector $\mathbf{k}_4$ is determined by the phase-matching condition $\Delta\mathbf{k}=0$ together with the relation $\mathbf{k}_2=-\mathbf{k}_1$ for the pump wave vectors and the reality condition which follows (7.20). It is not difficult to verify that in such a case the only possible configuration is $\mathbf{k}_4=-\mathbf{k}_3$, *i.e.*, the signal wave retraces exactly the path taken by the probe. This is another example of automatic phase matching, since this direction for the signal wave is phase-matched regardless of the dispersion of the medium. The coupled equations (7.21) for the probe and signal waves may be written as

$$\frac{\mathrm{d}}{\mathrm{d}z}\hat{E}_4 = iq^*\hat{E}_3^*>$$
$$\frac{\mathrm{d}}{\mathrm{d}z}\hat{E}_3^* = iq\,\hat{E}_4\,, \quad (7.45)$$

where $q^*=(3\omega/8nc)\chi^{(3)}(-\omega;\omega,-\omega,\omega)\hat{E}_1\hat{E}_2$. If the probe input to the medium at $z=0$ is denoted by $\hat{E}_{30}$, then the 'reflected' signal which emerges is $\hat{E}_4(z=0)=r_c\hat{E}_{30}^*$, where the reflectivity is

$$r_c = -i(q^*/|q|)\tan(|q|L) \quad (7.46)$$

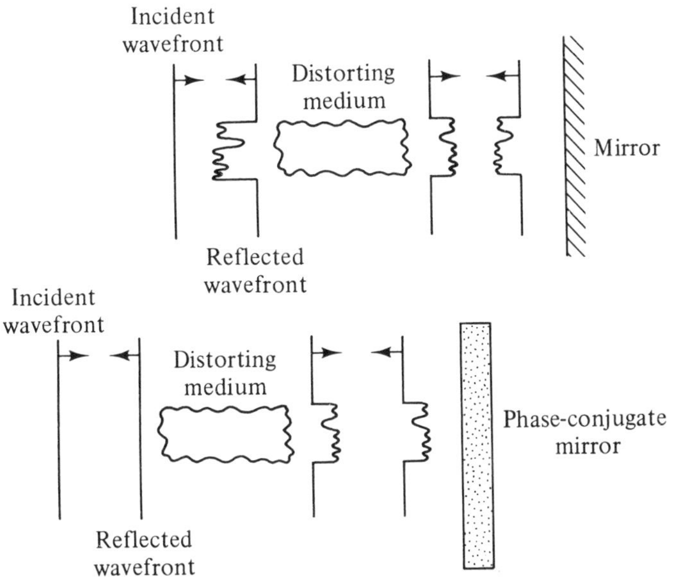

Fig. 7.5 Schematic comparison between the reflection properties of an ordinary mirror and that of a phase-conjugate mirror. In the latter case, the reflected wave is the 'time-reversed' replica of the incident one. Thus, phase aberrations which are inflicted on the signal wavefront during its propagation towards the phase-conjugate mirror are corrected during the reverse propagation of the reflected wave.

and L is the interaction distance (Yariv, 1978). The important feature is that the reflected signal is the phase conjugate of the probe wave. This peculiar property is quite unlike that of an ordinary mirror, as illustrated in Fig. 7.5. Phase-conjugate reflection can be used to correct for wavefront aberrations that occur during propagation through inhomogeneous optical media, such as the atmosphere or a high-power laser amplifier (Fisher, 1983). Unlike an ordinary mirror, the intensity-reflection coefficient $|r_c|^2 = \tan^2(|q|L)$ is adjustable by altering the pump intensity, and reflectivities of greater than unity can be obtained. Similar to the parametric amplifier discussed in §7.2.2, an output signal can be obtained without a probe wave being applied; the process is initiated by noise. In the limit $|q|L \rightarrow \pi/2$ the reflectivity becomes very large and self-oscillation can occur. High reflectivities can be obtained at modest pump intensity by exploiting resonances of the nonlinear medium; aspects of this are discussed in §6.4.

The process of phase conjugation is sometimes also called 'time reversal'. This can be understood by writing the input probe field in the form:

$$
\begin{aligned}
E(t) &= \tfrac{1}{2}\{\hat{E}_{30}\exp[-i(\omega t - \mathbf{k}_3\cdot\mathbf{r})] + \text{ c.c. }\} \\
&= \tfrac{1}{2}\{\hat{E}_{30}^{*}\exp[-i(\omega\{-t\} + \mathbf{k}_3\cdot\mathbf{r})] + \text{ c.c. }\}
\end{aligned}
\tag{7.47}
$$

where, in the second line, the order of the complex-conjugate terms is merely reversed. When the emerging phase-conjugate signal field is now written as

$$
E(t) = \tfrac{1}{2}\{r_c\hat{E}_{30}^{*}\exp[-i(\omega t - \mathbf{k}_4\cdot\mathbf{r})] + \text{ c.c. }\}
\tag{7.48}
$$

and it is noted that $\mathbf{k}_4\cdot\mathbf{r} = -\mathbf{k}_3\cdot\mathbf{r}$, we see that (7.47) and (7.48) are identical except for the reflectivity r_c and the time reversal $t\rightarrow -t$.

Degenerate four-wave mixing is just one of several nonlinear-optical processes with the ability to generate a phase-conjugate signal. Others include stimulated Brillouin scattering and Raman scattering, three-wave mixing and photon echoes. The first observation of phase conjugation was made by Zel'dovich *et al* (1972) using stimulated Brillouin scattering. Fisher (1983) provides a comprehensive review.

7.5 Self-phase modulation and solitons

As an example of a time-dependent process, we now consider the propagation of an optical pulse in a medium which exhibits a nonlinear refractive index. We assume here that the optical frequency is far removed from resonances, so that the refractive index n follows a Kerr-law dependence on intensity, given by (6.62): $n = n_0 + n_2|\hat{E}|^2$. The nonlinear-refraction coefficient n_2 is related to the susceptibility $\chi^{(3)}$ by (6.63). If the intensity $|\hat{E}|^2$ exhibits spatial and temporal variations, the refractive index will also be space- and time-dependent. Consider, for example, a beam with a Gaussian radial distribution of intensity. The induced index change will follow this radial intensity distribution. If the sign of n_2 is positive (as, for example, in the case of fused silica and other glasses), the refractive index increases with intensity and the medium acts as a lens giving rise to self-focusing; for negative n_2 the effect is self-defocusing. If the self-focusing effect is strong enough, it may overcome the spreading of the beam due to diffraction, a phenomenon known as 'self-trapping'. At yet higher intensities, self-focusing may cause the collapse of the beam to dimensions on the order of a wavelength, and can result in catastrophic optical damage in solids. Such dynamic effects are of great complexity (Shen, 1975). However, we shall simplify the following discussion by assuming the nonlinearity to be sufficiently weak ($n_2|\hat{E}|^2 \ll n_0$) that these transverse effects may be neglected. Instead we are concerned with the effect of time variations of the intensity, which give rise to phase changes in the optical carrier wave.

In the plane-wave limit, the propagation equation for a pulse $\hat{E}(z,t)$ in a Kerr-law medium is given by (7.25) expressed in the form:

$$i\left(\frac{\partial}{\partial z}+\frac{dk}{d\omega}\frac{\partial}{\partial t}\right)\hat{E}-\frac{1}{2}\frac{d^2k}{d\omega^2}\frac{\partial^2}{\partial t^2}\hat{E}=-\frac{\omega n_2}{c}|\hat{E}|^2\hat{E}, \qquad (7.49)$$

where loss is neglected and, as is well known in wave theory, the group velocity v_g is $(dk/d\omega)^{-1}$. We now investigate the solutions of this equation. To begin with, let us consider the most simple situation in which the second-order dispersion $d^2k/d\omega^2$ is negligible. In this regime, the effect of the nonlinear term on the right-hand side of (7.49) is known as self-phase modulation. To examine this case, we can further simplify (7.49) by eliminating $\partial/\partial t$ with the substitution of a reduced time scale $\xi \equiv t - z/v_g$. A particular value of ξ therefore represents a fixed position in a coordinate frame which travels at the group velocity in the direction of propagation. With $\hat{E}$ expressed as a function of z and ξ, the propagation equation (7.49) becomes:

$$\frac{\partial\hat{E}}{\partial z}=\frac{i\omega n_2}{c}|\hat{E}|^2\hat{E}. \qquad (7.50)$$

It is straightforward to verify that the solution of (7.50) is

$$\hat{E}(z,\xi)=\hat{E}(0,\xi)\exp[i\frac{\omega n_2}{c}|\hat{E}(0,\xi)|^2 z]. \qquad (7.51)$$

This is a pulse whose intensity is unaltered during propagation and which experiences a phase shift ϕ due to self-phase modulation:

$$\phi=\frac{\omega n_2}{c}|\hat{E}|^2 L \qquad (7.52)$$

after a distance L. Since the intensity $|\hat{E}|^2$ varies throughout the pulse ($|\hat{E}|^2$ is a function of ξ), the various parts of the pulse undergo different phase shifts, leading to a frequency shift or 'chirp' which is directly proportional to the distance travelled. Suppose, for example, the field $\hat{E}(z,t)$ is initially in the form of a chirp-free (constant-phase) pulse $\hat{E}(0,t)=\hat{E}_0\,\mathrm{sech}(t/\tau)$, where τ is a measure of the pulse duration. After such a pulse has travelled a distance L in the medium, the difference between its instantaneous optical frequency (at time $t+L/v_g$) and the carrier frequency ω is obtained by differentiating the phase shift (7.50) with respect to t:

$$\delta\omega=-\frac{d\phi}{dt}=-\frac{\omega n_2}{c}|\hat{E}_0|^2 L\frac{d}{dt}\mathrm{sech}^2(t/\tau). \qquad (7.53)$$

This is illustrated in Fig. 7.6. Assuming n_2 is positive, the instantaneous

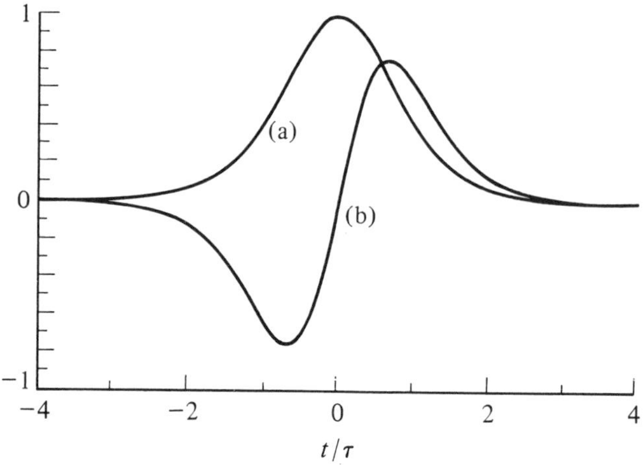

Fig. 7.6 (a) The function $\mathrm{sech}^2(t/\tau)$ representing pulse intensity, and (b) the function $-\mathrm{d}\,\mathrm{sech}^2(t/\tau)/\mathrm{d}t$ representing the corresponding frequency chirp arising from self-phase modulation in a Kerr-law nonlinear medium.

frequencies in the leading half of the pulse are lowered, whereas those in the trailing half are raised. The total frequency excursion is $2|\delta\omega_{\max}| = 1.54\Delta\phi_{\max}/\tau$, where $\Delta\phi_{\max} = \omega n_2|\hat{E}_0|^2L/c$ is the phase change at the peak of the pulse. This spectral broadening due to self-phase modulation becomes significant when the frequency excursion approaches and exceeds the initial bandwidth of the pulse, which is equal to $1.12/\tau$ (Kaiser, 1988).

We now consider the influence of group-velocity dispersion, which arises when the term $\mathrm{d}^2k/\mathrm{d}\omega^2$ in (7.49) is not zero; then the group velocity $v_g \equiv (\mathrm{d}k/\mathrm{d}\omega)^{-1}$ varies with frequency. The various frequency components within the bandwidth of the pulse travel at different group velocities. It is not difficult to see that, if the medium exhibits a group-velocity dispersion $\mathrm{d}v_g/\mathrm{d}\omega$ that is positive, the leading half of the pulse shown in Fig. 7.6 is delayed slightly and the trailing half is advanced, leading to pulse narrowing. (A value $\mathrm{d}v_g/\mathrm{d}\omega > 0$ implies that $\mathrm{d}v_g/\mathrm{d}\lambda$ and $\mathrm{d}^2k/\mathrm{d}\omega^2 \equiv \mathrm{d}(v_g^{-1})/\mathrm{d}\omega$ are both negative.) If, however, $\mathrm{d}v_g/\mathrm{d}\omega$ is negative ($\mathrm{d}v_g/\mathrm{d}\lambda > 0$ and $\mathrm{d}^2k/\mathrm{d}\omega^2 > 0$), then we see enhanced pulse-broadening. This enhanced effect reinforces the pulse-spreading phenomenon that is well known in the linear optics of dispersive media because any pulse must contain a spread of frequencies determined by its Fourier transform.

Acting in combination with group-velocity dispersion, the nonlinear refractive index can give rise to many interesting and useful pulse-propagation phenomena. In recent years, there has been an increasing interest in studying and utilising such effects in optical fibres

(Winful, 1986; Agrawal, 1989). This may seem surprising since standard optical fibres are fabricated from glasses whose nonlinear-refraction coefficients are several orders smaller than those of other materials in which nonlinear refractive index and self-phase modulation effects have been studied in the past. (For example, the value of $n_2 = 2.3 \times 10^{-22}$ $m^2\ V^{-2}$ for fused silica is about 2 orders smaller than for CS_2, and 4 orders smaller than for GaAs at frequencies well below the bandgap.) However, the tight transverse confinement of the optical field within the core (with dimensions of a few microns) means that high intensities $|\hat{E}|^2$ can be achieved at modest input power. Moreover, the very low transmission loss in good-quality fibres allows the effects of $n_2|\hat{E}|^2$ to become significant when integrated over long path lengths (distances varying from centimetres to tens of kilometres are typical). Monomode optical fibres are especially interesting media in which to study these effects; such waveguides support only one transverse mode of the electric field. In the weakly nonlinear regime which we are considering, the only effect of the nonlinearity is to modify the propagation constant (magnitude of the wave vector) for the guided mode. In other words, despite the transverse distribution of intensity, all parts of the travelling wavefront in a monomode fibre experience the same phase shift. Propagation in a monomode fibre is therefore described rather well by the uniform plane-wave equation (7.49), in which $\hat{E}$ is now interpreted as the maximum field on the axis (Adams, 1981; Snyder and Love, 1983; Neumann, 1988).

Fig. 7.7 shows the propagation loss of a good-quality monomode fibre as a function of wavelength, together with a plot of group-velocity dispersion (notice that it also common to refer to the dispersion of the group delay per unit distance, v_g^{-1}, which has the opposite sign). The material group-velocity dispersion for silica fibres passes through zero at a wavelength close to 1300 nm. The total group-velocity dispersion of a fibre consists mainly of the material group-velocity dispersion plus an additional component, called the waveguide dispersion, which depends on the structure, dimensions and refractive-index profile of the fibre. A third, and usually much smaller, component of the total dispersion arises from the fact that the refractive-index profile is itself wavelength dependent; this is termed profile dispersion. By fabricating fibres of different designs it is possible to manipulate the waveguide and profile dispersion so that the wavelength λ_0 at which the total group-velocity dispersion is zero can be adjusted to some extent (Adams, 1981).

We now return to consider the wave equation (7.49) which, as already mentioned, provides a good description of the propagation of the

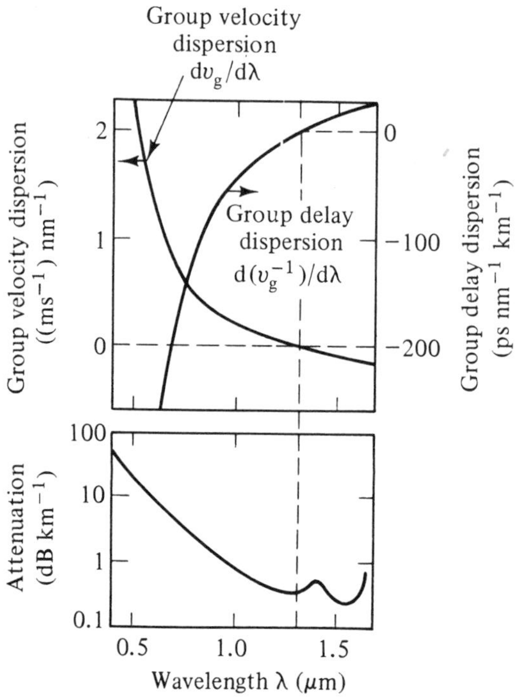

Fig. 7.7 Typical loss and group-velocity dispersion as functions of wavelength for good-quality monomode fibres based on silica glass. The dispersion characteristic shown is that of a 'standard' fibre having zero group-velocity dispersion at a wavelength $\lambda_0 \simeq 1300$ nm. By manipulating the waveguide design, other types of fibre can be fabricated with λ_0 positioned anywhere in the range approximately 1250 – 1600 nm.

pulse envelope in a monomode optical fibre. When the second-order dispersion term is retained but higher-order terms are neglected, (7.49) may be written in the form:

$$i\frac{\partial u}{\partial \varsigma} + |u|^2 u = \pm \frac{1}{2}\frac{\partial^2 u}{\partial s^2}, \tag{7.54}$$

where the parameters are expressed in the normalised form:

$$\begin{aligned} u &= \tau\sqrt{n_2\omega/c\,|\mathrm{d}^2k/\mathrm{d}\omega^2|^2}\,\hat{E} \\ s &= (t - z/v_g)/\tau \\ \varsigma &= |\mathrm{d}^2k/\mathrm{d}\omega^2|\,z/\tau^2 . \end{aligned} \tag{7.55}$$

Here s is a measure of the position in a moving coordinate frame which travels at the group velocity v_g, and τ is any convenient time scale (usually taken to be the initial pulse duration). The dispersion term which appears in the right-hand side of (7.54) takes the sign of $\mathrm{d}^2k/\mathrm{d}\omega^2$ (which is the opposite of the sign of $\mathrm{d}v_g/\mathrm{d}\omega$). In this normalised form, (7.54) is the celebrated nonlinear Schrödinger equation. The solutions of this seemingly simple equation display extraordinary and interesting properties, many of which have been observed experimentally in optical fibres (Agrawal, 1989).

The type of propagation behaviour that occurs depends critically on whether $dv_g/d\omega$ is positive or negative (*i.e.,* for silica fibres, whether the wavelength λ is respectively greater or less than λ_0 – see Fig. 7.7). For $dv_g/d\omega>0$ pulse-narrowing can occur, as described earlier, and it also becomes possible to observe pulses that propagate either without any change of shape or have shapes that vary periodically with distance. Such extraordinary pulses exist as solitary wave solutions (solitons) of the nonlinear wave equation. The propagation behaviour of solitons is fundamentally different from that of low-intensity 'linear' pulses which, by virtue of their finite bandwidth and group-velocity dispersion, must ultimately increase in width during propagation.

Solitary wave solutions of the nonlinear Schrödinger equation have been found both analytically and numerically (Zakharov and Shabat, 1973; Dodd *et al,* 1982; Blow and Doran, 1987; Agrawal, 1989). In the regime where $dv_g/d\omega>0$, an exact soliton is obtained when the pulse $u(\varsigma,s)$ has the initial form:

$$u(\varsigma,s) = A \operatorname{sech} s \tag{7.56}$$

and $A=N$, where N is an integer number ($N\geq1$). The amplitude A takes the normalised units defined by (7.55). Of course, there is no constraint; we are free to choose any initial boundary condition that we wish. To begin with, however, we focus attention on this particular case of (7.56) with $A=N$. (We consider the situation when the initial pulse is different from this later.) The solutions of the wave equation show that the whole energy of such an input pulse propagates subsequently as a single entity with the properties of a soliton. The soliton with $N=1$ (often called the 'fundamental' soliton) propagates as $u(\varsigma,s)= \operatorname{sech} s \exp(-i\varsigma/2)$. In other words, apart from a rotating phase, the fundamental soliton is a pulse which propagates without change. A simple intuitive explanation for this behaviour is that the frequency chirp generated by self-phase modulation acts together with the group-velocity dispersion to compensate exactly for the 'linear' pulse-broadening that the pulse would suffer in the absence of nonlinearity. This picture in which self-phase modulation and group-velocity dispersion are considered as separately identified processes is, however, vastly oversimplified and conceals a wealth of structure and subtleties that can be seen in the mathematical solutions of the nonlinear Schrödinger equation. The fundamental soliton is the only one which

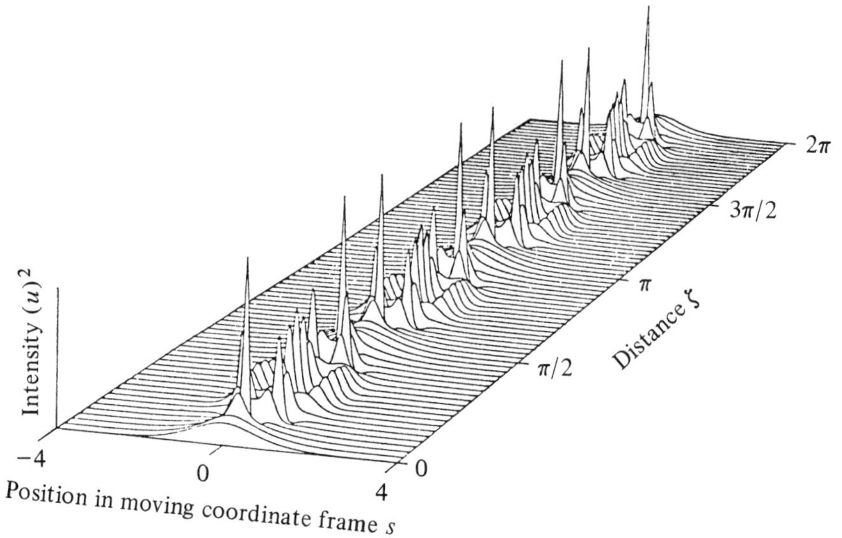

Fig. 7.8 Computer-generated solutions of the nonlinear Schrödinger equation showing the evolution of an $N=4$ soliton as it propagates over a distance equal to four soliton periods. (After Blow and Doran, 1987.)

propagates without change of shape. Higher-order solitons (integers $N>1$) undergo oscillations in shape with a period of $\pi/2$ on the distance scale ς. The pulse changes become more complicated in structure for increasing N value; for $N=2$ the pulse simply compresses during propagation for $\varsigma=0$ to $\pi/4$, and then reexpands to its original form at $\varsigma=\pi/2$ without developing further structure. The evolution is symmetric about $\varsigma=\pi/4$ for all solitons. To illustrate the periodic evolution of solitons, Fig. 7.8 shows the propagation of an $N=4$ soliton. Also visible is the periodic compression of the pulse which, in this case, is at its narrowest at $\varsigma \simeq 3\pi/40$, $17\pi/40$, $23\pi/40$, $37\pi/40$,... . Such behaviour has been observed experimentally (Mollenauer *et al*, 1980). It is found, both theoretically and experimentally, that single solitons behave as robust entities that are stable with respect to small perturbations of the propagation equation, such as loss and higher-order dispersion. Solitons can pass through each other and emerge unscathed apart from a change in phase.

The transformations (7.55) allow the parameters in the nonlinear wave equation (7.52) to be expressed in practical units. To set the scale of things, a pulse having a duration (full width at half maximum intensity) of 7 ps ($\tau=4$ ps) and peak power 1 W would generate a fundamental soliton in a standard monomode fibre at $\lambda=1.5\,\mu$m ($n_0=1.45$, core area = $100\,\mu\text{m}^2$, $\mathrm{d}^2k/\mathrm{d}\omega^2 = -(\lambda^2/2\pi c)\mathrm{d}(v_g^{-1})/\mathrm{d}\lambda = -1.8\times10^{-26}\ \text{s}^2\ \text{m}^{-1}$

from Fig. 7.7). The soliton period $\varsigma=\pi/2$ corresponds to a distance $\pi\tau^2/2|d^2k/d\omega^2|$, which in this case is 1.4 km.

Rarely in an experiment is the initial pulse launched into a fibre precisely of the form (7.56). For example, the pulse envelope may not have precisely a hyperbolic secant shape; its amplitude may not be exactly an integer number (in the normalised units); or it may be chirped in frequency. In this case, provided the initial pulse amplitude exceeds a particular threshold value, a certain proportion of the pulse energy will contribute to a soliton and the remainder will gradually disperse by the process of group-velocity dispersion during propagation. Not only will the soliton contain less energy than the original pulse, it will also have a different width (either narrower or wider). For an initial pulse of the form $u(0,s)=A\,\mathrm{sech}\,s$, a soliton is obtained provided the amplitude $A>\frac{1}{2}$. As mentioned previously, the special aspect of the initial condition $A=N$ with integers $N\geq 1$ is that *all* the energy of the initial pulse propagates as a soliton; none of the energy disperses.

Hasegawa and Tappert (1973) were the first to propose the use of solitons for long-distance optical-fibre communications to overcome one of the primary limitations of such systems in the linear regime, namely pulse-broadening (due to group-velocity dispersion), which limits the achievable bandwidth. A further limitation of such systems, however, is the inevitable attenuation over long transmission paths (see Fig. 7.7). The effects of loss in soliton propagation were considered by Blow and Doran (1985) using a numerical analysis of the nonlinear Schrödinger equation modified to incorporate loss as a perturbation. They concluded that, for the typical values of loss that occur in optical fibres in the wavelength region $\sim 1-1.5\ \mu$m (Fig. 7.7), a soliton will broaden during propagation, but that the asymptotic dispersion at long distances is lower than that of an equivalent pulse in the linear regime. Thus the soliton retains aspects of its special character even in the presence of loss. Since loss causes pulse-broadening in soliton systems, periodic amplification has been proposed as a means of both maintaining adequate pulse energy and eliminating broadening. Kodama and Hasegawa (1982) showed theoretically that if a soliton pulse propagating in a lossy fibre could be amplified periodically at distances determined by the group-velocity dispersion, the shape of the soliton would be almost unchanged even after several hundred amplifications. A scheme for achieving this using stimulated Raman amplification within the fibre itself has been demonstrated recently (Mollenauer and Smith, 1988). Soliton propagation over an effective path length as great as 6000 km was observed with only minor increase in the pulse width from its initial value

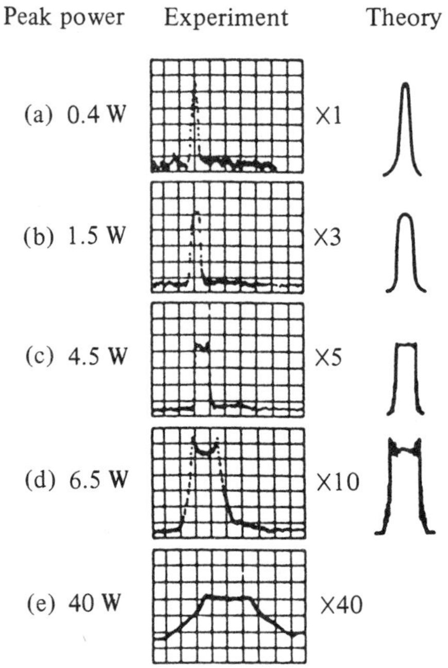

Fig. 7.9 Experimentally observed and calculated shapes of 1.06 μm optical pulses (initially 150 ps width) of various peak powers after transmission through 20 km monomode fibre, showing enhanced pulse-broadening. The horizontal scale is 500 ps per division. Vertical scaling factors are shown. (After Nelson *et al*, 1983.)

of 50 ps. (By contrast, an equivalent input pulse in the linear regime would broaden to ~6 ns during propagation over such a distance.)

We now return to the solutions of the nonlinear Schrödinger equation (7.54), and consider the situation when $dv_g/d\omega<0$. As noted previously, pulses undergo enhanced broadening in this regime (*i.e.*, the combination of nonlinearity and group-velocity dispersion produces a greater degree of pulse-broadening for an intense pulse compared to a low-intensity one). An example is shown in Fig. 7.9. The temporal broadening is accompanied by pronounced spectral broadening with a positive frequency chirp. For example, the almost-rectangular pulse shown in Fig. 7.9(c) contains a near-linear frequency sweep which describes essentially all of the pulse energy (Grischkowsky and Balant, 1982). By passing this chirped pulse through a linear dispersive delay line with the opposite sign of group-velocity dispersion ($dv_g/d\omega>0$), such as a pair of suitably aligned diffraction gratings, the pulse can be compressed in time to as short as one-tenth to one-hundredth of its initial duration. This technique is now widely employed as a method of optical pulse compression (Shank *et al*, 1982; Tomlinson *et al*, 1984) and pulse durations as short as a few femtoseconds can be achieved.

In the regime $dv_g/d\omega<0$ in which pulses undergo enhanced broadening, we can again obtain soliton solutions of the nonlinear Schrödinger equation, but of a quite different type from those discussed earlier. The soliton solutions are now characterised by the *absence* of light and are known as 'dark' solitons. A dark soliton consists of a long, steady and intense optical wave containing a momentary dark 'pulse' with an appropriate phase shift superimposed. These entities are solitons by virtue of their scattering and stability properties, and have been observed experimentally (Weiner *et al*, 1988).

We have cited as examples of time-dependent processes only a few of the many interesting and useful nonlinear-optical phenomena that occur in fibres (for extensive reviews, see Winful (1986), Cotter (1987) and Agrawal (1989)). In a more general sense, the current study of such phenomena in optical fibres is making an important contribution to the rapidly-evolving field of nonlinear physics.

8

Dynamic optical nonlinearities in semiconductors

Semiconductors contain free carriers. That is the characteristic feature which makes them different from the other systems which we have considered hitherto. The optical nonlinearities discussed in previous chapters arose from *bound* charges. Similar effects arise from bound charges in semiconductors but we do not consider them here. The optical nonlinearities which arise from free carriers in semiconductors are particularly important for applications because of the high degree of control which we have over the free-carrier densities and therefore on the performance of devices which make use of them.

When the free-carrier densities are changed by optical excitation we are concerned with *real* transitions (see §6.6). The resulting nonlinear processes proceed *via* a real exchange of energy from the optical field to the medium, and are often referred to as 'dynamic nonlinearities' (Miller *et al,* 1981a; Oudar, 1985); this is the nomenclature used here.

Another feature of semiconductors which has become of particular significance for applications in recent years is the ability to fabricate multiple 'quantum well' structures, in which the carriers are confined in one direction in repeated layers of the order of 5 nm wide. Within the layers the carrier motion is two-dimensional, which drastically affects their behaviour. This topic belongs more naturally to the next chapter. In this chapter, we confine our attention to the nonlinear-optical properties of bulk semiconductors.

In §§8.1 and 8.2 we outline the one-electron band structure and the behaviour of phonons in Group IV and III – V semiconductors. Then in §8.3 we give a relatively extensive account of the linear-optical properties of semiconductors. This is because a significant contribution to $\chi^{(1)}(-\omega;\omega)$ depends on the free-carrier concentrations. Many of the nonlinear effects which concern us can be described by replacing the free-carrier concentrations in thermal equilibrium (which enter into $\chi^{(1)}(-\omega;\omega)$) by the values resulting from optical pumping. We outline the theory of these processes in §8.4. A detailed and very readable account of the elementary physics of semiconductors can be found in the book by Kittel (1986).

8.1 The electronic band structure of Group IV and III – V semiconductors

To keep our discussion within a reasonable compass we confine our attention to the Group IV semiconductors Si and Ge and the III – V semiconductors GaAs and InSb. In the one-electron approximation each electron is pictured as moving in a potential field $V(\mathbf{r})$ with the periodicity of the crystal lattice. The one-electron energy eigenfunctions $\psi(\mathbf{r})$ and their associated energies E are determined by the Schrödinger equation

$$-\frac{\hbar^2}{2m}\nabla^2\psi(\mathbf{r}) + V(\mathbf{r})\,\psi(\mathbf{r}) = \mathrm{E}\,\psi(\mathbf{r})\,, \tag{8.1}$$

in which we ignore spin-orbit coupling terms for simplicity. In this equation, and throughout this chapter and the next, we use the conventional symbol $\mathbf{r}=(x,y,z)$ to denote an electron position vector. In §4.6 we used $\boldsymbol{\mu}$ for this purpose to avoid confusion, because we had already written the dipole moment of an individual molecule as $e\mathbf{r}$. In the present chapter and the next one, however, we are never concerned with molecules and the conventional notation is therefore to be preferred.

The periodicity of $V(\mathbf{r})$ ensures that $\psi(\mathbf{r})$ may be chosen to have the form of a Bloch function (Kittel, 1986):

$$\psi_{n\mathbf{k}}(\mathbf{r}) = \exp(i\mathbf{k}\cdot\mathbf{r})\,u_{n\mathbf{k}}(\mathbf{r})\,, \tag{8.2}$$

where $u_{n\mathbf{k}}(\mathbf{r})$ has the crystal periodicity. We suppose that $\psi_{n\mathbf{k}}(\mathbf{r})$ is normalised over a macroscopic crystal volume V and write the energy associated with it as $\mathrm{E}_n(\mathbf{k})$. In (8.2), $\mathbf{k}=(k_x, k_y, k_z)$ is a wave vector in reciprocal space. Now, since a factor $\exp(i\mathbf{G}\cdot\mathbf{r})$ may always be taken out of $u_{n\mathbf{k}}(\mathbf{r})$, where $\mathbf{G}$ is an arbitrary reciprocal lattice vector, we see that $\mathbf{k}$ is undefined to within a reciprocal lattice vector. To put this another way: both $\psi_{n\mathbf{k}}(\mathbf{r})$ and $\mathrm{E}_n(\mathbf{r})$ are periodic functions of $\mathbf{k}$ with the periodicity of the reciprocal lattice. It follows that we may describe all the states by restricting $\mathbf{k}$ to a unit cell in reciprocal space and using the index n (the 'band' index) to label the different values of energy which are allowed for any particular $\mathbf{k}$ (Kittel, 1986).

The unit cell in reciprocal space may be chosen at will. However, it is conventional to use the 'Brillouin zone', which is the interior of the region obtained by inserting planes which bisect all the reciprocal lattice vectors drawn from the origin. The Group IV and III – V semiconductors all have the same Brillouin zone which is sketched in Fig. 8.1. Inside the

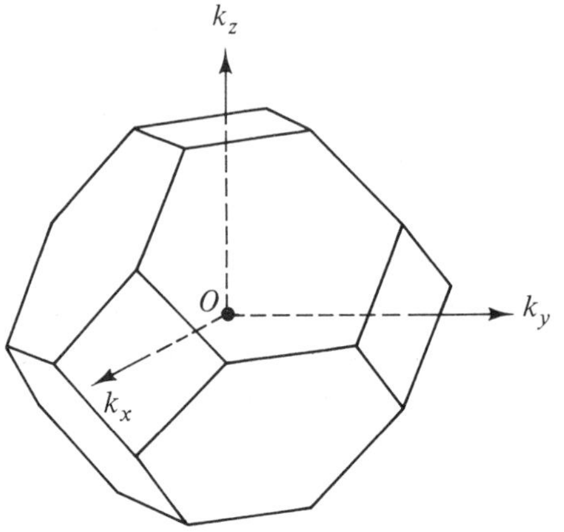

Fig. 8.1 The Brillouin zone of Ge, Si, InSb and GaAs.

Brillouin zone the energy $E_n(\mathbf{k})$ in band n is a continuous function of the wave vector $\mathbf{k}$ in an infinite crystal. However, it is conventional to make $\mathbf{k}$ quasi-continuous by applying periodic boundary conditions to $\psi_{n\mathbf{k}}(\mathbf{r})$ over a parallelopiped of crystal unit cells with the macroscopic crystal volume V. Then a familiar calculation (Kittel, 1986) gives $V/8\pi^3$ for the density of $\mathbf{k}$ values in $\mathbf{k}$-space. The density of states per unit energy range per unit volume of the crystal in band n is therefore given by:

$$N_n(E) = 2\sum_{\mathbf{k}}\delta[E - E_n(\mathbf{k})] = \frac{2}{8\pi^3}\int d\mathbf{k}\,\delta[E - E_n(\mathbf{k})]\,, \tag{8.3a}$$

where the factor of 2 allows for spin degeneracy and the integration is over the Brillouin zone.

We make one final remark about the behaviour of $\psi_{n\mathbf{k}}(\mathbf{r})$ and $E_n(\mathbf{k})$ which is often useful: both quantities exhibit time-reversal symmetry. Thus, since $V(\mathbf{r})$ is real we see from (8.1) that $\psi_{n\mathbf{k}}^*(\mathbf{r})$ is a Bloch function with the same energy as $\psi_{n\mathbf{k}}(\mathbf{r})$ but it obviously has wave vector $-\mathbf{k}$. It follows that the Bloch functions may be chosen so that

$$\psi_{n,-\mathbf{k}}(\mathbf{r}) = \psi_{n\mathbf{k}}^*(\mathbf{r}) \tag{8.4a}$$
$$E_n(-\mathbf{k}) = E_n(\mathbf{k})\,. \tag{8.4b}$$

We make most of our calculations for the 'standard' band structure which is shown in Fig. 8.2(a). It consists of a conduction band with $E_c(\mathbf{k}) = \hbar^2\mathbf{k}^2/2m_e$ (where m_e is the 'electron' effective mass and is positive) and a valence band with $E_v(\mathbf{k}) = -E_g - \hbar^2\mathbf{k}^2/2m_h$, where m_h is the 'hole' effective mass, which is also positive, and E_g is the forbidden energy gap. Each of the two energy bands is therefore a parabolic function of $\mathbf{k}$, with a curvature determined by the effective mass. The standard band structure is a gross oversimplification but it can be made to cover many problems in semiconductor physics by appropriate choice of m_e and m_h. We find immediately by transforming to polar coordinates in (8.3a) that, for the standard band structure,

$$N_c(E) = \frac{\sqrt{(2m_e)^3 E}}{2\pi^2\hbar^3}, \quad E>0, \tag{8.3b}$$

and

$$N_v(E) = \frac{\sqrt{(2m_h)^3(-E_g-E)}}{2\pi^2\hbar^3}, \quad E<-E_g, \tag{8.3c}$$

where the energy E is measured from the edge of the conduction band.

The semiconductors with features in their band structures which are closest to the standard form are GaAs and InSb (see Fig. 8.2(b)). They are both 'direct gap' semiconductors which have a minimum in the conduction band vertically above a maximum in the valence band at $\mathbf{k}=0$. Consequently, as we discuss in §8.3.1, interband absorption proceeds by 'direct' (*i.e.*, vertical) transitions on the band diagram. This feature of the band structure of GaAs and InSb is properly modelled by the standard band structure. However, these semiconductors have *three* valence bands: the heavy hole, light hole and split-off bands in order of decreasing energy in Fig. 8.2(b). We are mainly concerned with the heavy hole band which makes the most significant contribution to $N_v(E)$.

In the cases of Si and Ge the valence band structure near $\mathbf{k}=0$ is similar, but the minima in the conduction band are located along high symmetry directions away from $\mathbf{k}=0$, as is indicated for Si in Fig. 8.2(c). Consequently, as we discuss in §9.3.2, interband absorption at frequencies just above $E_g/\hbar$ proceeds by 'indirect' (*i.e.*, diagonal) transitions involving the simultaneous absorption of a phonon. Si and Ge are examples of an 'indirect gap' semiconductor. In Si the conduction band minima lie in the (100) directions 85% of the way to the zone boundary (see Fig. 8.2(c)). All six valleys are distinct. In Ge the conduction band minima are in the (111) directions on the boundary of the Brillouin zone. There are only four distinct valleys in this case because, as we have already remarked, points in $\mathbf{k}$-space which are separated by a reciprocal lattice vector describe the same electron states. The off-centre location of the conduction band valleys in these materials means that the constant-energy surfaces are ellipsoids, instead of spheres as they are in the standard band structure. However, (8.3b) remains valid for $N_c(E)$ when E is close to the conduction band edge provided that an appropriate 'density-of-states' effective mass is substituted for m_e. The valence bands

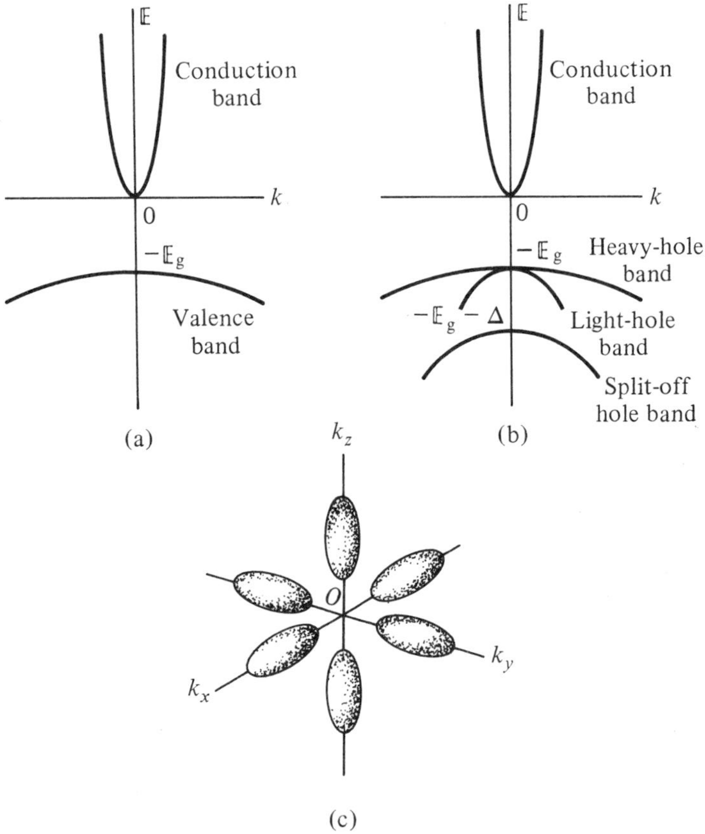

Fig. 8.2 Typical semiconductor band structures. (a) The 'standard' band structure (see text). (b) The band structures of GaAs and InSb. (c) Constant energy ellipsoids for the conduction band of Si.

do not have spherical constant-energy surfaces in any actual semiconductor. However, that complication is again not very significant for dynamical nonlinear-optical effects and we ignore it.

We remarked in §4.6 that electrons are Fermi particles. In the one-electron approximation the many-electron wave functions are therefore Slater determinants which are antisymmetrised combinations of products of one-electron wave functions (Inkson, 1983). In the Slater determinant each one-electron wave function occurs just once. That is to say, the electrons obey the Pauli exclusion principle and the occupation number of any particular one-electron state is therefore 0 or 1. The thermal average of a one-electron operator therefore has the general form given in (4.134), where u_a is the wave function for one-electron state a and $f_0(a)$ is the thermal average of the occupation number for that state, which is given by the Fermi-Dirac function (4.135). In §4.6, a

denotes an arbitrary one-electron state label. In the present chapter it is to be identified with the combination of the band index n and wave number $\mathbf{k}$ which label the Bloch functions. Then the Fermi-Dirac distribution (4.135) depends only on $E_n(\mathbf{k})$ and temperature T, and may be written in the form

$$f_0[(E_n(\mathbf{k})] = \{\exp[E_n(\mathbf{k}) - E_F]/kT + 1\}^{-1}. \tag{8.5}$$

The chemical potential E_F in (8.5) is often referred to as the Fermi level in the context of semiconductor physics. It is the same in all bands and is determined by the electron density $N_n^{(e)}$ in any band through the equation

$$N_n^{(e)} = \int dE\, N_n(E) f_0(E). \tag{8.6a}$$

In intrinsic (undoped) material the electrons which are thermally promoted to the conduction band leave an equal number of holes behind in the valence band. Hence (8.6a) may be written in the alternative form:

$$\int_0^{\infty} dt\, N_c(E) f_0(E) = \int_{-\infty}^{-E_g} dE\, N_v(E)\,[1 - f_0(E)]. \tag{8.6b}$$

8.2 Phonons in Group IV and III – V semiconductors

The crystal structure is that of diamond for Ge and Si and zinc blende for GaAs and InSb. In both cases there are two atoms in each unit cell and the phonon dispersion curves therefore have both acoustic and optical branches. The elemental semiconductors are homopolar and their optical vibrations do not produce an electric field. Symmetry considerations show that the transverse and longitudinal optic phonons are therefore degenerate in frequency when the phonon wave vector $\mathbf{q}$ vanishes. Conversely the binary semiconductors are heteropolar and have an electric field associated with their longitudinal optical (LO) vibrations. This increases the restoring force over and above that pertaining to transverse optical (TO) vibrations, and the LO phonons have a higher frequency than the TO phonons for small $\mathbf{q}$. These considerations are important in determining how the free carriers lose their energy to the phonons when they are pumped up by optical fields. It is the optical phonons with energies in the order of 30 meV which dominate the energy relaxation of 'hot' carrier distributions (Butcher, 1986).

Phonons are bosons. That is to say, for a noninteracting phonon system, the wave function which describes all the harmonic atomic vibrations is a symmetrised combination of product wave functions for the individual normal modes. There is therefore no restriction on the number of times a wave function for a particular normal mode may appear. Thus the number of phonons of a particular type present can take any nonnegative integer value. The thermal average number of phonons with wave vector $\mathbf{q}$ in a particular branch s of the phonon dispersion relation depends only on the frequency $\omega_s(\mathbf{q})$ of the phonon and the temperature T, and is given by the Planck distribution function:

$$n_{\text{ph}}\left(\omega_s(\mathbf{q})\right) = \{\exp[\hbar\omega_s(\mathbf{q})/kT] - 1\}^{-1} . \tag{8.7}$$

We note that the chemical potential vanishes for phonons because there is no limitation on their number, and consequently (8.6) has no analogue for phonons.

8.3 Linear-optical properties

8.3.1 The linear susceptibility

We emphasised in §4.6 that the susceptibility tensors for electrons in crystals are most conveniently expressed in terms of the matrix elements of the one-electron momentum operator $\mathbf{p} = -i\hbar\nabla$. In the following analysis it is sometimes convenient to use the Dirac bra–ket notation of (3.7a) for these matrix elements. We may easily verify that $\langle n\mathbf{k}|\mathbf{p}|n'\mathbf{k}'\rangle = 0$ unless $\mathbf{k}'=\mathbf{k}$, by Fourier expanding the two Bloch functions which enter into the integral expressions for the matrix element. When $\mathbf{k}'\neq\mathbf{k}$ the integral is a sum of integrals of plane waves with non-zero wave vectors. These all vanish because of the periodic boundary conditions satisfied by each plane wave. Hence we may write

$$\langle n\mathbf{k}|\mathbf{p}|n'\mathbf{k}\rangle = \mathbf{p}_{nn'}(\mathbf{k})\,\delta_{\mathbf{k}\mathbf{k}'} , \tag{8.8a}$$

where the Knonecker delta $\delta_{\mathbf{k}\mathbf{k}'}$ is defined by (3.4), and

$$\mathbf{p}_{nn'}(\mathbf{k}) = \int_V d\mathbf{r}\, \psi_{n\mathbf{k}}{}^*(\mathbf{r})\,(-i\hbar\nabla)\,\psi_{n'\mathbf{k}}(\mathbf{r}) \tag{8.8b}$$

in which V is the crystal volume and the integrand is periodic.

The linear-optical properties of the semiconductor are determined by $\chi^{(1)}_{\mu\alpha}(-\omega;\omega)$ which is given by (4.136) with $n=1$. In the present context we write $n\mathbf{k}$ for the state label a and $n'\mathbf{k}'$ for the state label b. Thus we obtain

$$\begin{aligned}\chi^{(1)}_{\mu\alpha}(-\omega;\omega) &= -\frac{N_0^{(e)}e^2}{\varepsilon_0 m\omega^2} I_{\mu\alpha} \\ &+ \frac{1}{\varepsilon_0 V}\frac{2e^2}{m^2\omega^2}\sum_{nn'}\sum_{\mathbf{k}} f_0[\mathrm{E}_n(\mathbf{k})] \\ &\times\left[\frac{p^{\mu}_{nn'}(\mathbf{k})\,p^{\alpha}_{n'n}(\mathbf{k})}{\mathrm{E}_{n'n}(\mathbf{k}) - \hbar\omega - i\hbar\Gamma_{n'n}(\mathbf{k})} + \frac{p^{\alpha}_{nn'}(\mathbf{k})\,p^{\mu}_{n'n}(\mathbf{k})}{\mathrm{E}_{n'n}(\mathbf{k}) + \hbar\omega + i\hbar\Gamma_{n'n}(\mathbf{k})}\right],\end{aligned} \tag{8.9}$$

where $N_0^{(e)}$ is the total electron density in the conduction and valence bands in thermal equilibrium. (Lower-energy bands contribute only to the bound-electron part of $\chi^{(1)}_{\mu\alpha}(-\omega;\omega)$ which does not concern us here.) The factor of 2 in the second line of (8.9) allows for the spin degeneracy. We have used the notation

$$\mathrm{E}_{n'n}(\mathbf{k}) = \mathrm{E}_{n'}(\mathbf{k}) - \mathrm{E}_{n}(\mathbf{k}) \tag{8.10}$$

for the energy difference of a 'vertical' transition at wave vector $\mathbf{k}$ from band n to band n'. We have also incorporated the real damping parameter $\Gamma_{n'n}(\mathbf{k}) = \Gamma_{nn'}(\mathbf{k})$ into the resonant denominators in accordance with the prescription given in §4.5.1. The invariance of $\Gamma_{n'n}(\mathbf{k})$ under an interchange of band indices ensures that the resonant frequencies are not split by the damping.

All the crystals with which we are concerned have cubic rotational symmetry and are therefore isotropic in their linear-optical properties (see Chapter 5). We may therefore simplify (8.9) by setting $\alpha = \mu$ and leaving μ understood and arbitrary. Then, writing $\chi^{(1)}_{\mu\mu}(-\omega;\omega) = \chi^{(1)}(-\omega;\omega)$ and $p^{\mu}_{nn'}(\mathbf{k}) = p_{nn'}(\mathbf{k})$, we may combine together the two resonant terms in (8.9) by interchanging n and n' in the second one. Thus we obtain

$$\chi^{(1)}(-\omega;\omega) = -\frac{N_0^{(e)}e^2}{\varepsilon_0 m\omega^2} + \frac{2e^2}{\varepsilon_0 V m^2 \omega^2} \times \sum_{nn'}\sum_{\mathbf{k}} \{f_0[\mathrm{E}_n(\mathbf{k})] - f_0[\mathrm{E}_{n'}(\mathbf{k})]\} \frac{|p_{nn'}(\mathbf{k})|^2}{\mathrm{E}_{n'n}(\mathbf{k}) - \hbar\omega - i\hbar\Gamma_{n'n}(\mathbf{k})}, \tag{8.11}$$

where we have exploited the Hermitian property of $\mathbf{p}$ (see §3.2.3).

8.3.2 Intraband free-carrier effects

It is instructive to consider the case of completely free electrons. Then the summations over band indices in (8.11) are irrelevant because there is only one band. Consequently the difference of Fermi functions is identically zero and the second term in (8.11) vanishes. We see immediately that the dielectric constant $\varepsilon(\omega) = 1 + \chi^{(1)}(-\omega;\omega)$ takes the familiar form $1 - \omega_p^2/\omega^2$, where $\omega_p = \sqrt{N_0^{(e)}e^2/\varepsilon_0 m}$ is the plasma frequency of the free-electron gas (Kittel, 1986). We may go a little further in this direction by noting that one ω in the denominator of the first term of (8.11) has the character of a resonant denominator with zero resonance frequency. We should therefore follow our usual practice (see §4.5.1) and augment this ω by adding on i/τ where $1/\tau$ is a damping factor of the same type as we have introduced in all the optical resonant denominators. The other ω remains unaltered because it simply reflects the relationship (4.145) between conductivities and susceptibilities. Thus we find, from (4.145) with $n=1$, that the linear conductivity is given by

$$\begin{aligned}\sigma^{(1)}(-\omega;\omega) &= -i\omega\varepsilon_0\,\chi^{(1)}(-\omega;\omega)\\ &= \frac{iN_0^{(e)}e^2}{m(\omega+i/\tau)}\\ &= \sigma_0/(1-i\omega\tau)\,,\end{aligned} \tag{8.12}$$

where $\sigma_0=N_0^{(e)}e^2\tau/m$ is the d.c. conductivity of a free-electron gas with a mean free time τ between collisions. We recognise (8.12) as the Drude formula which provides a reasonably good description of the linear-optical behaviour of good metals at infrared frequencies (Kittel, 1986).

Equation (8.12) also provides a reasonably good description of the infrared behaviour of free carriers near an extremum of either the conduction or valence band of a semiconductor, provided that we replace the bare electron mass by the electron or hole effective mass. For anisotropic bands an appropriate average mass must be used, and when there are free carriers in more than one band (*e.g.*, electrons in the conduction band and holes in the valence band) we must sum their contributions to the free-carrier part of $\sigma^{(1)}(-\omega;\omega)$. These results are obvious if one accepts the conceptual significance of the effective mass in the physics of semiconductors. They may also be derived directly from (8.11) by making $\omega\to 0$ and using the formulae for the effective masses in terms of momentum matrix elements and band separations (Butcher and McLean, 1963).

We consider in more detail an n-type semiconductor in which $\sigma^{(1)}(-\omega;\omega)$ is dominated by the contribution from the electrons in the conduction band. When the collision time τ is much larger than the optical period, so that $\omega\tau>>1$, we see that (8.12) reduces to

$$\sigma^{(1)}(-\omega;\omega)\simeq\sigma_0\,(1+i\omega\tau)/(\omega\tau)^2\,. \tag{8.13}$$

The complex dielectric constant of the material is $\varepsilon=n_0^2-\sigma^{(1)}(-\omega;\omega)/i\omega\varepsilon_0$, where n_0^2 is the square of the refractive index in the absence of free carriers (*i.e.*, the background dielectric constant due to the bound electrons). We find immediately that the complex refractive index $\sqrt{\varepsilon}=n+i\kappa$ in the presence of the free electrons has real and imaginary parts:

$$n\simeq n_0-n_0(\omega_{pe}/\omega)^2/2 \tag{8.14a}$$

and

$$\kappa\simeq n_0(\omega_{pe}/\omega)^2/2\omega\tau\,, \tag{8.14b}$$

where $\omega_{pe}=\sqrt{N_0^{(e)}e^2/m_e n_0^2\varepsilon_0}$ is the plasma frequency. The intensity absorption coefficient is

$$\alpha = 2\kappa\omega/c$$
$$= n_0(\omega_{pe}/\omega)^2/c\tau\,. \tag{8.15}$$

The absorption process described by (8.15) is usually referred to as free-carrier absorption. The equation exhibits its well-known ω^{-2}-dependence on frequency and inverse dependence on τ (Kittel, 1986). The change of refractive index due to the free carriers is given by $n-n_0$ from (8.14a). It also has an ω^{-2}-dependence on frequency. However, it is independent of τ when $\omega\tau >> 1$ because $n-n_0$ simply reflects the behaviour of the plasma dielectric constant $n_0^2(1-\omega_{pe}^2/\omega^2)$ in the limit $\tau\rightarrow\infty$.

8.3.3 Interband absorption in the one-electron approximation

At optical frequencies our main concern is with the interband terms appearing in the sum in (8.11). When $\hbar\omega$ is resonant with the energy difference between a conduction band state with energy $E_c(\mathbf{k})$ and a valence band state with energy $E_v(\mathbf{k})$ the contribution of the transition between them to $\chi^{(1)}(-\omega;\omega)$ behaves like an unsaturated resonance of a two-level system. Indeed, what we are currently concerned with *is* an assembly of two-level systems which are distinguished from one another by the wave vector **k**. In particular, we see that the resonant contribution to the susceptibility is proportional to the population (density) difference in thermal equilibrium: $N_0^{(e)}\{f_0[E_v(\mathbf{k})]-f_0[E_c(\mathbf{k})]\}$. This observation provides the key to understanding dynamic optical nonlinearities in semiconductors. The population differences may be altered drastically by intense optical fields, as we discuss in detail in §8.4. Our immediate task, however, is to complete the treatment of interband absorption in the linear regime.

We see from our earlier discussion of free-carrier absorption that the intensity absorption coefficient may be expressed in two ways:

$$\alpha = \frac{\omega}{n_0 c}\mathrm{Im}\,\chi^{(1)}(-\omega;\omega) \tag{8.16a}$$

$$= \frac{1}{n_0\varepsilon_0 c}\mathrm{Re}\,\sigma^{(1)}(-\omega;\omega)\,, \tag{8.16b}$$

where $\omega>0$ and n_0 is the background refractive index due to bound electrons. To evaluate $\mathrm{Im}\,\chi^{(1)}(-\omega;\omega)$ from (8.11) we confine our attention to the standard semiconductor band structure shown in Fig. 8.2(a) and to values of $\hbar\omega$ just above the forbidden energy gap. Then the resonant term in the double sum in (8.11) when $\omega>0$ is the one with $n'=c$ and $n=v$. We ignore all other contributions and write $f_0[E_c(\mathbf{k})]=0$ and

$f_0[E_v(\mathbf{k})]=1$, which are usually adequate approximations at room temperature and below. The **k**-dependence of both $p_{cv}(\mathbf{k})$ and $\Gamma_{cv}(\mathbf{k})$ is not important. We set $\mathbf{k}=0$ in both quantities and write $p_{cv}(0)=p_{cv}$ and $\Gamma_{cv}(0)=\Gamma_{cv}$. Then we have

$$\alpha = \frac{\omega}{n_0 c}\frac{2e^2|p_{cv}|^2}{\varepsilon_0 V m^2\omega^2}\sum_{\mathbf{k}}\frac{\hbar\Gamma_{cv}}{(E_g+\hbar^2k^2/2m_r-\hbar\omega)^2+(\hbar\Gamma_{cv})^2}, \quad (8.17)$$

where k denotes $|\mathbf{k}|$, and $m_r=m_e m_h/(m_e+m_h)$ is the reduced mass of the electrons and holes.

To evaluate (8.17) we let $V\rightarrow\infty$ and transform the sum to an integral, as indicated in (8.3a). In most semiconductors the damping parameter Γ_{cv} is small enough for us to take the limit $\Gamma_{cv}\rightarrow 0$ in (8.17). In that case we may write

$$\lim_{\Gamma_{cv}\rightarrow 0}\frac{\hbar\Gamma_{cv}}{(E_g+\hbar^2k^2/2m_r-\hbar\omega)^2+(\hbar\Gamma_{cv})^2} = \pi\,\delta(E_g+\hbar^2k^2/2m_r-\hbar\omega). \quad (8.18)$$

Consequently, the sum in (8.17) may be evaluated in the same way as we evaluated $N_c(E)$ in (8.3b), to yield the final result:

$$\alpha = \frac{\pi e^2|p_{cv}|^2}{n_0\varepsilon_0 c m^2\omega}\int_{E_g}^{\infty} dE_t\, N_{cv}(E_t)\,\delta(E_t-\hbar\omega) \quad (8.19a)$$

$$= \frac{\pi e^2|p_{cv}|^2}{n_0\varepsilon_0 c m^2\omega}N_{cv}(\hbar\omega), \quad (8.19b)$$

where $E_t=E_g+\hbar^2k^2/2m_r$ is the energy of the interband transition and, when $E_t>E_g$,

$$N_{cv}(E_t) = \frac{1}{2\pi^2}\sqrt{(2m_r/\hbar^2)^3\,(E_t-E_g)} \quad (8.20)$$

is the 'joint' density of states (per unit energy range per unit volume) for the valence and conduction bands with spin degeneracy included. The proportionality of α to $\sqrt{\hbar\omega-E_g}$ is the characteristic feature of interband absorption in direct gap semiconductors in the one-electron approximation. In the next section we show that the continuous absorption actually starts with a step when Coulomb interaction is taken into account.

The evaluation of the contribution to the refractive index due to transitions between the valence and conduction bands requires better approximations than we have used to obtain α, if a divergence is to be avoided in the relevant integral (see §8.4.3). Nevertheless, we see immediately by inspection of (8.11) that the resonant contribution to $\mathrm{Re}\chi^{(1)}(-\omega;\omega)$ is positive when $\hbar\omega<E_g$ so that $\alpha=0$. As is to be expected,

this result agrees with what we found in the discussion of a resonant two-level system given in Chapter 6. It means that, when $\hbar\omega < E_g$, the effect of the interband transitions is to increase the refractive index. Thus they act in the opposite direction to the *intraband* transitions (*i.e.,* free-carrier effects) which we discussed in §8.3.2. This is because the resonance frequency of the intraband transitions is zero.

When $\hbar\omega < E_g$ the absorption coefficient vanishes in our model. However, α may in fact be positive in this frequency range for a variety of reasons: level-broadening, combined photon and phonon absorption, and Coulomb interaction between electrons and holes. Coulomb interaction produces very significant modifications of (8.19), which we treat in detail in the next section. Here we look briefly at 'indirect' absorption which occurs in Si and Ge because the minima in the conduction band are located away from $\mathbf{k}=0$. In consequence they cannot be reached from the valence band maximum by vertical transitions. In the absence of electron-phonon interactions, α would therefore remain zero as the photon energy $\hbar\omega$ is increased until the energy for which vertical transitions at $\mathbf{k}=0$ becomes possible. The linear-optical absorption in indirect gap semiconductors does rise steeply at this point, but there is a weak absorption tail at lower photon energies which involves the electron-phonon interaction. The wave vectors of the phonons involved must be close to the separation of the band extrema so as to secure overall momentum conservation. Consequently, the phonon frequency is fixed at a value which we denote by ω_{ph}.

To calculate the contribution to α involving phonon absorption we have to integrate the absorption due to the excitation of electrons with an initial energy E in the valence band to a final energy $E+\hbar\omega+\hbar\omega_{ph}$ in the conduction band, for all values of E between $-E_g$ (the top of the valence band) and a minimum value $-\hbar\omega-\hbar\omega_{ph}$ below which energy conservation becomes impossible.

The integrand is the product of the density of states at these two energies (Smith, 1959). Hence this contribution to α is proportional to the integral

$$I = \int_{-\hbar\omega-\hbar\omega_{ph}}^{-E_g} dE \sqrt{-E-E_g}\sqrt{E+\hbar\omega+\hbar\omega_{ph}}$$

$$= (\hbar\omega+\hbar\omega_{ph}-E_g)^2 \int_0^1 dx \sqrt{x(1-x)}\,, \tag{8.21a}$$

where we have made the change of variable

$$x = -(E+E_g)/(\hbar\omega+\hbar\omega_{ph}-E_g)\,. \tag{8.21b}$$

The integral which remains in (8.21a) is independent of ω. Hence I and therefore the contribution to α involving phonon absorption are proportional to $(\hbar\omega+\hbar\omega_{ph}-E_g)^2$ times the Planck distribution n_{ph} given by (8.7) with $\omega_s(q)=\omega_{ph}$. A similar calculation gives the contribution to α involving phonon emission, which is proportional to $(\hbar\omega-\hbar\omega_{ph}-E_g)^2$ times $n_{ph}+1$. Consequently, indirect absorption edges are much more temperature dependent than the direct ones.

8.3.4 The exciton line spectrum

Throughout the analysis given in §8.3.3 we ignored the Coulomb interaction between an electron excited into the conduction band and the hole which it leaves behind in the valence band. The absorption edge in semiconductors is considerably modified when this interaction is taken into account. In particular, the electron and hole can form bound 'exciton' states which produce a line spectrum in α when $\hbar\omega<E_g$.

To investigate the exciton line spectrum we follow Elliott (1957) and extend the envelope-function formalism given in Appendix 6 to handle an exciton wave function $\psi(\mathbf{r}_e,\mathbf{r}_h)$ involving electron and hole position vectors $\mathbf{r}_e$ and $\mathbf{r}_h$. We confine our attention to a standard semiconductor band structure (see §8.1) in which the electron and hole effective masses are m_e and m_h, respectively. Then, in the spirit of the envelope-function approximation, we write

$$\psi(\mathbf{r}_e,\mathbf{r}_h) = \sum_{\mathbf{k}\mathbf{k}'} a_{\mathbf{k}\mathbf{k}'}\,\psi_{c\mathbf{k}}(\mathbf{r}_e)\,\psi_{v\mathbf{k}'}(\mathbf{r}_h)\,, \tag{8.22}$$

where $\mathbf{k}$ labels the normalised conduction-band Bloch functions near $\mathbf{k}=0$, and $\mathbf{k}'$ labels the normalised valence-band Bloch functions near $\mathbf{k}'=0$. The envelope function $\phi(\mathbf{r}_e,\mathbf{r}_h)$ is defined by

$$\phi(\mathbf{r}_e,\mathbf{r}_h) = \sum_{\mathbf{k}\mathbf{k}'} a_{\mathbf{k}\mathbf{k}'}\exp[i(\mathbf{k}\cdot\mathbf{r}_e + \mathbf{k}'\cdot\mathbf{r}_h)] \tag{8.23}$$

and $a_{\mathbf{k}\mathbf{k}'}$ (which determines $\psi(\mathbf{r}_e,\mathbf{r}_h)$ through (8.22)) is given by V^{-2} times the double Fourier transform of $\phi(\mathbf{r}_e,\mathbf{r}_h)$:

$$a_{\mathbf{k}\mathbf{k}'} = V^{-2}\int_V d\mathbf{r}_e \int_V d\mathbf{r}_h\, \phi(\mathbf{r}_e,\mathbf{r}_h)\exp[-i(\mathbf{k}\cdot\mathbf{r}_e + \mathbf{k}'\cdot\mathbf{r}_h)]\,. \tag{8.24a}$$

When $\psi(\mathbf{r}_e,\mathbf{r}_h)$ is normalised to unity we may show easily that $\phi(\mathbf{r}_e,\mathbf{r}_h)$ is normalised to V^2, *i.e.*,

$$\int_V d\mathbf{r}_e \int_V d\mathbf{r}_h\, |\phi(\mathbf{r}_e,\mathbf{r}_h)|^2 = V^2 \tag{8.24b}$$

(see the remark at the end of Appendix 6). These equations are the analogues of (A6.1)−(A6.3) and (A6.11) for an electron-hole pair. To complete the formalism we suppose that the Coulomb interaction

$$V_C(\mathbf{r}_e - \mathbf{r}_h) = -\frac{e^2}{4\pi\varepsilon_0\varepsilon|\mathbf{r}_e - \mathbf{r}_h|} \tag{8.25}$$

varies slowly over a unit cell in the crystal. Then an approximate equation for $\phi(\mathbf{r}_e, \mathbf{r}_h)$ can be readily obtained by simple extensions of the arguments given in Appendix 6; it is

$$\left[-\frac{\hbar^2}{2m_e}\nabla_e^2 - \frac{\hbar^2}{2m_h}\nabla_h^2 + V_C(\mathbf{r}_e - \mathbf{r}_h)\right]\phi(\mathbf{r}_e, \mathbf{r}_h) = (\mathrm{E} - \mathrm{E}_g)\phi(\mathbf{r}_e, \mathbf{r}_h). \tag{8.26}$$

We recognise (8.26) as the Schrödinger equation for a hydrogen atom with the bare electron mass replaced by m_e and the bare proton mass replaced by m_h. This equation is consequently easily solved by transforming to centre-of-mass and relative coordinates $\mathbf{R}$ and $\mathbf{r}$:

$$\begin{aligned} \mathbf{R} &= (m_e\mathbf{r}_e + m_h\mathbf{r}_h)/m_t \\ \mathbf{r} &= \mathbf{r}_e - \mathbf{r}_h, \end{aligned} \tag{8.27}$$

where m_t is the total mass $m_e + m_h$. Then $\phi(\mathbf{R}, \mathbf{r})$ and E are labelled by the wave vector $\mathbf{K}$ associated with the motion of the centre of mass and the hydrogenic quantum numbers n, l and m_l associated with the relative motion. Thus we find that

$$\phi_{\mathbf{K}nlm_l}(\mathbf{R}, \mathbf{r}) = \sqrt{V}\,\exp(i\mathbf{K}\cdot\mathbf{R})\,\xi_{nlm_l}(\mathbf{r}) \tag{8.28}$$

which is normalised in accordance with (8.24b), and

$$\mathrm{E}_{nlm_l}(\mathbf{K}) = \mathrm{E}_g + \hbar^2K^2/2m_t - \mathrm{E}_{\mathrm{Ry}}^*/n^2, \tag{8.29}$$

where

$$\mathrm{E}_{\mathrm{Ry}}^* = \hbar^2/2m_r(a_0^*)^2 \tag{8.30a}$$

is the effective Rydberg. In (8.28), $\xi_{nlm_l}(\mathbf{r})$ denotes the wave function of the hydrogen orbital state with principal quantum number n, total orbital angular momentum quantum number l, and azimuthal quantum number m_l. It is normalised to unity. In (8.29) $m_r = m_e m_h/m_t$ is the reduced mass of the electron and hole, and

$$a_0^* = 4\pi\hbar^2\varepsilon_0\varepsilon/e^2m_r \tag{8.30b}$$

is the radius of the Bohr orbit associated with the ground state of the relative motion with $n = 1$, $l = 0$ and $m_l = 0$. This is the lowest s-state; we shall see that the s-states are particularly significant for exciton absorption. We find by inspection of the normalised s-states given in many textbooks (*e.g.*, Condon and Shortley (1957)) that

$$|\xi_{n00}(0)|^2 = [\pi(a_0^*)^3n^3]^{-1}. \tag{8.31}$$

We make use of this result in what follows.

We made the point in §8.3.1 that, in the cases of interest to us, the operator which controls interband absorption is the one-electron momentum component p in any chosen direction. The matrix element of p between the ground state (in which the valence band is full and the conduction band is empty) and the pair state $\psi(\mathbf{r}_e, \mathbf{r}_h)$ (in which there is one hole in the valence band and one electron in the conduction band) is (Elliott, 1957)

$$\begin{aligned} <\psi|p|0> &= \sum_{\mathbf{k}\mathbf{k}'} a_{\mathbf{k}\mathbf{k}'} \int d\mathbf{r}_e \, \psi_{c\mathbf{k}}(\mathbf{r}_e) p \, \psi_{v\mathbf{k}'}(\mathbf{r}_e) \\ &= \sum_{\mathbf{k}\mathbf{k}'} a_{\mathbf{k}\mathbf{k}'} \, p_{cv}(-\mathbf{k}) \, \delta_{\mathbf{k}',-\mathbf{k}} \\ &\simeq p_{cv} \sum a_{\mathbf{k},-\mathbf{k}} \, . \end{aligned} \qquad (8.32)$$

In the second line we have used (8.8) and the identity (8.4a). In the third line we have written $p_{cv}(-\mathbf{k}) \simeq p_{cv}(0) \equiv p_{cv}$ and have thus neglected the **k**-dependence of this matrix element. To evaluate the sum in (8.32) we go back to (8.23), set $\mathbf{r}_h = \mathbf{r}_e$ and integrate over $\mathbf{r}_e$. The result is

$$\begin{aligned} \sum_{\mathbf{k}} a_{\mathbf{k},-\mathbf{k}} &= V^{-1} \int_V d\mathbf{r}_e \, \phi(\mathbf{r}_e, \mathbf{r}_e) \\ &= \sqrt{V} \, \xi_{nlm_l}(0) \, \delta_{\mathbf{K},0} \, , \end{aligned} \qquad (8.33)$$

where we have used (8.28). Thus we have, finally,

$$<\psi|p|0> = p_{cv} \sqrt{V} \, \xi_{nlm_l}(0) \, \delta_{\mathbf{K},0} \, . \qquad (8.34)$$

We see that the matrix element vanishes unless $\mathbf{K}=0$. This is because we are working in the electric-dipole approximation. The photon momentum is taken to be zero and momentum conservation ensures that a stationary exciton is produced when a photon is annihilated. We also see that the matrix element vanishes unless the hydrogen function considered is an s-function because all other hydrogen functions vanish at the origin (Condon and Shortley, 1957).

To evaluate the exciton line spectrum we use the formula (8.16b) for α and note that $\mathrm{Re}\,\sigma^{(1)}(-\omega;\omega)$ may be expressed in the form

$$\mathrm{Re}\,\sigma^{(1)}(-\omega;\omega) = 2\,W/|\mathbf{E}_\omega|^2 \, , \qquad (8.35)$$

where W is the power absorbed per unit volume of the crystal when it is subject to a uniform simple harmonic electric field of the form

$$\mathbf{E}(t) = \tfrac{1}{2}\left[\mathbf{E}_\omega \exp(-i\omega t) + \mathbf{E}_{-\omega} \exp(i\omega t)\right] , \qquad (8.36)$$

with $\mathbf{E}_{-\omega} = \mathbf{E}_\omega^*$ (*cf.* §2.4.1). Equation (8.35) is simply a rearrangement of the familiar formula for the power dissipation density in such a field. It is useful because we may readily express W in terms of the matrix element of **p** in the direction of $\mathbf{E}_\omega$ by using Fermi's Golden Rule, as follows.

We see from (4.147) that the linear perturbation of the one-electron Hamiltonian is

$$\begin{aligned} H_\mathrm{I} &= \frac{e}{m}\mathbf{A}(t)\cdot\mathbf{p} \\ &= \frac{e}{m}\frac{1}{2i\omega}[\mathbf{E}_\omega\cdot\mathbf{p}\exp(-i\omega t) - \mathbf{E}_{-\omega}\cdot\mathbf{p}\exp(i\omega t)], \end{aligned} \tag{8.37}$$

where we have used the first line of (4.146) to evaluate the vector potential from the electric field given in (8.36). The first term in (8.37) is responsible for photon absorption and, at absolute-zero temperature, we have

$$W = \frac{\hbar\omega}{V}\sum_\psi 2\pi/\hbar\left[\frac{e^2}{(2m\omega)^2}|\mathbf{E}_\omega|^2\,|<\psi|p|0>|^2\right]\delta(\hbar\omega-\mathrm{E}_\psi), \tag{8.38}$$

where the summation is over all excited pair states ψ with energy E_ψ. The summand in (8.38) is just Fermi's Golden Rule formula for the transition rate from the ground state to ψ in the presence of the perturbation H_I. When (8.38) is substituted into (8.35) we find, with the aid of (8.29) and (8.34), that (8.16b) gives

$$\alpha = \frac{\pi}{n_0\varepsilon_0\omega c}(e/m)^2\,|p_{\mathrm{cv}}|\,2\sum_n|\xi_{n00}(0)|^2\,\delta(\hbar\omega-\mathrm{E_g}+\mathrm{E}^*_{\mathrm{Ry}}/n^2), \tag{8.39}$$

since only s-states with $\mathbf{K}=0$ contribute to the sum. The factor of 2 in front of the summation sign makes allowance for the spin degeneracy.

We see from (8.31) that the absorption spectrum is the hydrogen s-state line spectrum with the line at $\hbar\omega=\mathrm{E_g}-\mathrm{E}^*_{\mathrm{Ry}}/\hbar^2$ weighted by a factor n^{-3}. As $\hbar\omega$ approaches the series limit $\mathrm{E_g}$, the values of n involved in (8.39) increase indefinitely. We may therefore evaluate α when $\hbar\omega\to\mathrm{E_g}$ by treating n as a continuous variable and replacing the sum by an integral. With the aid of (8.31) we find that

$$\begin{aligned} \lim_{\hbar\omega\to\mathrm{E_g}}\alpha &= \lim_{\hbar\omega\to\mathrm{E_g}}\frac{2(e/m)^2}{n_0\varepsilon_0\omega c(a_0^*)^3}|p_{\mathrm{cv}}|^2\int_{\hbar\omega-\mathrm{E_g}}^{\hbar\omega-\mathrm{E_g}+\mathrm{E}^*_{\mathrm{Ry}}}\mathrm{d}x\,\frac{\delta(x)}{\mathrm{E}^*_{\mathrm{Ry}}} \\ &= \frac{(e/m)^2}{n_0\varepsilon_0(\mathrm{E_g}/\hbar)c(a_0^*)^3\mathrm{E}^*_{\mathrm{Ry}}}|p_{\mathrm{cv}}|^2, \end{aligned} \tag{8.40}$$

where the integration variable $x=\hbar\omega-\mathrm{E_g}+\mathrm{E}^*_{\mathrm{Ry}}/n^2$.

Since α approaches the finite value (8.40) when $\hbar\omega\to\mathrm{E_g}$ from below, it is to be expected that the continuous absorption spectrum which lies in the frequency range $\hbar\omega>\mathrm{E_g}$ starts with a step. This is in complete contrast with the square-root behaviour exhibited in (8.20) for the case when Coulomb interaction is ignored. This expectation is confirmed by extending the sum over bound exciton states in (8.39) to include the contributions due to transitions to ionised exciton states. Finally, when $\hbar\omega >> E_g$, we may ignore the Coulomb interaction in (8.26). Then we see by inspection that the normalised envelope eigenfunctions are most conveniently labelled by electron and hole wave vectors $\mathbf{k}_e$ and $\mathbf{k}_h$, so that

$$\phi_{\mathbf{k}_e\mathbf{k}_h}(\mathbf{r}_e, \mathbf{r}_h) = \exp[i(\mathbf{k}_e\cdot\mathbf{r}_e + \mathbf{k}_h\cdot\mathbf{r}_h)], \tag{8.41}$$

which has the energy eigenvalue

$$E(\mathbf{k}_e, \mathbf{k}_h) = E_g + \frac{\hbar^2}{2m_e}\mathbf{k}_e^2 + \frac{\hbar^2}{2m_h}\mathbf{k}_h^2 . \tag{8.42}$$

Hence, it follows from (8.24) that $a_{\mathbf{k}\mathbf{k}'} = \delta_{\mathbf{k}\mathbf{k}_e}\delta_{\mathbf{k}'\mathbf{k}_h}$, so that for these pair wave functions we find, from (8.32) and the first line of (8.33), that

$$<\psi_{\mathbf{k}_e\mathbf{k}_h}|p|0> = p_{cv}\,\delta_{\mathbf{k}_e,-\mathbf{k}_h} . \tag{8.43}$$

When this result is substituted into (8.38) we find, with the aid of (8.35) and allowing for spin degeneracy, that (8.16b) reduces to our earlier result (8.19). We trust this happy outcome increases the reader's confidence in the pair-state formalism given here.

We have concentrated throughout this exciton analysis on a standard semiconductor band structure involving direct transitions between spherically symmetrical nondegenerate energy bands. Indirect transitions are treated by Elliott (1957), and Dresselhaus (1956) discusses exciton states for degenerate energy bands. Both topics are outside the scope of this book. We have also confined our attention to absorption from the ground state. At temperature T we may make a crude correction for the variation of the one-electron state occupation probabilities by multiplying α for $T=0$ by the difference of the Fermi functions at the band edges.

8.4 Nonlinear-optical properties

We are concerned with the nonlinearities which occur when intense optical radiation pumps carriers from the valence band to the conduction band. In thermal equilibrium the occupancy of a state $\mathbf{k}$ is determined by the Fermi-Dirac function (8.5). This function involves two parameters, T and E_F, which are the same for all bands. In the presence of optical pumps, the occupation probability of a state in any particular band is usually thermalised by rapid (usually sub-picosecond) *intravalley* scattering processes. It retains the form of the Fermi-Dirac function but E_F is now different in different bands and is referred to as the quasi-Fermi level. In principle T may also differ from the lattice temperature but, in the context of nonlinear optics, it is usually assumed that the coupling between the carriers and the lattice is strong enough to prevent such 'hot carrier' effects (Miller *et al*, 1981a; Banyai and Koch, 1986; de Rougemont and Frey, 1988; Koch *et al*, 1988). They are, however, important for low frequency and d.c. fields (see, for example, Butcher (1986)).

The regime which we have just described is the one which concerns us throughout this section; it is often called the 'band-filling' regime. It describes the physical picture which is often the most appropriate for discussing optical nonlinearities in semiconductors. We note in passing that an essentially different regime may be envisaged: If there were *no* intravalley scattering processes then we would simply be concerned with an assembly of saturable, resonant, two-level systems distinguished from one another by the common wave vector of the states connected optically in the valence and conduction bands. This is sometimes called the 'state-filling' regime (Miller *et al* 1981a). State-filling effects may be observed when ultrashort (sub-picosecond) optical pulses are used to excite the semiconductor; in that case it may be possible to establish nonthermal (*i.e.*, non-Fermi) distributions of electrons and holes in the conduction and valence bands, and to observe them in such a short time that the carriers have not yet scattered out of the states into which they were injected (see, for example, Tang and Erskine (1983), Shah (1985), Oudar (1985), Lin *et al* (1987), Knox *et al* (1986)). State-filling effects may be discussed in terms of the two-level model presented in §6.2, and we do not consider these further here.

In §8.4.1 we describe an elementary calculation of the quasi-Fermi levels in the conduction and valence bands resulting from optical pumping. The results are used in §8.4.2 to discuss intraband free-carrier nonlinearities and in §8.4.3 to discuss nonlinearities arising from interband absorption. Finally, in §8.4.4, we discuss dynamic nonlinear effects in terms of saturated absorption involving the exciton line spectrum by following a semiphenomenological theory of Koch *et al* (1988) which allows for screening of the Coulomb interaction between the electrons and holes.

8.4.1 Calculation of the quasi-Fermi levels in the presence of an optical pump

We denote the quasi-Fermi levels in the conduction and valence bands by E_{Fc} and E_{Fv} respectively. To calculate them in intrinsic (*i.e.* undoped) material we need to know only the electron density $N^{(e)}$. The hole concentration $N^{(h)}$ in the valence band is equal to $N^{(e)}$ because of charge conservation. Hence we may use appropriate modifications of (8.6) which are:

$$N^{(e)} = \int_0^\infty \mathrm{dE}\, N_c(\mathrm{E})/\{\exp[(\mathrm{E}-\mathrm{E_{Fc}})/kT]+1\}\,, \tag{8.44a}$$

$$N^{(h)} = \int_{-\infty}^{-\mathrm{E}_g} \mathrm{dE}\, N_v(\mathrm{E})/\{\exp[(\mathrm{E_{Fv}}-\mathrm{E})/kT]+1\}$$

$$= N^{(e)}\,. \tag{8.44b}$$

The modifications of these equations which are necessary in doped semiconductors are easily incorporated.

The calculation of $N^{(e)}$ may be taken to various degrees of sophistication. Here we use a simple and commonly-used approach which assumes a monochromatic optical beam with intensity I and employs the elementary rate equation:

$$\mathrm{d}N^{(e)}/\mathrm{d}t = \alpha I/\hbar\omega - (N^{(e)}-N_0^{(e)})/\tau_R\,, \tag{8.45}$$

where τ_R is the electron-hole recombination time, $N_0^{(e)}$ is the value of $N^{(e)}$ in thermal equilibrium and α is the absorption coefficient which depends on I (Miller *et al,* 1981a,b). We have assumed uniform illumination so as to remove carrier-diffusion effects from (8.45). In the steady state, $\mathrm{d}N^{(e)}/\mathrm{d}t=0$ and (8.45) yields the simple result:

$$N^{(e)} = N_0^{(e)} + \tau_R\alpha I/\hbar\omega\,. \tag{8.46}$$

We may employ (8.44) and (8.46) to calculate the quasi-Fermi levels when α is known. At low intensities we may use the linear value calculated in §8.3. When α depends on I, as we discuss in subsequent sections of this chapter, (8.44) and (8.46) must be solved self-consistently. This is also necessary when τ_R depends on $N^{(e)}$.

8.4.2 Intraband free-carrier effects

We give only a brief discussion of optical nonlinearities due to intraband free-carrier effects because interband absorption and exciton effects are more important. For simplicity, we calculate the coefficient n_2^I which is introduced in Appendix 5 to describe the intensity dependence of the refractive index n in the presence of a weak monochromatic light beam. We see from Appendix 5 that n_2^I is the derivative of n with respect to I when $I\to 0$. In most semiconductors the effective mass of the electrons is considerably less than that of the holes. We may then concentrate our attention on the electrons in the conduction band because they dominate the intraband effects. To calculate n_2^I we have only to replace $N_0^{(e)}$ by $N^{(e)}$ from (8.46) in the square of the plasma frequency which appears in (8.14a). Then we find that n is linear in I when $I\to 0$, with the result:

$$n_2^I = -\frac{e^2\tau_R\alpha}{2m_e n_0\varepsilon_0\hbar\omega^3}, \tag{8.47}$$

where α is the absorption coefficient in the linear regime. By substituting typical values of the parameters in (8.47) ($m_e=0.1m$, $n_0=3.5$, $\alpha=10^6$ m^{-1}, $\hbar\omega=0.75$ eV and $\tau_R=10$ ns) we find that $n_2^I=2.9\times10^{-3}$ cm^2 kW^{-1} and, using (A5.13), the effective value of $\chi^{(3)}_{esu}\sim10^{-3}$.

8.4.3 Interband absorption effects

In the linear regime the absorption coefficient α is given by (8.19a) for a standard semiconductor band structure. It is convenient to write that equation down again here because we need to modify it in two ways in the nonlinear regime. Thus we have:

$$\alpha = \frac{\pi e^2|p_{cv}|^2}{n_0\varepsilon_0 cm^2\omega}\int_{E_g}^{\infty} dE_t\, N_{cv}(E_t)\,\delta(E_t-\hbar\omega), \tag{8.48}$$

where

$$E_t = E_g + \hbar^2k^2/2m_r \tag{8.49a}$$

is the energy increase in an interband transition between states with $|\mathbf{k}|=k$ and, for $E_t>E_g$,

$$N_{cv}(E_t) = \frac{1}{2\pi^2}\sqrt{(2m_r/\hbar^2)^3(E_t-E_g)} \tag{8.50}$$

is the joint density of states of the conduction and valence bands.

In deriving (8.48) we set $f_0[E_v(\mathbf{k})]-f_0[E_c(\mathbf{k})]=1$. The most important modification is to replace this factor of 1 in the integrand of (8.48) by

$$F(E_t) = f_v[E_v(\mathbf{k})]-f_c[E_c(\mathbf{k})], \tag{8.51}$$

where $f_v[E_v(\mathbf{k})]$ and $f_c[E_c(\mathbf{k})]$ are Fermi-Dirac functions of the form (8.5) with quasi-Fermi levels E_{Fv} and E_{Fc} respectively. As is indicated by the notation $F(E_t)$, we may express $E_v(\mathbf{k})=-E_g-\hbar^2k^2/2m_h$ and $E_c(\mathbf{k})=\hbar^2k^2/2m_e$ in terms of E_t by solving (8.49a) for k^2. Thus we find that

$$E_c(\mathbf{k}) = m_h(E_t-E_g)/m_t \tag{8.49b}$$

and

$$E_v(\mathbf{k}) = -E_g-m_e(E_t-E_g)/m_t, \tag{8.49c}$$

where $m_t=m_e+m_h$. The second important modification is to recognise that the δ-function in (8.48) is simply an approximation to the normalised line shape when the line width parameter $\Gamma_{cv}\to0$. Since Γ_{cv} plays an important rôle in the nonlinear regime, we must replace the δ-function by the Lorentzian line-shape function:

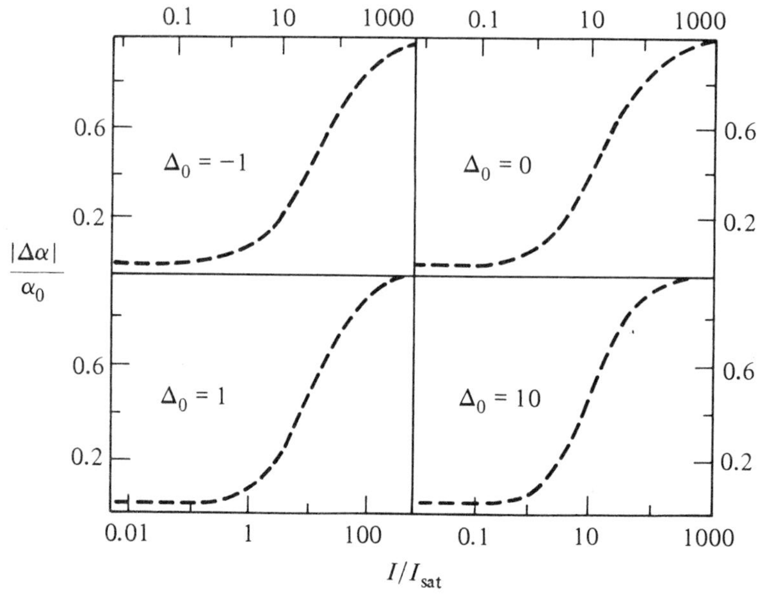

Fig. 8.3 Plots of $|\Delta\alpha|/\alpha_0$ (where $\Delta\alpha$ is the decrease in absorption coefficient α, and α_0 is the value of α in the linear regime) against I/I_{sat} for four different values of $\Delta_0 = (\hbar\omega - E_g)/\hbar\Gamma_{cv}$. (After de Rougemont and Frey, 1988.)

$$g(E_t - \hbar\omega) = \frac{\hbar\Gamma_{cv}/\pi}{(E_t - \hbar\omega)^2 + (\hbar\Gamma_{cv})^2} \tag{8.52}$$

from which it was derived in §8.4.3. With these modifications (8.48) becomes

$$\alpha = \frac{\pi e^2 |p_{cv}|^2}{n_0 \varepsilon_0 c m^2 \omega} \int_{E_g}^{\infty} dE_t\, N_{cv}(E_t)\, F(E_t)\, g(E_t - \hbar\omega)\,. \tag{8.53}$$

We may calculate α from (8.53) when E_{Fv} and E_{Fc} are known. These quasi-Fermi levels are determined in terms of α by (8.44) and (8.46). A self-consistent solution of these three equations is therefore necessary. It has been carried out numerically by several authors; for example, de Rougemont and Frey (1988) performed a calculation for a semiconductor at $T = 293$ K using the following typical values for the parameters: $E_g = 1.8$ eV, $m_e/m = 0.13$, $m_h/m = 0.61$ and $\hbar\Gamma_{cv} = 6.6$ meV, and using Fermi-Dirac statistics. Their calculation generalises the pioneering work of Miller *et al* (1981b) who used Boltzmann statistics. Some of the results calculated by de Rougemont and Frey (1988) are given in Fig. 8.3; the four curves each assume a different value of the detuning parameter $\Delta_0 = (\hbar\omega - E_g)/\hbar\Gamma_{cv}$. The curves show the variation of the magnitude of the change in α, denoted by $\Delta\alpha$, divided by its linear

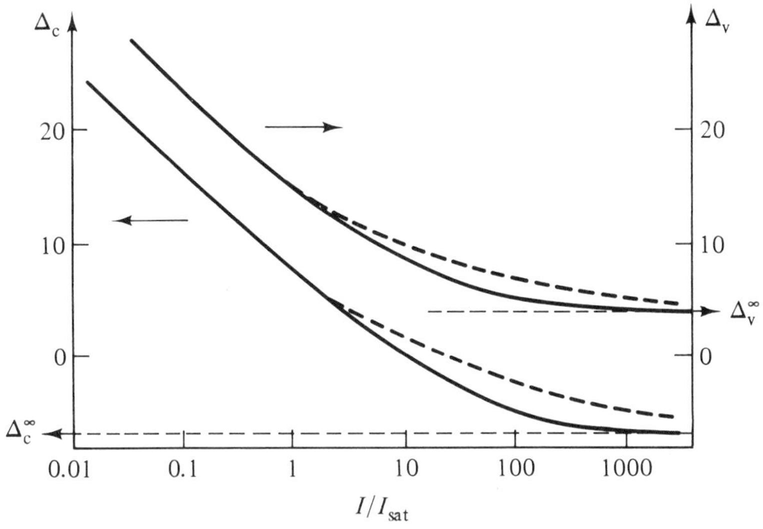

Fig. 8.4 Intensity-dependence of the quasi-Fermi levels in a typical semiconductor. The notation is $\Delta_c = 1 - \mathbb{E}_{Fc}/\hbar\Gamma_{cv}$ and $\Delta_v = (\mathbb{E}_g + \mathbb{E}_{Fv})/\hbar\Gamma_{cv}$ with Δ_c^∞ and Δ_v^∞ indicating the fully saturated values of these quantities. (After de Rougemont and Frey, 1988.)

value α_0 (*i.e.*, the value of α measured at vanishly small light intensities) as a function of I/I_{sat}, where I_{sat} is the saturation intensity on resonance which we introduced in the treatment of a two-level saturable absorber given in §6.3. It is a measure of the optical intensity required to produce marked saturation, and we discuss it in detail below. For the moment we simply note that, as is to be expected from §6.3.2, $|\Delta\alpha|/\alpha_0 \to 1$ when $I/I_{sat} \to \infty$ for all Δ_0. Moreover, the saturation behaviour occurs at larger values of I as $\hbar\omega$ (*i.e.*, Δ_0) is reduced because this reduces α. Further insight into the saturation of α for large values of I/I_{sat} is provided by the full lines in Fig. 8.4 (de Rougemont and Frey, 1988) which show the variation of $\Delta_c = \Delta_0 - \mathbb{E}_{Fc}/\hbar\Gamma_{cv}$ and $\Delta_v = (\mathbb{E}_g + \mathbb{E}_{Fv})/\hbar\Gamma_{cv}$ with I/I_{sat} when $\Delta_0 = 1$. When $I/I_{sat} \to 0$ both quasi-Fermi levels lie in the middle of the gap at the energy $-\mathbb{E}_g/2$. Hence $\Delta_v \simeq \Delta_c \simeq \mathbb{E}_g/2\hbar\Gamma_{cv} = 136$. When $I/I_{sat} \to \infty$, on the other hand, we see from Fig. 8.4 that $\Delta_v \to 4$ and $\Delta_c \to 7$, *i.e.*, $\mathbb{E}_{Fc} \to -\mathbb{E}_g + 4\hbar\Gamma_{cv}$ and $\mathbb{E}_{Fc} \to 8\hbar\Gamma_{cv}$. Thus the quasi-Fermi level associated with the conduction band actually moves into it, which is the reason for the 'band-filling' terminology used to describe the physics we are now discussing. The occupation of the states at the bottom of the conduction band is accompanied by an emptying of the states near the top of the valence band. However, for the parameters used here, the quasi-Fermi level associated with the valence band never moves right into it.

This is because the large heavy hole mass ($m_h = 0.61m$) implies a high density of states in the valence band, which means that E_{Fv} moves, with increasing carrier density, much more slowly than E_{Fc}. The band filling (and emptying) exhibited in Fig. 8.4 is responsible for the shift of the absorption edge to higher frequencies at high intensities; this is known as the 'Moss-Burnstein shift' (Miller *et al*, 1981a). The dashed lines in Fig. 8.4 show results obtained when allowance is made for Auger recombination (Smith, 1959) by replacing $(\tau_R)^{-1}$ by $(\tau_R)^{-1} + A(N^{(e)})^2$ in (8.45), with $A\tau_R = 1.5\times10^{-47}$ m^6. This additional recombination mechanism is significant only when $N^{(e)}$ is large, and it then delays somewhat the final approach to complete saturation.

There is a component of the total refractive index n which is associated with the interband absorption. We denote this component by n_{cv}, and write $n = n_0 + n_{cv}$ where, as previously in this chapter, n_0 denotes the background refractive index in the absence of interband absorption effects. In the band-filling regime, both n and α are determined by $\overline{\chi}_{cv}$ via the complex refractive index

$$\begin{aligned} n + i\kappa &= \sqrt{n_0^2 + \overline{\chi}_{cv}} \\ &\simeq n_0 + \overline{\chi}_{cv}/2n_0 , \end{aligned} \tag{8.54}$$

where $\overline{\chi}_{cv}$ denotes the contribution to the intensity-dependent susceptibility $\overline{\chi}(\omega;I)$ resulting from band-filling effects alone. The first line of (8.54) is derived from the previous expression (6.65a) for $n + i\kappa$ in terms of $\overline{\chi}(\omega;I)$ by assuming that band-filling effects make the dominant intensity-dependent contribution, *i.e.*, $1 + \overline{\chi} \simeq n_0^2 + \overline{\chi}_{cv}$. The second line in (8.54) follows from the reasonable assumption that $\overline{\chi}_{cv} << n_0^2$. Since $\alpha = 2\kappa\omega/c$ and $n_{cv} = n - n_0$, we see that $\alpha - i\,2\omega n_{cv}/c$ is given by $-i\omega\overline{\chi}_{cv}/n_0 c$. In the calculation of α leading to (8.53) we retained only the imaginary part of $\overline{\chi}_{cv}$. This is the origin of the line-shape factor $g(E_t - \hbar\omega)$ in the integrand of (8.53). To calculate $\alpha - i\,2\omega n_{cv}/c$ it is only necessary to replace $g(E_t - \hbar\omega)$ in (8.53) by $-i/\pi\,[(E_t - \hbar\omega) - i\hbar\Gamma_{cv}]$, which contains the full complex resonant denominator and correctly incorporates the real part of $\overline{\chi}_{cv}$. The resulting integral for n_{cv} is, however, divergent for this Lorentzian line-shape function. The fault lies in our inadequate treatment of the wings of the line. Many stochastic factors are involved in determining the shape of the wings with the result that they are Gaussian. To allow for this, de Rougemont and Frey (1988) multiplied the right-hand side of (8.52) by $\exp[-(E_t - \hbar\omega)^2/(20\hbar\Gamma_{cv})^2]$, because the Gaussian tails are estimated to have a line width in the order of 20 times the Lorentzian central peak. The results for Re $\overline{\chi}_{cv}$ are shown

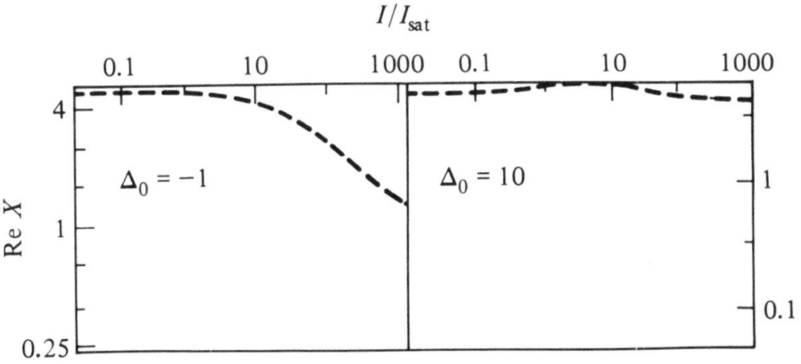

Fig. 8.5 Plots of $\mathrm{Re}X = \mathrm{Re}\,\overline{\chi}_{cv}\,\omega/\alpha_0 c n_0 = 2\omega n_{cv}/c\alpha_0$, where $\alpha_0(\omega)$ is the value of the interband absorption coefficient in the linear regime, plotted against I/I_{sat} for two values of the detuning parameter $\Delta_0 = (\hbar\omega - \mathrm{E_g})/\hbar\Gamma_{cv}$. (After de Rougemont and Frey, 1988.)

by the dashed lines in Fig. 8.5 for $\Delta_0 = -1$ and 10. The values shown are the real part of the function $\overline{\chi}_{cv}\,\omega/\alpha_0 c n_0 = 2n_{cv}\,\omega/c\alpha_0 + i\alpha/\alpha_0$, where $\alpha_0(\omega)$ is the low-intensity interband absorption coefficient.

It remains for us to discuss I_{sat} in detail and make some numerical estimates. In §6.3, I_{sat} is defined by (6.79). We are concerned with the on-resonance case, for which

$$(I_{sat})^{-1} = \frac{2e^2 T_1 T_2}{\varepsilon_0 n_0 c \hbar^2} |\mathbf{r}_{ab} \cdot \mathbf{e}|^2 . \tag{8.55}$$

To put this formula in semiconductor notation we write $T_1 = \tau_R$ and $T_2 = (\Gamma_{cv})^{-1}$. The correspondence between the lower state a and the upper state b of the two-level system in §6.3 with the conduction band and valence band labels c and v is $a \rightarrow$v and $b \rightarrow$c. Hence $|\mathbf{r}_{ab} \cdot \mathbf{e}| \rightarrow |x_{cv}|$, where x is a cartesian coordinate of an electron. Now, the Heisenberg operators for p ($\equiv p_x$) and x obey the elementary classical relation $p = m\,\mathrm{d}x/\mathrm{d}t$. When we set $t = 0$ in this equation and take the matrix element between the states at the band extrema, we find that $|p_{cv}| = m|x_{cv}|\mathrm{E_g}/\hbar$. In Appendix 7 we use $\mathbf{k}\cdot\mathbf{p}$ theory (Kane, 1982) to investigate the relationship between m_e, m_h, $\mathrm{E_g}$ and p_{cv} in the standard band structure. In particular we find from (A7.12) and the above result that we may write

$$I_{sat}^{-1} = \frac{2e^2 \tau_R}{\varepsilon_0 n_0 c \hbar^2 \Gamma_{cv}} |x_{cv}|^2 , \tag{8.56}$$

where

$$|x_{cv}|^2 = \hbar^2/2m_e \mathrm{E_g} . \tag{8.57}$$

For the parameter values used by de Rougemont and Frey (*i.e.* those listed above, together with $|x_{cv}|=0.4$ nm, $\tau_R=10$ ns and $n_0=3.5$), we find that $I_{sat}=1.25$ kW cm^{-2}. Hence $I/I_{sat}\sim 10$ when I takes a typical value in the order of 10 kW cm^{-2}. We see from Fig. 8.5 that, with $\Delta_0=-1$, $\mathrm{Re}\,\overline{\chi}_{cv}\,\omega/\alpha_0 c n_0=2\omega n_{cv}/c\alpha_0$ decreases by roughly one when I/I_{sat} increases from 0 to 10. Hence the decrease of n_{cv} is in the order of 5×10^{-2} when $\alpha_0=10^6$ m^{-1} and $n_2^I\sim5\times10^{-3}$ cm^2 kW^{-1}, which has the same order of magnitude as the value calculated in the previous section for the contribution to n_2^I due to free carrier effects. Much larger values of n_2^I associated with interband absorption may be obtained by using a semiconductor with a smaller band gap and a smaller electron effective mass. Thus, for InSb, $E_g=0.17$ eV and $m_e=0.015m$, which are both smaller by an order of magnitude than the values assumed by de Rougemont and Frey. It follows from (8.57) that, in this material, $|x_{cv}|$ increases by an order of magnitude to 40 nm. Hence, other things being equal, we may expect n_2^I to increase by two orders of magnitude to 0.5 cm^2 kW^{-1}. Miller *et al* (1981a,b) quote experimental values in this order for InSb at 77 K. Their size is due to the very large values of the effective dipole moment associated with the electrons because of small values of m_e and E_g which enter into the relationship (8.57) and to the relative ineffectiveness of electron-hole recombination which leads to large values of $\tau_R\sim50$ ns.

8.4.4 Exciton effects in the nonlinear regime

As described in §8.3.4, the Coulomb interaction between electrons and holes can have a substantial influence on the linear absorption spectrum of semiconductors in the region of the band edge. Similarly, the presence of these excitonic states can have a profound effect on the nonlinear-optical properties. The most important process is that due to screening; the photogeneration of free carriers in the semiconductor results in an electric field which screens – or reduces the effectiveness – of the field component binding the electron and hole in an exciton. The effect of screening is thus to 'ionise' the exciton, causing the sharp resonance lines in the absorption spectrum to vanish and so produce a large optical nonlinearity.

The nonlinearity exhibited by an exciton in a semiconductor is much greater than that of a hydrogen atom. This can be understood – at least in part – by considering the relative magnitudes of the internal electric fields which bind the particles (electron and hole in the exciton, or electron and proton in the hydrogen atom). We can expect nonlinearity to be observed when the optical field is not insignificant in comparison with the internal Coulombic field $e/4\pi\varepsilon_0(a_0{}^*)^2$, where $a_0{}^*$ is the effective Bohr radius given by (8.30b). For an exciton in a semiconductor, typical values are $m_r = 0.1m$ and $\varepsilon = 5$, so that the internal field is some 2500 times smaller than that of a hydrogen atom. In fact the nonlinearity due to excitons may be even stronger than suggested by this elementary calculation of the internal field. As noted above, the electric field produced by the optically-generated free carriers in the semiconductor is usually more effective than the optical field itself in screening the internal field of the exciton.

A first-principles treatment of excitonic effects in the nonlinear regime is beyond the scope of this book. We outline a partly phenomenological theory which has been developed recently by Banyai and Koch (1986) and Koch *et al* (1988) and which provides a good description of the saturated interband absorption observed in GaAs. The starting point is the formula (8.39) for excitonic absorption in the linear regime. It is convenient to write that equation down again here because we need to modify it in several ways to make it useful in the nonlinear regime. Thus we have

$$\alpha = \frac{\pi}{n_0\varepsilon_0\omega c}(e/m)^2\,|p_{cv}|\,2\sum_n|\xi_{n00}(0)|^2\,\delta(\hbar\omega - \mathrm{E_g} + \mathrm{E}^*_{\mathrm{Ry}}/n^2), \quad (8.58)$$

where $\mathrm{E}^*_{\mathrm{Ry}}$ is the effective Rydberg and $\xi_{n00}(r)$ is the hydrogenic s-state function of the distance r between the electron and hole with principal quantum number n.

In deriving (8.58) we set $f_0[\mathrm{E_v}(\mathbf{k})] - f_0[\mathrm{E_c}(\mathbf{k})] = 1$. An important modification is to introduce a factor to take account of the shift of the quasi-Fermi levels as the intensity increases. This is the analogue of $F(\mathrm{E_t})$ in (8.51) and (8.53). Koch *et al* carry out this task approximately by multiplying the right-hand side of (8.58) by $F(\hbar\omega)$, where $F(\hbar\omega)$ is given exactly by (8.51). We may easily manipulate this equation into the form

$$F(\hbar\omega) = G\tanh[(\hbar\omega - \mathrm{E_{Fc}} + \mathrm{E_{Fv}})/2kT], \quad (8.59)$$

where

$$G = \frac{\exp[(\mathrm{E_c}(\mathbf{k}) - \mathrm{E_{Fc}})/kT] + \exp[(\mathrm{E_v}(\mathbf{k}) - \mathrm{E_{Fv}})/kT]}{\{\exp[(\mathrm{E_c}(\mathbf{k}) - \mathrm{E_{Fc}})/kT] + 1\}\{\exp[(\mathrm{E_v}(\mathbf{k}) - \mathrm{E_{Fv}})/kT] + 1\}} \quad (8.60)$$

and we have used the identity $\mathrm{E_c}(\mathbf{k}) - \mathrm{E_v}(\mathbf{k}) = \hbar\omega$. For low intensities the first exponential in the numerator of G is much larger than 1 and the second exponential is much less than 1. Consequently, $G \simeq 1$. This will remain true at high intensities provided that $\mathrm{E_{Fc}}$ and $\mathrm{E_{Fv}}$ do not penetrate too far into the conduction and valence bands, respectively. Koch *et al* therefore approximate G by 1 at all intensities.

The second modification of (8.58) is to replace the hydrogenic s-functions ξ_{n00} and their energies by the spherical symmetrical solutions of the corresponding problem for a *screened* Coulomb potential. Ideally one would like to use the Yukawa potential: $V_{\mathrm{Y}}(r) = -\exp(-2\kappa r)/4\pi\varepsilon_0 n_0^2 r$, where κ^{-1} is a screening length which is calculated below in (8.62). However, the determination of the eigenfunctions and eigenenergies is difficult for this potential, and Koch *et al* therefore replace it by the Hulthen potential:

$$V_{\mathrm{H}}(r) = \frac{2\kappa}{4\pi\varepsilon_0 n_0^2}\,[\exp(2\kappa r) - 1]^{-1}. \tag{8.61}$$

This is identical to the Yukawa potential when $2\kappa r \ll 1$, but has a different asymptotic behaviour. Then $\xi_{n00}(0)$ and $\mathrm{E}_{\mathrm{Ry}}^{*}/n^2$ in (8.58) are replaced by the corresponding quantities $\phi_{n00}(0)$ and E_n in the Hulthen potential problem, which are well known (Banyai and Koch, 1986).

The screening constant κ is evaluated in the Thomas-Fermi approximation (Kittel, 1986) for a degenerate electron-hole gas. Thus

$$\kappa^2 = \frac{e^2}{4\pi\varepsilon_0}(\mathrm{d}N^{(\mathrm{e})}/\mathrm{dE}_{\mathrm{Fc}} + \mathrm{d}N^{(\mathrm{h})}/\mathrm{dE}_{\mathrm{Fv}}), \tag{8.62}$$

where

$$N^{(\mathrm{e})} = \int_0^\infty \mathrm{dE}\, N_{\mathrm{c}}(\mathrm{E}) f_{\mathrm{c}}(\mathrm{E}) \tag{8.63a}$$

and

$$N^{(\mathrm{h})} = \int_{-\infty}^{-\mathrm{E_g}} \mathrm{dE}\, N_{\mathrm{v}}(\mathrm{E})\,[1 - f_{\mathrm{v}}(\mathrm{E})] \tag{8.63b}$$

are the electron and hole densities in which the Fermi functions $f_{\mathrm{c}}(\mathrm{E})$ and $f_{\mathrm{v}}(\mathrm{E})$ involve the quasi-Fermi levels E_{Fc} and E_{Fv}, respectively.

The last modification to (8.58) is to replace the normalised line-shape function $\delta(\hbar\omega - \mathrm{E_g} + \mathrm{E}_n)$ by $g(\hbar\omega - \mathrm{E_g} + \mathrm{E}_n)$, where

$$g(x) = (\pi\hbar\Gamma_{\mathrm{cv}})^{-1}\,\mathrm{sech}(x/\hbar\Gamma_{\mathrm{cv}}). \tag{8.64}$$

The expression finally obtained for α is thus:

$$\alpha = \frac{2\pi e^2}{n_0\varepsilon_0\omega c m^2}\,|p_{\mathrm{cv}}|^2 \tanh\left[\frac{\hbar\omega - \mathrm{E}_{\mathrm{Fc}} + \mathrm{E}_{\mathrm{Fv}}}{2kT}\right] \times \sum_n |\phi_{n00}(0)|^2\, g(\hbar\omega - \mathrm{E_g} + \mathrm{E}_n). \tag{8.65}$$

In this equation we include only the contribution from the exciton line spectrum. Koch *et al* (1988) add on a contribution from the continuous spectrum in the form (8.53), with $F(\mathrm{E_t})$ replaced by $F(\hbar\omega)$ in (8.59) with

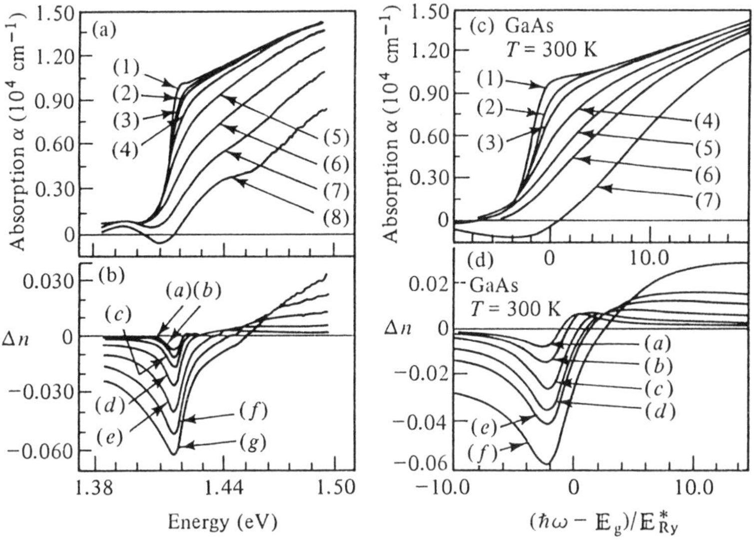

Fig. 8.6 Experimental and theoretical absorption and refractive-index spectra for GaAs at room temperature.

(a) Experimental absorption spectra with intensity increasing from curve (1) (linear regime) to curve (8).

(b) Corresponding incremental refractive-index changes Δn obtained through a Kramers-Kronig transformation of the differences between curves (2)–(8) and curve (1) in Fig. 8.4(a).

(c) Calculated absorption spectra for electron concentrations increasing from 10^{15} cm^{-3} for curve (1) (linear regime) to 1.5×10^{18} cm^{-3} for curve (7).

(d) Corresponding incremental refractive-index changes Δn obtained through a Kramers-Kronig transformation of the difference between curves (2)–(7) and curve (1) in Fig. 8.4(c). Details are given in the text.

(After Koch *et al*, 1988.)

$G = 1$, and with $g(\mathrm{E_t} - \hbar\omega)$ replaced by $g(x)$ in (8.64) with $x = \mathrm{E_t} - \hbar\omega$.

We see by inspection that α is determined by (8.62)–(8.65) when $\mathrm{E_{Fc}}$ and $\mathrm{E_{Fv}}$ are known. The quasi-Fermi levels are determined in terms of α in (8.44) and (8.46). All these equations must be solved self-consistently to determine the intensity dependence of α. In comparing the results of their calculations with experiment, Koch *et al* (1988) avoid this difficult calculation. Instead they present calculated values of α for different values of $N^{(e)}$ which are shown in Fig. 8.6(c). The curves (1)–(7) are for: (1) $N^{(e)} = 10^{15}$ cm^{-3} (linear regime); (2) 8×10^{16}, (3) 2×10^{17}, (4) 5×10^{17}, (5) 8×10^{17}, (6) 10^{18} and (7) 1.5×10^{18} cm^{-3}. The calculations were made for GaAs at room temperature using the parameter values: $\mathrm{E_g} = 1.42$ eV, the exciton Rydberg $\mathrm{E}^*_{\mathrm{Ry}} = 4.2$ meV, $m_\mathrm{e} = 0.066m$, $m_\mathrm{h} = 0.52m$, $n_0^2 = 13.5$ and $\hbar\Gamma_{\mathrm{cv}} = 4.2$ meV (Banyai and Koch, 1986).

Curve (1) is appropriate to the linear regime. It exhibits the shoulder typical of the absorption edge predicted by the linear analysis given in §8.3. With increasing $N^{(e)}$ (*i.e.*, increasing intensity) the edge broadens out and α decreases sufficiently rapidly when $\hbar\omega < E_g$ to become negative in curves (6) and (7). Thus optical gain can be achieved at sufficiently high intensities. It comes about because of population inversion, which we see from (8.65) occurs when $E_{Fc} > E_{Fv} + \hbar\omega$, so that $\tanh[(\hbar\omega - E_{Fc} + E_{Fv})/2kT]$ changes sign.

Measured values of α are shown for comparison in Fig. 8.6(a). They were obtained for continuous (c.w.) excitation powers of: (1) 0 mW; (2) 0.2, (3) 0.5, (4) 1.3, (5) 3.2, (6) 8, (7) 20 and (8) 50 mW incident on a spot with a diameter $\sim 15\,\mu$m. The qualitative and quantitative agreement with the theoretical curves in Fig. 8.6(c) is excellent. To make the comparison complete it is necessary to use (8.44) and (8.46) to relative the optical intensity I to $N^{(e)}$. Banyai and Koch (1986) show that agreement to better than 10% is obtained when τ_R is taken to be 3.4 ns.

One advantage of using the electron density as a parameter in the theoretical curves for α is that the Kramers-Kronig relations given in Appendix 8 may be used to calculate the increment Δn of the refractive index from its linear value, from the corresponding change $\Delta\alpha$ of the absorption coefficient. (Miller *et al* (1981a) emphasise that the relations are, strictly speaking, not valid when I is used as a parameter.) Curves $(a)-(f)$ in Fig. 8.6(d) show results obtained in this way after subtracting curve (1) from the curves (2)$-$(6) in Fig. 8.6(c). Curves $(a)-(g)$ in Fig. 8.6(b) show results obtained in the same way from the experimental absorption data in Fig. 8.6(a). (Direct measurement of Δn shows that this procedure is valid in the immediate neighbourhood of the absorption edge in spite of the fact that the intensity I is the parameter labelling the experimental curves.)

Finally, it is useful to consider how well the predictions of the simpler theory of interband absorption neglecting excitonic effects, as discussed in §8.4.3, agree with those of the exciton theory for photon energies in the tail of the absorption band. We see from Fig. 8.6(b) that $\Delta n = -0.025$ for curve (f) when $\hbar\omega$ is 40 meV below the band gap. The intensity in this case is that for curve (7) in Fig. 8.6(a), *i.e.*, 20 mW in a spot of diameter 15 μm, which is of the order of 11.3 kW cm^{-2}. Hence the experimental value of $n_2^I = -0.025/11.3 = -2.2\times10^{-3}$ cm^2 kW^{-1}. We see from (A5.13) that this value of n_2^I corresponds to an effective value of $\chi^{(3)}_{esu} \sim 7\times10^{-4}$, which is consistent with the relatively large energy gap and electron effective mass in GaAs. The theoretical value is similar and has the same order of magnitude as the value estimated in §8.4.3 from the numerical calculations of de Rougemont and Frey.

9

The optical properties of artificial materials

In the earlier days of nonlinear optics, the materials used for experiments and devices were mainly inorganic dielectric crystals, vapours, liquids and bulk semiconductors. The search for 'good' nonlinear-optics media was made amongst the known materials. However in more recent years, with the growing interest in optical devices and applications, attention has focused increasingly on new artificial solid-state materials which may offer higher nonlinearity; in particular, those that will allow nonlinear-optical devices to operate efficiently at relatively low power levels, such as the outputs from semiconductor-diode lasers. Organic materials offer great scope since modern methods of synthesis allow considerable flexibility in the design of materials at the molecular level. As mentioned in §4.4.3, the macroscopic nonlinear-optical properties of many organic crystals are given by the tensor sum of the properties of the constituent molecules, with due regard to local-field factors and molecular orientation. It is this feature of organic materials that allows a 'molecular-engineering' approach to the optimisation of macroscopic properties. Several materials with large second-order nonlinearity have been successfully fabricated (Chemla and Zyss, **1**, 1987). Some of these newer organic optical materials also exhibit other desirable properties, such as a greater resistance to optical damage. These have applications in devices such as compact optical-frequency doublers and parametric-amplifiers and -oscillators. However, materials with large *third*-order nonlinearity are perhaps of greater interest currently, since the nonlinear refractive-index effect can be exploited for switching, optical bistability, phase conjugation and other types of signal processing (Gibbs, 1985). Some recent advances have been made in synthesising organic materials with large third-order nonlinearity (Chemla and Zyss, **2**, 1987), but so far, the greatest progress has been made with inorganic semiconductors.

There are many semiconducting materials and many doping regimes. Bulk semiconductors therefore provide a rich variety of optical behaviour. Over the last decade the range of possibilities has been greatly extended by fabricating new kinds of artificial semiconductor structures, such as 'quantum wells', 'quantum wires' and 'superlattices'. These are made possible by means of the sophisticated crystal-growth technology that has been developed recently. Semiconductors have the additional potential advantage of allowing optical and electronic devices to be fabricated together on the same material substrate. We focus mainly on the linear- and nonlinear-optical properties of artificial semiconductors in this chapter, and thus review the background to the current intensive research.

Most attention has been given to 'multiple quantum wells', which are made by growing alternate layers of two different semiconductors (Chemla and Miller, 1985; Miller *et al,* 1985). The pure material GaAs and the alloy $Ga_{1-x}Al_xAs$ are often used and we discuss this case specifically. However, many other combinations yield interesting behaviour: *e.g.,* Si and $Ge_{1-x}Si_x$ and the $CdTe/Cd_{1-x}Mn_xTe$ and $InAs_xP_{1-x}/InP$ systems. Modern crystal-growing techniques give typical layer thicknesses of the order of 10 nm and interfaces which are smooth on an atomic scale (Chang and Ploog, 1985). The name 'multiple quantum wells' refers to the alternating succession of potential hills and valleys (*i.e.,* potential energy wells) for electrons and holes which these structures produce. The carriers are confined in the potential wells and exhibit quasi-two-dimensional behaviour which we describe in §9.1.1. The linear- and nonlinear-optical properties of multiple quantum wells are described in §§9.1.2 – 9.1.5.

The linear-optical properties of metal colloids and 'colour filter' glasses containing semiconductor crystallites have been studied for many years. The discovery that such colour filters exhibit relatively large band-edge-resonant third-order nonlinearity (Jain and Lind, 1983) with relaxation times as short as ~10 ps (Yao *et al,* 1985; Cotter, 1986) has stimulated much current investigation (Hache *et al,* 1986; Schmitt-Rink *et al,* 1987; Flytzanis *et al,* 1986; Cotter *et al,* 1988). The linear dimensions of the conducting particles in these colour filter glasses and metal colloids may again be of the order of a few nanometres and the carriers in them are then confined in all three directions. It has therefore become customary to talk of 'quantum dots' or 'zero-dimensional' structures in this case (Schmitt-Rink *et al,* 1987; Flytzanis *et al,* 1986). It is also possible to fabricate 'quantum wires' or 'one-dimensional' semiconductor structures; however, there has been little evaluation of their nonlinear-optical properties and we do not discuss them further here.

Multiple quantum wells and quantum dots are treated in §§9.1 and 9.2 respectively. The nomenclature used for these structures is strictly speaking appropriate only when the dimensions of confinement are smaller than the mean free path of the carriers. Then the description of the optical behaviour involves the energy eigenvalues and eigenfunctions resulting from the carrier confinement. We focus on this interesting case and do not treat the opposite extreme in which the optical properties of each part of the system are adequately described by their bulk characteristics. In the latter case the overall behaviour of the whole heterogeneous system may be evaluated simply by using Maxwell's equations and the electromagnetic boundary conditions at the interfaces. This approach fails in quantum-well and quantum-dot structures because the quantum-mechanical states of the carriers are strongly affected by the interfaces.

9.1 Quantum wells

9.1.1 Electronic structure

A single quantum well may be fabricated by sandwiching a thin layer of GaAs between two extended layers of the alloy $Al_xGa_{1-x}As$ with $x \sim 0.3$. The energy gap of the alloy is larger than that of GaAs and the conduction- and valence-band edges are spatially modulated, as illustrated in Fig. 9.1(a). We concentrate on the effect of this modulation on the behaviour of the electron in one well for the 'standard' band structure shown in Fig. 8.2(a). The complications which arise when a more realistic band structure is used are not essential to our discussion.

Let z be the spatial coordinate in the direction of growth and suppose that it is measured from one side of the well. The conduction-band edge shown in Fig. 9.1(a) is the potential energy function $V(z)$ for an electron moving in the conduction band. We describe the electronic motion by means of the envelope-function approximation outlined in Appendix 6. The Schrödinger equation for the envelope function $\phi(\mathbf{r})$ and the energy E is obtained from (A6.9) by writing $m^* = m_e$ and $V_{ext}(\mathbf{r}) = V(z)$:

$$-\frac{\hbar^2}{2m_e}\nabla^2\phi(\mathbf{r}) + V(z)\,\phi(\mathbf{r}) = \mathrm{E}\,\phi(\mathbf{r}). \tag{9.1}$$

We see that the electrons remain free in the x and y directions so that (9.1) has a solution of the form:

$$\phi_{n\mathbf{k}}(\mathbf{r}) = \exp[i\,(k_x x + k_y y)]\,Z_n(z), \tag{9.2}$$

where $\mathbf{k} = (k_x, k_y)$ and n is a quantum number which is specified below. We see that $\phi_{n\mathbf{k}}(\mathbf{r})$ is correctly normalised to V over the crystal volume V (see the remark at the end of Appendix 6) when $Z_n(z)$ is normalised to

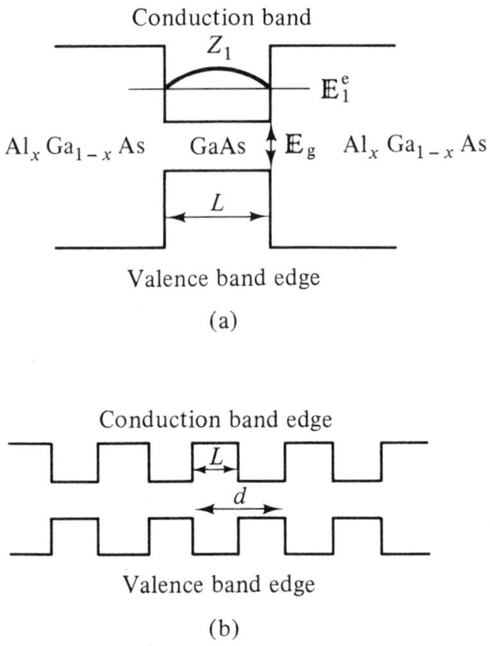

Fig. 9.1 Energy diagrams with distance as the abscissa, showing:
(a) A single quantum well for electrons and holes (with the ground-state electron envelope function $Z_1(z)$ superimposed).
(b) A multiple quantum well.

L_z, the length of the crystal in the z direction.

To calculate $Z_n(z)$ we substitute (9.2) in (9.1) and find that

$$-\frac{\hbar^2}{2m_e}\frac{d^2Z_n(z)}{dz^2} + V(z)\,Z_n(z) = \left(E - \frac{\hbar^2 k^2}{2m_e}\right)Z_n(z), \tag{9.3}$$

where $k = \sqrt{k_x^2 + k_y^2}$. For simplicity we solve (9.3) with $V(z) = 0$ inside the well and suppose that $V(z) \to \infty$ at the edges of the well ($z = 0$ and $z = L$). Then

$$Z_n(z) = \sqrt{2L_z/L}\,\sin(n\pi z/L), \tag{9.4a}$$

where n is a positive integer, and $E = E_n^e(\mathbf{k})$ with

$$E_n^e(\mathbf{k}) = E_n^e + \hbar^2 k^2/2m_e \tag{9.4b}$$

in which

$$E_n^e = \hbar^2 (n\pi/L)^2/2m_e. \tag{9.4c}$$

We exhibit $Z_1(z)$ and E_1^e in Fig. 9.1(a).

The assumption that $V(z) \to \infty$ at the edges of the well is crude. The height of the potential barrier there is typically only a few tenths of

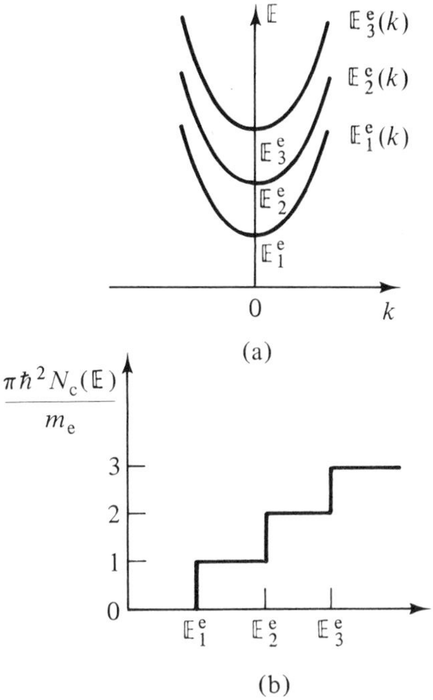

Fig. 9.2 (a) Dispersion curves of the first three electron sub-bands of a quantum well. (b) The density of states of the electron sub-bands of a quantum well.

an electron volt. However, the main point is that the confinement produces an energy-band structure of the form (9.4b) in which $\mathbb{E}_n^e$ has discrete values which are determined by $V(z)$. For each n we have a parabolic relation between $\mathbb{E}_n^e(\mathbf{k})$ and k, with a minimum at $\mathbb{E}_n^e$, as shown schematically in Fig. 9.2(a). We refer to the energy eigenvalue $\mathbb{E}_n^e(\mathbf{k})$ and its associated eigenfunction as the 'electron sub-band n'. A similar treatment may be given for holes in the valence band of the standard band structure. The hole energies $\mathbb{E}_n^h(\mathbf{k})$ (measured downwards from the conduction-band edge) take the form (9.4b) with $\mathbb{E}_n^e$ replaced by $\mathbb{E}_n^h = \mathbb{E}_g + \hbar^2 (n\pi/L)^2/2m_h$, where $\mathbb{E}_g$ is the energy gap in the well (as shown in Fig. 9.1(a)). The corresponding envelope functions are again given by (9.2) with $Z_n(z)$ defined by (9.4a), *i.e.*, they are identical to the envelope functions for electrons. In the case of holes we speak of 'hole sub-band n'.

In very wide quantum wells (large L) the values of $n\pi/L$ form a quasi-continuum. When we set $n\pi/L = k_z$, (9.4b) and (9.4c) simply reproduce the relation between energy and the wave vector (k_x, k_y, k_z) for free electrons moving in three dimensions. In this case the sub-band structure is not important. However, when $L = 10$ nm and we set $m_e = 0.067m$ (the value for GaAs), we find from (9.4c) that $\mathbb{E}_1^e = 55$ meV.

Consequently significant tuning of the sub-band minima is possible in these structures, with the result that the forbidden energy gap increases by $\hbar^2k^2/2m_r \sim 50-100$ meV, where $m_r = m_e m_h/(m_e + m_h)$ is the reduced mass of the electrons and holes.

In each sub-band $\mathbf{k} = (k_x, k_y)$ is a two-dimensional vector. The density of states per unit area $N_n(\mathrm{E})$ is easily evaluated for the nth electron sub-band by making some obvious changes in the three-dimensional calculations given in §8.1. The result is very simple: $N_n(\mathrm{E}) = 0$ when $\mathrm{E} < \mathrm{E}_n$, and $N_n(\mathrm{E}) = m_e/\pi\hbar^2$ when $\mathrm{E} > \mathrm{E}_n$, where we have allowed for spin degeneracy. The density of states for all the electron sub-bands is therefore given by the 'staircase' function:

$$N_c^{QW}(\mathrm{E}) = \frac{m_e}{\pi\hbar^2} \sum_n \theta(\mathrm{E} - \mathrm{E}_n^e), \tag{9.5a}$$

where $\theta(x) = 0$ when $x < 0$, $= 1$ when $x > 1$. We plot $N_c^{QW}(\mathrm{E})$ against E in Fig. 9.2(b). A similar calculation for holes gives

$$N_v^{QW}(\mathrm{E}) = \frac{m_h}{\pi\hbar^2} \sum_n \theta(\mathrm{E} - \mathrm{E}_n^h). \tag{9.5b}$$

Finally, the joint density of states for the electron-hole pair states derived from electron and hole sub-bands with the same sub-band index n is

$$N_{cv}^{QW}(\mathrm{E}) = \frac{m_r}{\pi\hbar^2} \sum_n \theta(\mathrm{E} - \mathrm{E}_n^e - \mathrm{E}_n^h). \tag{9.5c}$$

We take the same n in both sets of sub-bands because, as shown in §9.1.2, this is the case for the allowed optical transitions.

Experimental optical studies are usually made on a large number of periodically repeated quantum wells which are separated by alloy barrier regions thick enough to make tunnelling between the walls negligible. This is the multiple quantum-well structure which concerns us here. It is also possible to make the barriers so thin that there is appreciable tunnelling between the wells. This is the 'superlattice' structure in which the discrete quantum-well levels E_n^e and E_n^h broaden out into bands. Until now, superlattices have been more important in studies of electron transport, although it seems likely that in future some superlattice structures will be designed specifically to produce desirable linear- and nonlinear-optical properties (Adams, 1986).

9.1.2 Single-particle absorption spectrum

When Coulomb interaction between electrons and holes is neglected we may calculate the interband absorption coefficient α from (8.11) and (8.16a). The carriers in each quantum well make identical contributions to α. We may therefore confine our attention to one well and multiply the result by the number of quantum wells in the crystal volume V; to do this, V in (8.11) is replaced by the volume of one period of the multiple quantum well V_{QW}. Moreover, we may identify n and n' in (8.11) with hole and electron sub-band labels so as to pick out the resonant terms. Thus we obtain the multiple quantum-well analogue of (8.17) and (8.18):

$$\alpha = \frac{2e^2}{n_0\varepsilon_0 c V_{QW} m^2\omega} \sum_{nn'} \sum_{\mathbf{k}} |p_{n'n}(\mathbf{k})|^2$$
$$\times \pi\,\delta(\mathrm{E_g} + \mathrm{E}_n^{\mathrm{h}} + \mathrm{E}_{n'}^{\mathrm{e}} + \hbar^2k^2/2m_{\mathrm{r}} - \hbar\omega). \tag{9.6}$$

In (9.6) $p_{n'n}(\mathbf{k})$ is a matrix element of the momentum operator in the direction of the optical electric field which we take to be parallel to Ox. Then

$$p_{n'n}(\mathbf{k}) = \int_V \mathrm{d}\mathbf{r}\, \phi_{n'\mathbf{k}}{}^*(\mathbf{r})\, u_{\mathrm{c}}{}^*(\mathbf{r})\,(-i\hbar\,\partial/\partial x)\,\phi_{n\,\mathbf{k}}(\mathbf{r})\, u_{\mathrm{v}}{}^*(\mathbf{r})\,, \tag{9.7}$$

where $u_{\mathrm{c}}(\mathbf{r})$ and $u_{\mathrm{v}}(\mathbf{r})$ denote the Bloch functions at the extrema of the conduction and valence bands respectively. Since the envelope functions are slowly varying we may write:

$$\begin{aligned}
p_{n'n}(\mathbf{k}) &\simeq \int_V \mathrm{d}\mathbf{r}\, \phi_{n'\mathbf{k}}{}^*(\mathbf{r})\, \phi_{n\mathbf{k}}(\mathbf{r})\, u_{\mathrm{c}}{}^*(\mathbf{r})\,(-i\hbar\,\partial/\partial x)\, u_{\mathrm{v}}{}^*(\mathbf{r}) \\
&\simeq \sum_{\mathbf{R}} \phi_{n'\mathbf{k}}{}^*(\mathbf{R})\, \phi_{n\mathbf{k}}(\mathbf{R}) \int_{V_{\mathrm{c}}} \mathrm{d}\mathbf{r}\, u_{\mathrm{c}}{}^*(\mathbf{r})\,(-i\hbar\,\partial/\partial x)\, u_{\mathrm{v}}{}^*(\mathbf{r}) \\
&\simeq \sum_{\mathbf{R}} \phi_{n'\mathbf{k}}{}^*(\mathbf{R})\, \phi_{n\mathbf{k}}(\mathbf{R})\, p_{\mathrm{cv}}\, V_{\mathrm{c}}/V \\
&\simeq p_{\mathrm{cv}}\, V^{-1} \int_V \mathrm{d}\mathbf{r}\, \phi_{n'\mathbf{k}}{}^*(\mathbf{r})\, \phi_{n\mathbf{k}}(\mathbf{r}) \\
&= p_{\mathrm{cv}}\, \delta_{n'n}\,.
\end{aligned} \tag{9.8}$$

In the first line we have neglected $\partial\phi_{n\mathbf{k}}(\mathbf{r})/\partial x$ in comparison to $\partial u_{\mathrm{v}}(\mathbf{r})/\partial x$. In the second line the integral has been broken into contributions from unit cells with one corner at $\mathbf{R}$ and volume V_{c}. The envelope functions are treated as constants over each unit cell. The remaining integral over the unit cell volume V_{c} is just V_{c}/V times p_{cv} as defined in §8.3 in which the integration region was the entire crystal volume V. Taking p_{cv}/V outside the summation sign leaves a discrete approximation to the overlap integral in the fourth line. The final result follows from (9.2) which gives zero for the overlap integral when $n'\neq n$, and V when $n'=n$.

We see from (9.8) that optical transitions are only possible between sub-bands with the same sub-band index. When this result is substituted into (9.6) we may set $V_{\mathrm{QW}} = Ad$ to interpret the final formula; here A is the area of the structure in xy plane and d is its period. Thus we find that in a multiple quantum well α simply takes the bulk form (8.19b) with the three-dimensional joint density of states per unit volume $N_{\mathrm{cv}}(\hbar\omega)$ replaced by $N_{\mathrm{cv}}^{\mathrm{QW}}(\hbar\omega)/d$ where $N_{\mathrm{cv}}^{\mathrm{QW}}(\mathrm{E})$ is the staircase-like joint density of states (per unit energy range per unit area) defined in (9.5c).

9.1.3 Exciton line spectrum

The exciton line spectrum for a multiple quantum well is also easily evaluated by making some simple modifications of the bulk calculation given in §8.3.4. We may again use the envelope-function approximation and write the exciton wave function $\psi(\mathbf{r}_{\mathrm{e}}, \mathbf{r}_{\mathrm{h}})$ in one quantum well in the form (8.22). Equations (8.23)–(8.25) are valid as they stand, and in the exciton Schrödinger equation (8.26) it is only necessary to add the hole and electron confinement potentials $V_{\mathrm{h}}(z_{\mathrm{h}})$ and $V_{\mathrm{e}}(z_{\mathrm{e}})$ (as illustrated, for example, by the valence- and conduction-band edges in Fig. 9.1(a)) to the Coulomb potential $V_{\mathrm{c}}(\mathbf{r}_{\mathrm{e}} - \mathbf{r}_{\mathrm{h}})$. The equation retains its translational invariance in the xy plane. We may therefore usefully continue to make the transformation (8.27) to centre-of-mass and relative coordinates in the xy plane. However, the z coordinates z_{e} and z_{h} are best left as they stand. Then the form (8.28) for the exciton wave function is replaced by

$$\phi_{\mathbf{K}n}(\mathbf{R}, \mathbf{r}, z_{\mathrm{e}}, z_{\mathrm{h}}) = \sqrt{VNd}\,\exp(i\mathbf{K}\cdot\mathbf{R})\,\xi_n(\mathbf{r}, z_{\mathrm{e}}, z_{\mathrm{h}}), \tag{9.9}$$

where $\mathbf{R} = (X, Y)$, $\mathbf{r} = (x, y)$ and $\mathbf{K} = (K_x, K_y)$. In (9.9) n denotes the quantum numbers associated with the final factor $\xi_n(\mathbf{r}, z_{\mathrm{e}}, z_{\mathrm{h}})$ which we have supposed to be normalised to unity. The prefactor $\sqrt{VNd}$ then ensures that $\phi_{\mathbf{K}n}(\mathbf{R}, \mathbf{r}, z_{\mathrm{e}}, z_{\mathrm{h}})$ is normalised to V^2 as required by (8.24b). Here V, N and d denote respectively the crystal volume, the number of periods of the multiple quantum well in V, and the length of each one (*i.e.*, the period in the growth direction). Finally, we write the energy of the exciton state in the form

$$\mathrm{E}_n(\mathbf{k}) = \frac{\hbar^2}{2m_{\mathrm{t}}}\mathrm{K}^2 + \mathrm{E}_n, \tag{9.10}$$

where $m_{\mathrm{t}} = m_{\mathrm{e}} + m_{\mathrm{h}}$ is the total mass. The calculation of ξ_n and E_n is discussed below.

When (9.9) and (9.10) are substituted in the exciton Schrödinger equation we find that ξ_n and E_n are determined by

$$(H_{\mathrm{ex}} + H_{\mathrm{e}} + H_{\mathrm{h}})\,\xi_n = \mathrm{E}_n\,\xi_n, \tag{9.11}$$

where the Hamiltonian is composed of the following parts:

$$H_{\mathrm{ex}} = -\frac{\hbar^2}{2m_{\mathrm{r}}}\left(\frac{\partial^2}{\partial x^2} + \frac{\partial^2}{\partial y^2}\right) + V_{\mathrm{c}}(\mathbf{r}, z_{\mathrm{e}}, z_{\mathrm{h}}), \tag{9.12a}$$

$$H_{\mathrm{e}} = -\frac{\hbar^2}{2m_{\mathrm{e}}}\frac{\partial^2}{\partial z_{\mathrm{e}}^2} + V_{\mathrm{e}}(z_{\mathrm{e}}), \tag{9.12b}$$

and

$$H_{\mathrm{h}} = -\frac{\hbar^2}{2m_{\mathrm{h}}}\frac{\partial^2}{\partial z_{\mathrm{h}}^2} + V_{\mathrm{h}}(z_{\mathrm{h}}). \tag{9.12c}$$

The solution of this equation is usually effected by variational methods (Miller *et al*, 1985; Miller *et al*, 1981c; Bastard *et al*, 1982). We use a simple ansatz which is appropriate when $V_{\mathrm{e}}(z_{\mathrm{e}})$ and $V_{\mathrm{h}}(z_{\mathrm{h}})$ are square wells with infinite side walls:

$$\xi_n(\mathbf{r}, z_{\mathrm{e}}, z_{\mathrm{h}}) = (2/L)\sin(\pi z_{\mathrm{e}}/L)\sin(\pi z_{\mathrm{h}}/L)\, f_n(r) \tag{9.13}$$

where $r=\sqrt{x^2+y^2}$. The sinusoidal factors in (9.13) take account of confinement in the *ground* electron and hole subbands. The final factor $f_n(r)$ can be adjusted to approximate the cylindrically-symmetrical exciton states which are important for absorption. We suppose that $f_n(r)$ is normalised to unity in the *xy* plane.

To complete the calculation of the absorption coefficient for a multiple quantum well we use (9.9) and (9.13) to substitute for $\phi(\mathbf{r}_{\mathrm{e}}, \mathbf{r}_{\mathrm{h}})$ in (8.32) and (8.33). The result is

$$\langle\psi|p|0\rangle = p_{\mathrm{cv}}\frac{\sqrt{VNd}}{Nd}\, f_n(0)\,\delta_{\mathbf{K}\,0}\,. \tag{9.14}$$

The factor $(Nd)^{-1}$ arises because the integration when $\mathbf{K}=0$ gives the cross-sectional area of the multiple quantum well which cancels out of the V in the denominator of the first line of (8.33), to leave Nd. The rest of the three-dimensional analysis goes through as it stands except that we have to multiply by a factor of N to allow for absorption by all the quantum wells. The final expression for the absorption coefficient α is just (8.39) with $|\xi_{n\,00}(0)|^2$ replaced by $|f_n(0)|^2/d$ and $\mathrm{E_g} - \mathrm{E}^*_{\mathrm{Ry}}/n^2$ replaced by E_n. Both these quantities remain to be determined.

Insight into the effects of exciton confinement may be obtained by dropping H_{e} and H_{h} from (9.11) and setting $z_{\mathrm{e}}=z_{\mathrm{h}}$ in H_{ex}. This gives the Schrödinger equation for a strictly two-dimensional exciton which is easily solved and provides a zeroth-order approximation when the well width is much less than the three-dimensional Bohr radius $a_0{}^* = 4\pi\varepsilon_0\varepsilon\hbar^2/e^2m_{\mathrm{r}}$. We may verify that the ground-state wave function is (Chemla and Miller, 1985):

$$f_1(r) = \sqrt{2/\pi}\,(2/a_0{}^*)\exp(-2r/a_0{}^*). \tag{9.15a}$$

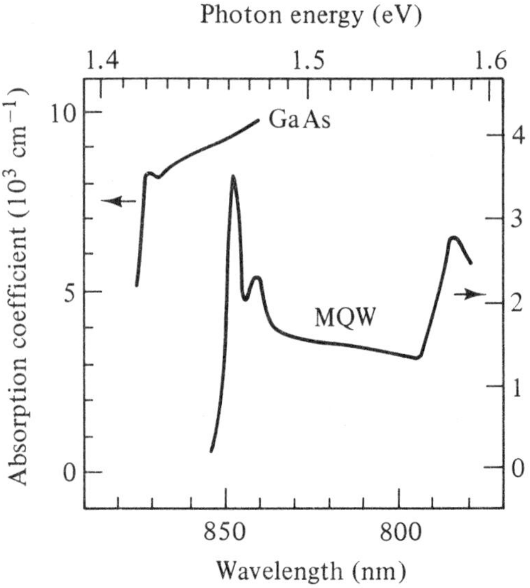

Fig. 9.3 Comparison of the room-temperature absorption spectra of a 3.2 μm thick bulk sample of GaAs and that of a multiple quantum well (MQW) structure consisting of 77 periods of 10.2 nm GaAs layers alternating with 20.7 nm $Al_xGa_{1-x}As$ layers. (After Chemla *et al,* 1985.)

The squared magnitudes of all the cylindrically-symmetrical states when $r = 0$ are given by (Bastard *et al,* 1982):

$$|f_n(0)|^2 = |f_1(0)|^2/(n - \tfrac{1}{2})^3 , \tag{9.15b}$$

and the corresponding energies are

$$\mathrm{E}_n = \mathrm{E_g} - \mathrm{E^*_{Ry}}/(n - \tfrac{1}{2})^2 \tag{9.16}$$

where $\mathrm{E^*_{Ry}}$ is the three-dimensional Rydberg (8.30a). In these equations n takes positive integer values. For the ground state $\mathrm{E_g} - \mathrm{E_1} = 4\mathrm{E^*_{Ry}}$, which is four times the binding energy of the three-dimensional exciton. The increase of binding energy arises because the wave function (9.15a) is compressed by comparison with its three-dimensional counterpart. We may readily verify, for example, that the peak of the radial charge distribution $2\pi r|f_1(r)|^2$ is at $a_0{}^*/4$, which is one-quarter of the corresponding three-dimensional quantity. The compression of the bound state wave function in two-dimensional also enhances the ratio of the 1s-exciton peak to the continuum absorption as compared with the three-dimensional case. As the well width increases the binding energy decreases from $4\mathrm{E_{Ry}}$ towards $\mathrm{E^*_{Ry}}$ and the exciton wave function transforms smoothly to its three-dimensional form when $L >> a_0{}^*$.

The dramatic effect of replacing bulk GaAs by multiple quantum

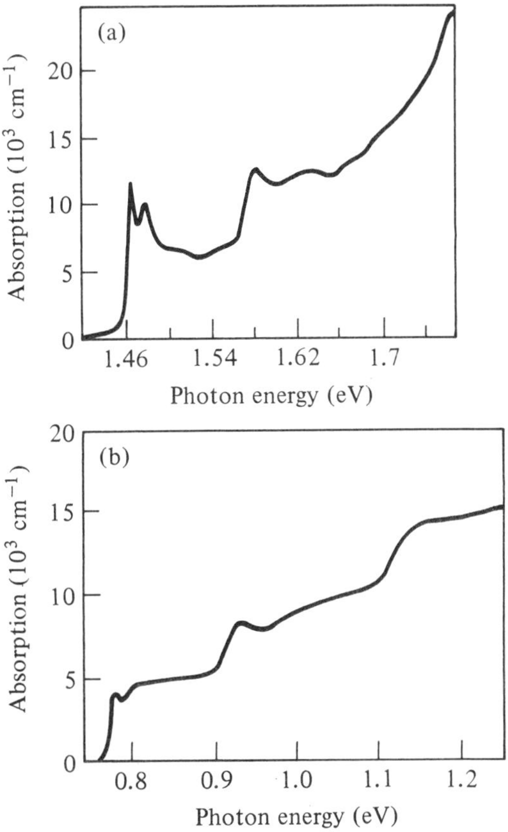

Fig. 9.4 Room-temperature absorption spectra of multiple quantum well structures of (a) GaAs/$Ga_{1-x}Al_xAs$ and (b) GaInAs/AlInAs with well thicknesses of 10.0 nm. (After Chemla and Miller, 1985.)

wells is illustrated by the room-temperature absorption data shown in Fig. 9.3. In the multiple quantum well (MQW) both the heavy-hole and light-hole exciton peaks are strong and clearly resolved. Such strong resonances are only seen in bulk material at low temperatures. In the multiple quantum well the reduced symmetry reduces the decay rate via optical phonon emission which is the principal decay channel. A major contribution to the remaining line width is inhomogeneous broadening arising from fluctuations of the well width. In Fig. 9.4 we show room-temperature absorption data for multiple quantum wells which illustrate better than Fig. 9.3 the staircase structure of the continuous absorption spectrum.

9.1.4 Quantum-confined Stark effect

When a d.c. electric field is applied to a semiconductor, a change

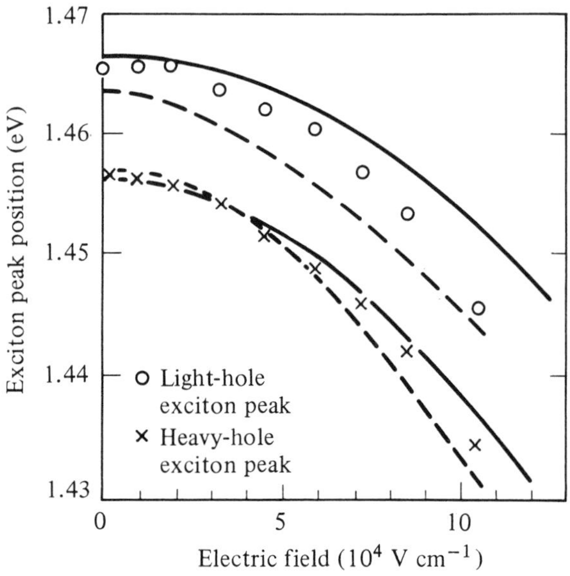

Fig. 9.5 Variation of the energies of the exciton peaks in a GaAs/$Ga_{1-x}Al_xAs$ multiple quantum-well structure with an electric field applied in the direction perpendicular to the layers. Experimental points are shown for both heavy- and light-hole excitons. The theoretical curves are calculated for a split of the energy gap difference between conduction and valence bands of 57:43% (full lines) and 85:15% (dashed lines). (After Miller *et al,* 1985.)

in the optical absorption spectrum may result; this is termed electroabsorption. The electroabsorption exhibited by bulk semiconductors is due to the Franz-Keldysh effect: there is a small shift of the absorption edge to lower frequencies in the presence of an electric field. The excitons play little part in this phenomenon because they are ionised by the field. In narrow quantum wells, however, for fields parallel to the growth direction (*i.e.,* perpendicular to the layers) the quantum confinement is strong enough to prevent significant ionisation. The field produces a considerable red shift of the exciton peak, a phenomenon known as the quantum-confined Stark effect (Miller *et al,* 1985). This may be employed in a number of optical devices, including modulators and low-energy switches (Miller *et al,* 1988a). The effect is illustrated by the room-temperature absorption data shown in Fig. 9.5. Very large fields cause ionisation and broaden the exciton peak out of existence, but shifts in the order of 20 meV can be achieved before this occurs. As we would expect: for fields perpendicular to the growth direction the situation is similar to that in bulk material. The shift of the exciton peak may be calculated by extending the Hamiltonian used in (9.11) by adding

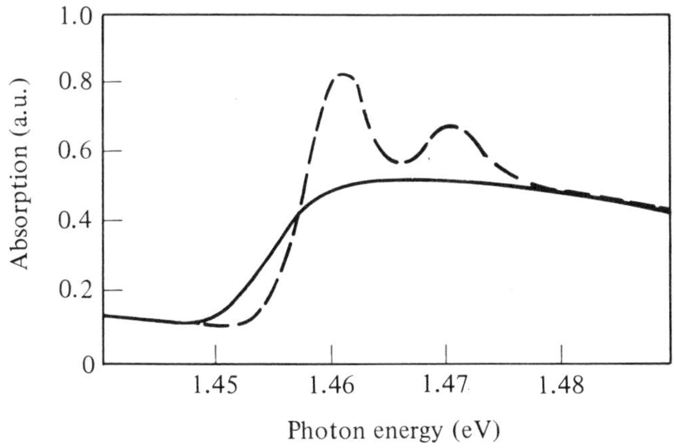

Fig. 9.6 Absorption spectra of a GaAs/$Ga_{1-x}Al_xAs$ multiple quantum-well structure measured in c.w. pump-and-probe experiments. Dashed line: pump off. Full line: pump on. (After Chemla and Miller, 1985.)

a term $eE(z_e - z_h)$ to account for the effect of the applied d.c. field E on both electrons and holes. For quantum wells ~10 nm wide the shift is dominated by the shift of E_1^e and E_1^h. Fig. 9.5 shows a comparison of experimental data with calculated values taken from Miller *et al* (1985). At the time of the calculation there was considerable uncertainty about how the energy-gap difference between GaAs and $Al_xGa_{1-x}As$ was split between conduction and valence bands and two possibilities are considered in the figure. The dashed lines which are for a 85:15% split should be ignored. The full lines are calculated for a 57:43% split which is now generally accepted as close to the correct value.

Recently it has been suggested that the electroabsorption exhibited by quantum dots (see §9.2) should be even greater than that in quantum wells (Miller *et al,* 1988b).

9.1.5 Nonlinear exciton absorption

We saw in Chapter 8 that the nonlinearity of the exciton absorption in bulk semiconductors comes about primarily through screening produced by free carriers. The same mechanism operates in multiple quantum wells. The theory (Schmitt-Rink *et al,* 1985) is beyond the scope of this book and we confine ourselves to an outline of the experimental situation (Chemla and Miller, 1985).

The result of a pump-and-probe experiment is shown in Fig. 9.6. The heavy- and light-hole peaks are both bleached out by a 800 W cm^{-2} pump beam of photons with energy 35 meV above the heavy-hole exciton peak. The pump generates 2×10^{12} cm^{-2} free electron-hole pairs in the quantum wells for which $L = 10$ nm, *i.e.*, a volume density of $2\times10^{18}\ \mathrm{cm}^{-3}$. We see from our discussion of excitons in the bulk (*cf.* Fig. 8.6) that radical changes of the absorption spectrum are to be expected at free-carrier densities of this order. The blue shift exhibited in Fig. 8.6 for bulk GaAs is absent from Fig. 9.6 because of the increased strength of the below-band-gap exciton absorption in a multiple quantum-well structure.

In these experiments the change of the refractive index n was also measured in a degenerate four-wave mixing experiment (Chemla *et al*, 1984). Writing n as $n_0 + n_2^I I$, where I is the pump intensity (see Appendix 5), gives $n_2^I = -0.2\ \mathrm{cm^2\ kW^{-1}}$. Using (A5.13) we find the effective value of $\chi^{(3)}_{\mathrm{esu}} \sim 10^{-1}$ which is some two orders of magnitude larger than the value estimated at the end of §9.4 for bulk excitons. It is, however, consistent with a hard-core picture of two-dimensional excitons in which each exciton has an area in the order of $(a_0{}^*)^2$ from which other excitons are excluded (Chemla and Miller, 1985). In this picture complete saturation occurs when the pump generates one free carrier per exciton area in a multiple quantum well.

9.2 Quantum dots

9.2.1 Electronic structure

We consider an insulating dielectric medium containing 'quantum dots' with a length scale in the order of 2.5 − 25 nm. The dots are made of semiconducting or metallic material and we suppose that they are large enough for the energy-band structure of the material in the dot to be relevant to its electronic structure. Then, as was the case with quantum wells, we may use the envelope-function approximation to investigate the electronic and optical properties of the system. Since the electrons in a dot are confined in all three spatial directions the linear absorption spectrum consists of a series of broadened lines. In what follows we consider a dot in the form of a semiconductor cube of side L with infinite potential barriers on all sides (Schmitt-Rink *et al*, 1987). Let V_{D} be the volume occupied by a dot and its surrounding insulating material (*i.e.*, V_{D}^{-1} is the number of dots per unit volume). Then the one-electron envelope functions for the electron states are

$$\phi_{nlm}(\mathbf{r}) = \sqrt{V_{\mathrm{D}}(2/L)^3}\ \sin(n\pi x/L)\sin(l\pi y/L)\sin(m\pi z/L) \quad (9.17)$$

which has the energy

$$\mathrm{E}^{\mathrm{e}}_{nlm} = (\hbar\pi/L)^2\ 2m_{\mathrm{e}}^{-1}\ (n^2+l^2+m^2)\,, \quad (9.18)$$

where n, l and m are all positive integers and the coordinate axes $Oxyz$

coincide with cube edges. The envelope functions for the hole states are also given by (9.17) but (measuring hole energies down from the conduction-band edge) the corresponding energy levels are

$$\mathrm{E}^{\mathrm{h}}_{nlm} = \mathrm{E}_{\mathrm{g}} + (\hbar\pi/L)^2\, 2m_{\mathrm{h}}^{-1}\, (n^2+l^2+m^2)\,. \tag{9.19}$$

We see that the confinement leads to a blue shift of the fundamental absorption edge equal to $\mathrm{E}^{\mathrm{e}}_{111} - \mathrm{E}^{\mathrm{h}}_{111} - \mathrm{E}_g = 3(\hbar\pi/L)^2/2m_r$ where $m_{\mathrm{r}} = m_{\mathrm{e}}m_{\mathrm{h}}/(m_{\mathrm{e}}+m_{\mathrm{h}})$ is the reduced mass of the electrons and holes. In a metallic quantum dot we are concerned only with the electron states.

9.2.2 Linear-optical properties of semiconducting quantum dots

The argument developed in §9.1.2 for quantum wells that optical transitions take place only between electron and hole states with the same quantum numbers is readily extended to quantum dots. We may also easily modify the calculation of α made there to deal with quantum dots. The answer is the bulk result (8.19b) with the joint density of states (per unit energy range per unit volume) replaced by

$$N_{\mathrm{cv}}^{\mathrm{QW}}(\hbar\omega) = V_{\mathrm{D}}^{-1} \sum_{nlm} \delta(\hbar\omega - \mathrm{E}^{\mathrm{e}}_{nlm} - \mathrm{E}^{\mathrm{h}}_{nlm})\,. \tag{9.20}$$

Equation (9.20) clearly exhibits the anticipated line spectrum. We also see that α is proportional to V_{D}^{-1}. It is useful to recall that the exciton line spectrum derived for a bulk semiconductor in §8.3.4 had a similar form with V_{D} replaced by $(a_0{}^*)^3$. Consequently, when $V_{\mathrm{D}} < (a_0{}^*)^3$ we may expect quantum-dot absorption lines whose strength is greater than that of the bulk-exciton lines. We concentrate on this interesting case. Then the confinement dominates the Coulomb interaction between an electron and hole. That is not to say that excitons do not form in a semiconductor quantum dot; they do, but their binding energy is small in comparison to the confinement energies.

In bulk material the exciton lines are broadened by exciton decay *via* optical phonon creation. In quantum dots the large confinement energies will block this mode of decay. The confinement of the carriers will also increase the coupling to short wavelength phonons, but for GaAs at least, the quantum-dot line width is estimated to be close to that of excitons in the bulk (Schmitt-Rink *et al,* 1987). In practice the absorption lines will be inhomogeneously broadened by fluctuations in the dot dimensions from one dot to another.

We may treat each isolated resonance as a two-level system of the type discussed in §6.2. For 5.6 nm GaAs spheres, Schmitt-Rink *et al* (1987) estimate that the first resonance occurs at $\hbar\omega$=2.1 eV and take the

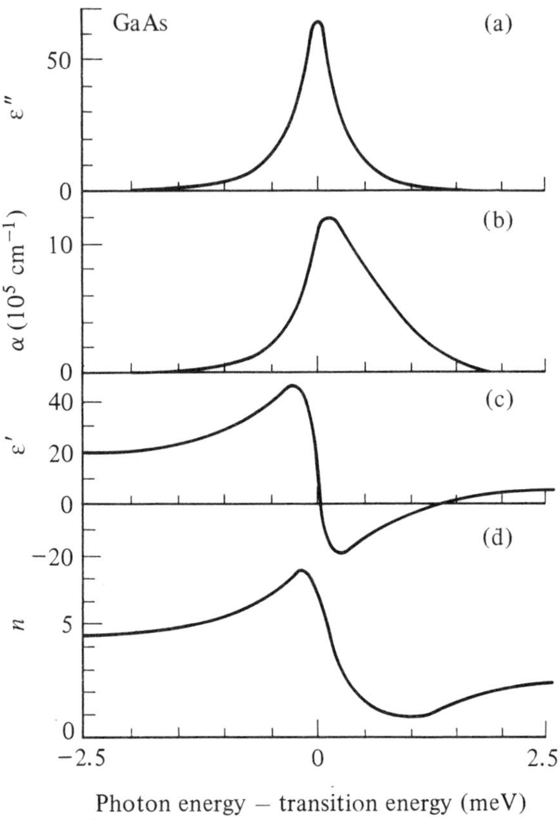

Fig. 9.7 Linear optical properties of a 5.6 nm diameter GaAs sphere.
(a) Imaginary part of the dielectric constant, ε''.
(b) Absorption coefficient, α.
(c) Real part of the dielectric constant, ε'.
(d) Refractive index, n.
Drawn for an artificial filling factor of unity. (After Schmitt-Rink *et al*, 1987.)

line width to be ~0.5 meV. They use a background dielectric constant of 12 in calculating the linear-optical properties exhibited in Fig. 9.7, which is drawn for an artificial filling factor of 1. We see that the predicted resonant absorption is now so large that the refractive index changes significantly in the neighbourhood of resonance.

9.2.3 Nonlinear-optical properties of semiconducting quantum dots

The one-particle states in quantum dots are separated by large energies and the spatial rearrangement of charges responsible for screening is strongly inhibited. Consequently, 'state filling' is the primary mechanism for optical nonlinearities (*cf.* the introductory remarks in

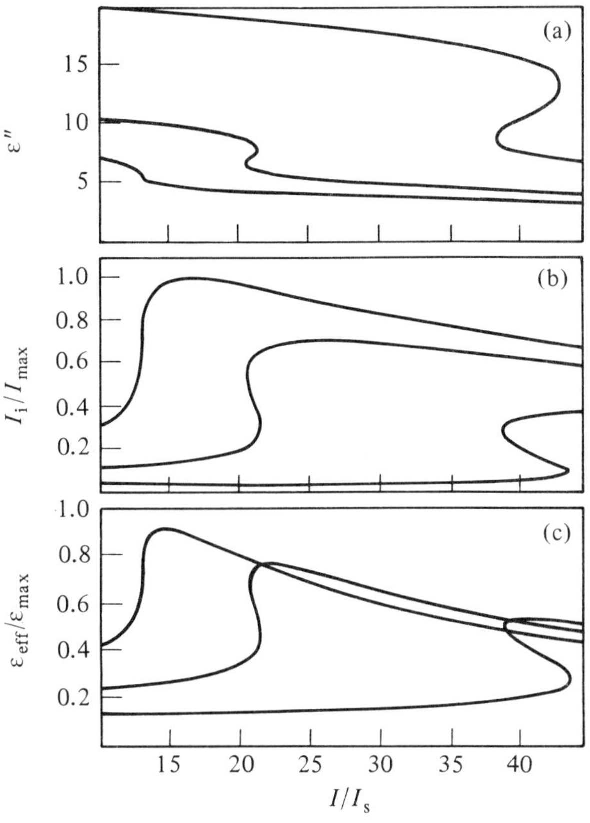

Fig. 9.8 Plots against incident intensity of
(a) the imaginary part of the dielectric constant, ε'',
(b) the normalised local intensity factor,
(c) the normalised real part of the dielectric constant.
Drawn for 5.6 nm diameter GaAs spheres with an artificial filling factor of unity. (After Schmitt-Rink *et al,* 1987.)

§8.4). Indeed, in the nonlinear regime, we may treat each resonance as a *saturable* two-level system of the type already discussed in §§6.2 and 6.3. Since one starts with strong absorption peaks at low intensities, quantum-dot materials offer the possibility of large changes of n and α before saturation occurs. Optical bistable behaviour may also be expected because saturation changes the effective dielectric constant inside a quantum dot and thus modifies the local-field corrections appropriate to the dot and its surroundings (Schmitt-Rink *et al,* 1987). The results of calculations made for the GaAs spheres considered in §9.2.2, on the assumption that they are embedded in free space, are shown in Fig. 9.8 as a function of the incident intensity (normalised to the

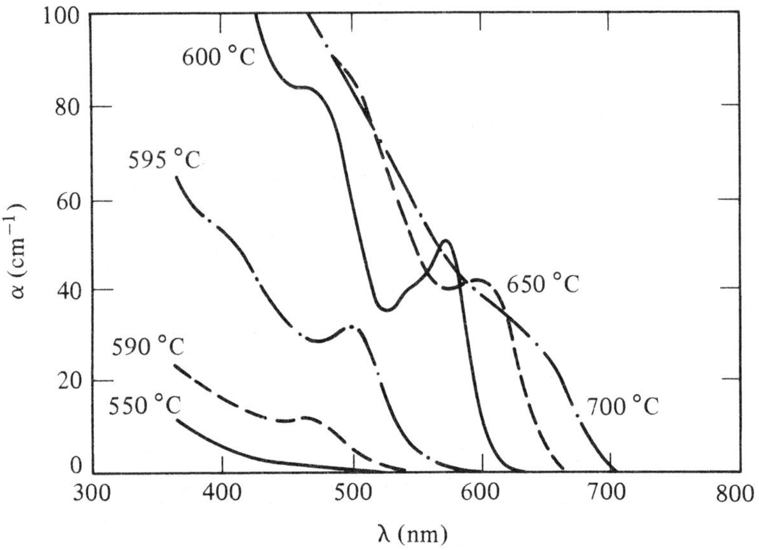

Fig. 9.9 Room-temperature absorption spectra of a glass containing CdSe microcrystals. The temperatures given refer to the heat treatment used for microcrystal nucleation and growth. Banding ascribed to quantum-confinement effects is seen in samples receiving lower-temperature heat treatments. The average measured microcrystal diameters (and distribution deviations) are as follows, in nm: 700 °C: 7.9 (2.2); 650 °C: 4.4 (1.2); 600 °C: 3.0 (−). (After Hall and Borrelli, 1988.)

saturation value I_{sat}) for detunings $\delta = -1.5$, -2.25 and -2.75. The quantities (a) ε'', (b) I_i/I_{max} and (c) $\varepsilon_{eff}/\varepsilon_{max}$ are respectively the imaginary part of the effective interior dielectric constant, the normalised ratio of the interior intensity to the exterior intensity, and the normalised real part of the effective interior dielectric constant. The S-shaped curves which are characteristic of optical bistability are evident.

Experimental studies of nonlinear-optical phenomena in semiconducting quantum dots are still at an early stage. The technological problems associated with the epitaxial growth of semiconductor particles with uniform dimensions on the scale of ~10 nm have yet to be resolved. However, glasses containing semiconductor microcrystals with such small dimensions can be fabricated easily, and these are the subject of much current investigation (Peyghambarian and Gibbs, 1988; references therein). These materials are prepared by heat treating glasses doped with Group II and Group VI elements, to effect within the glass the nucleation and growth of II–VI semiconductor crystals, such as CdS and CdSe. Because the crystallite growth is determined by a thermal process, the resulting particles have a distribution of sizes, so that the absorbing transitions observed in the glass are inhomogeneously broadened. Nevertheless, when the heat treatment conditions are chosen so as to prepare microcrystals with a mean diameter of $\lesssim 5$ nm (size distribution width ~ 1 nm) the effects of carrier confinement are clearly apparent in the sequence of peaks in the room-temperature absorption spectrum (shown in Fig. 9.9), as predicted by the expression (9.20) (Ekimov and Onushchenko, 1984; Hall and Borrelli, 1988). The Maxwell-Garnet equation (Genzel and Martin, 1973) is used to express the average dielectric function $\varepsilon_{av}(\omega)$ of the composite material (glass plus embedded semiconductor microcrystals):

$$\varepsilon_{av}(\omega) = \varepsilon_G \frac{\varepsilon_S(\omega)(1+2p) + 2\varepsilon_G(1-p)}{\varepsilon_S(\omega)(1-p) + \varepsilon_G(2+p)}, \tag{9.21}$$

where $\varepsilon_G \simeq 2.2$ is the dielectric constant of the host glass, p is the fraction of the total volume occupied by semiconductor microcrystals (typically, $p \sim 10^{-3}$), and $\varepsilon_S(\omega)$ is the complex dielectric function of the microcrystals.

The absorptive and refractive nonlinearity of glasses containing microcrystals of larger dimensions (~ 10 nm) has been measured and successfully explained (Peyghambarian and Gibbs, 1988) by calculating $\varepsilon_S(\omega)$ using the plasma theory of Banyai and Koch (1986), as described in §8.4.4. The measured values of the nonlinear refraction coefficient n_2^I are in the range $\sim 10^{-9} - 10^{-8}$ cm^2 kW^{-1} (which, using (A5.13), correspond to effective values of $\chi^{(3)}_{esu} \sim 10^{-10} - 10^{-9}$). These values are several orders of magnitude smaller than those measured in multiple quantum wells (see §9.1.5). Nevertheless, semiconductor-doped glass is of considerable interest for optical switching devices because the nonlinear refractive index exhibits a very fast recovery ($\lesssim 10$ ps), perhaps due to a large cross-section for the capture of free carriers by deep-lying traps (Cotter *et al*, 1988). In contrast, the recovery time of the nonlinear refractive index in quantum wells may be as long as several tens of nanoseconds (determined by direct interband carrier recombination) or ~ 100 ps in proton-bombarded material. The semiconductor-doped glass can also be fashioned into the form of planar- and fibre-waveguides.

The optical nonlinearity of smaller-sized semiconductor microcrystals is under investigation currently.

9.2.4 Metallic quantum dots

Flytzanis *et al* (1986) treat the optical behaviour of metallic spheres randomly dispersed in a dielectric matrix. The spheres support a surface plasmon mode, and both the dipole moments of the spheres and their contribution to the susceptibility tensors are resonantly enhanced when the frequency of the applied optical field approaches that of the surface plasmon. Hache *et al* (1986) use the surface plasmon resonance in phase-conjugation measurements of $\chi^{(3)}(-\omega;\omega,-\omega,\omega)$ for systems containing either Au or Ag spheres. They also calculate $\chi^{(3)}$ using the formulae in terms of momentum matrix elements set out in §4.6, together with the Maxwell-Garnet formula (9.21). To do so they assume that the electrons are free inside each sphere and that the potential barrier at the surface of the sphere is infinitely high. This model is similar to that used in §§9.2.1–9.2.3 to discuss cubic quantum dots, but has the additional complications inherent in the spherical geometry. Hache *et al* find that, like α (see §9.2.2), $\chi^{(3)}$ is proportional to the reciprocal of the sphere volume. For 5 nm spheres they estimate, very roughly, that $\chi^{(3)}_{\text{esu}} \sim 10^{-9}$ and 10^{-10} esu for Au and Ag respectively. The corresponding measured values are an order of magnitude larger in both cases.

APPENDIX 1

Universal constants

Constants used in text

Symbol	Value	Name
a_0	5.29177×10^{-11} m	Bohr radius
c	2.997925×10^{8} m s^{-1}	Velocity of light in vacuum
e	1.60218×10^{-19} C	Modulus of electronic charge
ε_0	8.85419×10^{-12} F m^{-1}	Permittivity of free space
$\hbar = h/2\pi$	1.05457×10^{-34} J s	Dirac constant
	(6.58212×10^{-16} eV s)	
k	1.38066×10^{-23} J K^{-1}	Boltzmann constant
m	9.10939×10^{-31} kg	Electron rest mass
μ_B	9.27402×10^{-24} J T^{-1}	Bohr magneton
μ_0	$4\pi\times10^{-7}$ H m^{-1}	Permeability of free space
D	$10^{-21}/c$ Cm	Debye unit of electric
	(10^{-18} esu)	dipole moment
E_{Ry}	2.17987×10^{-18} J	Rydberg energy
	(13.6056 eV)	

Relationships

$a_0 = 4\pi\varepsilon_0\hbar^2/me^2$ $\qquad$ $\mu_0\varepsilon_0 c^2 = 1$

$\mu_B = e\hbar/2m$ $\qquad$ $E_{Ry} = \hbar^2/2m(a_0)^2$

Benchmarks

$kT = 26$ meV at $T = 300$ K

$= 6.6$ meV at $T = 77$ K

$= 0.34$ meV at $T = 4$ K

1 eV $= 1.60\times10^{-19}$ J

$$\left.\begin{array}{l}\text{vacuum wavelength } \lambda = 1\ \mu\text{m}\\ \text{wavenumber } \overline{\nu} = 10{,}000\ \text{cm}^{-1}\end{array}\right\}\ \text{at } \hbar\omega = \hbar 2\pi c/\lambda = 1.24\ \text{eV}$$

wavenumber $\overline{\nu}$ (cm^{-1}) $= [100\,\lambda(\text{m})]^{-1}$

APPENDIX 2

Relations between esu and SI units for the susceptibilities

There are three main sources of numerical error in work with nonlinear susceptibilities: first, numerical factors that arise from permutation symmetry and the definition of the field; second, the choice of macroscopic or microscopic susceptibilities; and third, inconsistent units.

The numerical factors referred to above are embodied in the factor K which is defined by (2.55) and (2.56), and listed in Table 2.1. These factors must be taken into account when comparing susceptibilities with different frequency arguments. Here we put $K=1$ and consider the units of the macroscopic susceptibilities $\chi^{(n)}$. (The relationship between microscopic and macroscopic susceptibilities and the molecular hyper-polarizabilities are set out in §4.4.3. The units of the hyperpolarisabilities are given in §4.4.1.)

We use SI units throughout the main text, leading to a definition of $\chi^{(n)}$ with the following units:

$$\left.\begin{array}{lll} P^{(n)} = \varepsilon_0 \chi^{(n)} E^n & & \mathrm{C\,m^{-2}} \\ \chi^{(n)} & & (\mathrm{m\,V^{-1}})^{n-1} \\ E & & \mathrm{V\,m^{-1}} \end{array}\right\}\ \mathrm{SI} \tag{A2.1}$$

In the earlier literature it was common to use the cgs/esu system of units; in that case the definition corresponding to (A2.1) is

$$\left.\begin{array}{lll} P^{(n)} = \chi^{(n)} E^n & & \mathrm{statvolt\,cm^{-1}} \\ \chi^{(n)} & & (\mathrm{cm\,statvolt^{-1}})^{n-1} \\ E & & \mathrm{statvolt\,cm^{-1}} \end{array}\right\}\ \mathrm{esu} \tag{A2.2}$$

The statvolt $\mathrm{cm^{-1}}$ is the unit of electric-field strength in esu; it is equivalent in cgs units to $\mathrm{cm^{-1/2}\,g^{1/2}\,s^{-1}}$, and 1 statvolt $\mathrm{cm^{-1}} \equiv 3\times10^4\ \mathrm{V\,m^{-1}}$. The relation between the susceptibilities in A2.1 and A2.2 is:

$$\chi^{(n)}(\mathrm{SI})/\chi^{(n)}(\mathrm{esu}) = 4\pi/(10^{-4}c)^{n-1}, \tag{A2.3}$$

where $c = 3\times10^8$. The linear susceptibility $\chi^{(1)}$ is dimensionless in both systems of units. The relation between the polarisations in A2.1 and A2.2 is:

$$P^{(n)}(\mathrm{SI})/P^{(n)}(\mathrm{esu}) = 10^3/c\,. \qquad \text{(A2.4)}$$

Further relations between esu and SI units appear in Appendix 5.

APPENDIX 3

Tables of electric-dipole susceptibility tensors for isotropic media and the crystal classes

In the tables each tensor element is denoted by its subscripts in a cartesian coordinate system with axes oriented along the directions of the crystallographic axes (see §5.3.4). A bar denotes the negative.

TABLE A3.1 The form of the first-order susceptibility tensor $\chi^{(1)}_{\mu\alpha}$ for the seven crystal systems and for isotropic media. The number in parenthesis is the number of independent nonzero elements.

Triclinic	$\begin{bmatrix} xx & xy & zx \\ xy & yy & yz \\ zx & yz & zz \end{bmatrix}$	(6)
Monoclinic	$\begin{bmatrix} xx & 0 & zx \\ 0 & yy & 0 \\ zx & 0 & zz \end{bmatrix}$	(4)
Orthorhombic	$\begin{bmatrix} xx & 0 & 0 \\ 0 & yy & 0 \\ 0 & 0 & zz \end{bmatrix}$	(3)
Tetragonal **Trigonal** **Hexagonal**	$\begin{bmatrix} xx & 0 & 0 \\ 0 & xx & 0 \\ 0 & 0 & zz \end{bmatrix}$	(2)
Cubic **Isotropic**	$\begin{bmatrix} xx & 0 & 0 \\ 0 & xx & 0 \\ 0 & 0 & xx \end{bmatrix}$	(1)

TABLE A3.2 The form of the second-order susceptibility tensor $\chi^{(2)}_{\mu\alpha\beta}$ for those crystal classes which have no centre of symmetry (the tensor is identically zero for those classes which do not appear). The number in parenthesis is the number of independent nonzero elements.

Triclinic	Class 1	$\begin{bmatrix} xxx & xyy & xzz & xyz & xzy & xzx & xxz & xxy & xyx \\ yxx & yyy & yzz & yyz & yzy & yzx & yxz & yxy & yyx \\ zxx & zyy & zzz & zyz & zzy & zzx & zxz & zxy & zyx \end{bmatrix}$	(27)
Monoclinic	Class 2	$\begin{bmatrix} 0 & 0 & 0 & xyz & xzy & 0 & 0 & xxy & xyx \\ yxx & yyy & yzz & 0 & 0 & yzx & yxz & 0 & 0 \\ 0 & 0 & 0 & zyz & zzy & 0 & 0 & zxy & zyx \end{bmatrix}$	(13)

Class m

$$\begin{bmatrix} xxx & xyy & xzz & 0 & 0 & xzx & xxz & 0 & 0 \\ 0 & 0 & 0 & yyz & yzy & 0 & 0 & yxy & yyx \\ zxx & zyy & zzz & 0 & 0 & zzx & zxz & 0 & 0 \end{bmatrix} \qquad (14)$$

Orthorhombic Class 222

$$\begin{bmatrix} 0 & 0 & 0 & xyz & xzy & 0 & 0 & 0 & 0 \\ 0 & 0 & 0 & 0 & 0 & yzx & yxz & 0 & 0 \\ 0 & 0 & 0 & 0 & 0 & 0 & 0 & zxy & zyx \end{bmatrix} \qquad (6)$$

Class mm2

$$\begin{bmatrix} 0 & 0 & 0 & 0 & 0 & xzx & xxz & 0 & 0 \\ 0 & 0 & 0 & yyz & yzy & 0 & 0 & 0 & 0 \\ zxx & zyy & zzz & 0 & 0 & 0 & 0 & 0 & 0 \end{bmatrix} \qquad (7)$$

Tetragonal Class 4

$$\begin{bmatrix} 0 & 0 & 0 & xyz & xzy & xzx & xxz & 0 & 0 \\ 0 & 0 & 0 & xxz & xzx & \overline{xzy} & \overline{xyz} & 0 & 0 \\ zxx & zxx & zzz & 0 & 0 & 0 & 0 & zxy & \overline{zxy} \end{bmatrix} \qquad (7)$$

Class $\bar{4}$

$$\begin{bmatrix} 0 & 0 & 0 & xyz & xzy & xzx & xxz & 0 & 0 \\ 0 & 0 & 0 & \overline{xxz} & \overline{xzx} & xzy & xyz & 0 & 0 \\ zxx & \overline{zxx} & 0 & 0 & 0 & 0 & 0 & zxy & zxy \end{bmatrix} \qquad (6)$$

Class 422

$$\begin{bmatrix} 0 & 0 & 0 & xyz & xzy & 0 & 0 & 0 & 0 \\ 0 & 0 & 0 & 0 & 0 & \overline{xzy} & \overline{xyz} & 0 & 0 \\ 0 & 0 & 0 & 0 & 0 & 0 & 0 & zxy & \overline{zxy} \end{bmatrix} \qquad (3)$$

Class 4mm

$$\begin{bmatrix} 0 & 0 & 0 & 0 & 0 & xzx & xxz & 0 & 0 \\ 0 & 0 & 0 & xxz & xzx & 0 & 0 & 0 & 0 \\ zxx & zxx & zzz & 0 & 0 & 0 & 0 & 0 & 0 \end{bmatrix} \qquad (4)$$

Class $\bar{4}2m$

$$\begin{bmatrix} 0 & 0 & 0 & xyz & xzy & 0 & 0 & 0 & 0 \\ 0 & 0 & 0 & 0 & 0 & xzy & xyz & 0 & 0 \\ 0 & 0 & 0 & 0 & 0 & 0 & 0 & zxy & zxy \end{bmatrix} \qquad (3)$$

Cubic Class 432

$$\begin{bmatrix} 0 & 0 & 0 & xyz & \overline{xyz} & 0 & 0 & 0 & 0 \\ 0 & 0 & 0 & 0 & 0 & xyz & \overline{xyz} & 0 & 0 \\ 0 & 0 & 0 & 0 & 0 & 0 & 0 & xyz & \overline{xyz} \end{bmatrix} \qquad (1)$$

Class $\bar{4}3m$

$$\begin{bmatrix} 0 & 0 & 0 & xyz & xyz & 0 & 0 & 0 & 0 \\ 0 & 0 & 0 & 0 & 0 & xyz & xyz & 0 & 0 \\ 0 & 0 & 0 & 0 & 0 & 0 & 0 & xyz & xyz \end{bmatrix} \qquad (1)$$

Class 23

$$\begin{bmatrix} 0 & 0 & 0 & xyz & xzy & 0 & 0 & 0 & 0 \\ 0 & 0 & 0 & 0 & 0 & xyz & xzy & 0 & 0 \\ 0 & 0 & 0 & 0 & 0 & 0 & 0 & xyz & xzy \end{bmatrix} \qquad (2)$$

Trigonal Class 3

$$\begin{bmatrix} xxx & \overline{xxx} & 0 & xyz & xzy & xzx & xxz & \overline{yyy} & \overline{yyy} \\ \overline{yyy} & yyy & 0 & xxz & xzx & \overline{xzy} & \overline{xyz} & \overline{xxx} & \overline{xxx} \\ zxx & zxx & zzz & 0 & 0 & 0 & 0 & zxy & \overline{zxy} \end{bmatrix} \qquad (9)$$

Class 32

$$\begin{bmatrix} xxx & \overline{xxx} & 0 & xyz & xzy & 0 & 0 & 0 & 0 \\ 0 & 0 & 0 & 0 & 0 & \overline{xzy} & \overline{xyz} & \overline{xxx} & \overline{xxx} \\ 0 & 0 & 0 & 0 & 0 & 0 & 0 & zxy & \overline{zxy} \end{bmatrix} \qquad (4)$$

Class 3m

$$\begin{bmatrix} 0 & 0 & 0 & 0 & 0 & xzx & xxz & \overline{yyy} & \overline{yyy} \\ \overline{yyy} & yyy & 0 & xxz & xzx & 0 & 0 & 0 & 0 \\ zxx & zxx & zzz & 0 & 0 & 0 & 0 & 0 & 0 \end{bmatrix} \qquad (5)$$

Table A3.2 (continued)

Hexagonal Class 6

$$\begin{bmatrix} 0 & 0 & 0 & xyz & xzy & xzx & xxz & 0 & 0 \\ 0 & 0 & 0 & xxz & xzx & \overline{xzy} & \overline{xyz} & 0 & 0 \\ zxx & zxx & zzz & 0 & 0 & 0 & 0 & zxy & \overline{zxy} \end{bmatrix} \qquad (7)$$

Class $\bar{6}$

$$\begin{bmatrix} xxx & \overline{xxx} & 0 & 0 & 0 & 0 & 0 & \overline{yyy} & \overline{yyy} \\ \overline{yyy} & yyy & 0 & 0 & 0 & 0 & 0 & \overline{xxx} & \overline{xxx} \\ 0 & 0 & 0 & 0 & 0 & 0 & 0 & 0 & 0 \end{bmatrix} \qquad (2)$$

Class 622

$$\begin{bmatrix} 0 & 0 & 0 & xyz & xzy & 0 & 0 & 0 & 0 \\ 0 & 0 & 0 & 0 & 0 & \overline{xzy} & \overline{xyz} & 0 & 0 \\ 0 & 0 & 0 & 0 & 0 & 0 & 0 & zxy & \overline{zxy} \end{bmatrix} \qquad (3)$$

Class 6mm

$$\begin{bmatrix} 0 & 0 & 0 & 0 & 0 & xzx & xxz & 0 & 0 \\ 0 & 0 & 0 & xxz & xzx & 0 & 0 & 0 & 0 \\ zxx & zxx & zzz & 0 & 0 & 0 & 0 & 0 & 0 \end{bmatrix} \qquad (4)$$

Class $\bar{6}$m2

$$\begin{bmatrix} 0 & 0 & 0 & 0 & 0 & 0 & 0 & \overline{yyy} & \overline{yyy} \\ \overline{yyy} & yyy & 0 & 0 & 0 & 0 & 0 & 0 & 0 \\ 0 & 0 & 0 & 0 & 0 & 0 & 0 & 0 & 0 \end{bmatrix} \qquad (1)$$

TABLE A3.3 The form of the third-order susceptibility tensor $\chi^{(3)}_{\mu\alpha\beta\gamma}$ for all of the 32 crystal classes and isotropic media (incorporating corrections by Shang and Hsu (1987)).

Triclinic

For both classes (1 and $\bar{1}$) there are 81 independent nonzero elements.

Monoclinic

For all three classes (2, m and 2/m) there are 41 independent nonzero elements, consisting of:

3 elements with suffixes all equal,
18 elements with suffixes equal in pairs,
12 elements with suffixes having two *y*s, one *x* and one *z*,
4 elements with suffixes having three *x*s and one *z*,
4 elements with suffixes having three *z*s and one *x*.

Orthorhombic

For all three classes (222, mm2 and mmm) there are 21 independent nonzero elements, consisting of:

3 elements with suffixes all equal,
18 elements with suffixes equal in pairs.

Tetragonal

For the three classes 4, $\bar{4}$ and 4/m, there are 41 nonzero elements of which only 21 are independent. They are:

	$xxxx = yyyy$	$zzzz$	
$zzxx = zzyy$	$xyzz = \overline{yxzz}$	$xxyy = yyxx$	$xxxy = \overline{yyyx}$
$xxzz = yyzz$	$zzxy = \overline{zzyx}$	$xyxy = yxyx$	$xxyx = \overline{yyxy}$
$zxzx = zyzy$	$xzyz = \overline{yzxz}$	$xyyx = yxxy$	$xyxx = \overline{yxyy}$
$xzxz = yzyz$	$zxzy = \overline{zyzx}$		$yxxx = \overline{xyyy}$
$zxxz = zyyz$	$zxyz = \overline{zyxz}$		
$xzzx = yzzy$	$xzzy = \overline{yzzx}$		

For the four classes 422, 4mm, 4/mmm, and $\bar{4}$2m, there are 21 nonzero elements of which only 11 are independent. They are:

$$\begin{array}{lll} & xxxx = yyyy & zzzz \\ yyzz = xxzz & yzzy = xzzx & xxyy = yyxx \\ zzyy = zzxx & yzyz = xzxz & xyxy = yxyx \\ zyyz = zxxz & zyzy = zxzx & xyyx = yxxy \end{array}$$

Cubic

For the two classes 23 and m3, there are 21 nonzero elements of which only 7 are independent. They are:

$$\begin{array}{l} xxxx = yyyy = zzzz \\ yyzz = zzxx = xxyy \\ zzyy = xxzz = yyxx \\ yzyz = zxzx = xyxy \\ zyzy = xzxz = yxyx \\ yzzy = zxxz = xyyx \\ zyyz = xzzx = yxxy \end{array}$$

For the three classes 432, $\bar{4}$3m and m3m, there are 21 nonzero elements of which only 4 are independent. They are:

$$\begin{array}{l} xxxx = yyyy = zzzz \\ yyzz = zzyy = zzxx = xxzz = xxyy = yyxx \\ yzyz = zyzy = zxzx = xzxz = xyxy = yxyx \\ yzzy = zyyz = zxxz = xzzx = xyyx = yxxy \end{array}$$

Trigonal

For the two classes 3 and $\bar{3}$, there are 73 nonzero elements of which only 27 are independent. They are:

$$zzzz$$

$$xxxx = yyyy = xxyy + xyyx + xyxy \left\{ \begin{array}{l} xxyy = yyxx \\ xyyx = yxxy \\ xyxy = yxyx \end{array} \right.$$

$$\begin{array}{ll} yyzz = xxzz & xyzz = \overline{yxzz} \\ zzyy = zzxx & zzxy = \overline{zzyx} \\ zyyz = zxxz & zxyz = \overline{zyxz} \\ yzzy = xzzx & xzzy = \overline{yzzx} \\ yzyz = xzxz & xzyz = \overline{yzxz} \\ zyzy = zxzx & zxzy = \overline{zyzx} \end{array}$$

$$xxxy = \overline{yyyx} = yyxy + yxyy + xyyy \left\{ \begin{array}{l} yyxy = \overline{xxyx} \\ yxyy = \overline{xyxx} \\ xyyy = \overline{yxxx} \end{array} \right.$$

$$\begin{array}{l} yyyz = \overline{yxxz} = \overline{xyxz} = \overline{xxyz} \\ yyzy = \overline{yxzx} = \overline{xyzx} = \overline{xxzy} \\ yzyy = \overline{yzxx} = \overline{xzyx} = \overline{xzxy} \\ zyyy = \overline{zyxx} = \overline{zxyx} = \overline{zxxy} \\ xxxz = \overline{xyyz} = \overline{yxyz} = \overline{yyxz} \\ xxzx = \overline{xyzy} = \overline{yxzy} = \overline{yyzx} \\ xzxx = \overline{yzxy} = \overline{yzyx} = \overline{xzyy} \\ zxxx = \overline{zxyy} = \overline{zyxy} = \overline{zyyx} \end{array}$$

Table A3.3 (continued)

For the three classes 3m, $\bar{3}$m and 32 there are 37 nonzero elements of which only 14 are independent. They are:

$$zzzz$$

$$xxxx = yyyy = xxyy + xyyx + xyxy \begin{cases} xxyy = yyxx \\ xyyx = yxxy \\ xyxy = yxyx \end{cases}$$

$$\begin{array}{ll} yyzz = xxzz & xxxz = \overline{xyyz} = \overline{yxyz} = \overline{yyxz} \\ zzyy = zzxx & xxzx = \overline{xyzy} = \overline{yxzy} = \overline{yyzx} \\ zyyz = zxxz & xzxx = \overline{xzyy} = \overline{yzxy} = \overline{yzyx} \\ yzzy = xzzx & zxxx = \overline{zxyy} = \overline{zyxy} = \overline{zyyx} \\ yzyz = xzxz & \\ zyzy = zxzx & \end{array}$$

Hexagonal

For the three classes 6, $\bar{6}$ and 6/m, there are 41 nonzero elements of which only 19 are independent. They are:

$$zzzz$$

$$xxxx = yyyy = xxyy + xyyx + xyxy \begin{cases} xxyy = yyxx \\ xyyx = yxxy \\ xyxy = yxyx \end{cases}$$

$$\begin{array}{ll} yyzz = xxzz & xyzz = \overline{yxzz} \\ zzyy = zzxx & zzxy = \overline{zzyx} \\ zyyz = zxxz & zxyz = \overline{zyxz} \\ yzzy = xzzx & xzzy = \overline{yzzx} \\ yzyz = xzxz & xzyz = \overline{yzxz} \\ zyzy = zxzx & zxzy = \overline{zyzx} \end{array}$$

$$xxxy = \overline{yyyx} = yyxy + yxyy + xyyy \begin{cases} yyxy = \overline{xxyx} \\ yxyy = \overline{xyxx} \\ xyyy = \overline{yxxx} \end{cases}$$

For the four classes 622, 6mm, 6/mmm and $\bar{6}$m2, there are 21 nonzero elements of which only 10 are independent. They are:

$$zzzz$$

$$xxxx = yyyy = xxyy + xyyx + xyxy \begin{cases} xxyy = yyxx \\ xyyx = yxxy \\ xyxy = yxyx \end{cases}$$

$$\begin{array}{l} yyzz = xxzz \\ zzyy = zzxx \\ zyyz = zxxz \\ yzzy = xzzx \\ yzyz = xzxz \\ zyzy = zxzx \end{array}$$

Isotropic

There are 21 nonzero elements of which only 3 are independent. They are:

$$\begin{array}{l} xxxx = yyyy = zzzz \\ yyzz = zzyy = zzxx = xxzz = xxyy = yyxx \\ yzyz = zyzy = zxzx = xzxz = xyxy = yxyx \\ yzzy = zyyz = zxxz = xzzx = xyyx = yxxy \\ xxxx = xxyy + xyxy + xyyx \end{array}$$

Tables of **d**-tensors and d_{eff}-coefficients for the crystal classes

TABLE A4.1 The form of the **d**-tensor for the seven crystal systems. The coefficients $d_{\mu m}$ are defined by (5.23) in terms of the elements of the second-order susceptibility $\boldsymbol{\chi}^{(2)}$, where the index m runs over the values 1–6 with the correspondence given by (5.21), and the cartesian label μ is given the values 1, 2 and 3 representing the x, y and z directions respectively. The main entries are derived from the elements of Table A3.2 by invoking the permutation symmetry properties described in §5.1.2. A bar over an entry denotes the negative. The supplementary relations given on the right apply when Kleinman symmetry holds. An illustrative example of the derivation of the latter relations is given at the end of the table.

Triclinic	Class 1	$\begin{bmatrix} d_{11} & d_{12} & d_{13} & d_{14} & d_{15} & d_{16} \\ d_{21} & d_{22} & d_{23} & d_{24} & d_{25} & d_{26} \\ d_{31} & d_{32} & d_{33} & d_{34} & d_{35} & d_{36} \end{bmatrix}$	$\begin{cases} d_{12} = d_{26} \\ d_{13} = d_{35} \\ d_{14} = d_{25} = d_{36} \\ d_{15} = d_{31} \\ d_{16} = d_{21} \\ d_{23} = d_{34} \\ d_{24} = d_{32} \end{cases}$
Monoclinic	Class 2	$\begin{bmatrix} 0 & 0 & 0 & d_{14} & 0 & d_{16} \\ d_{21} & d_{22} & d_{23} & 0 & d_{25} & 0 \\ 0 & 0 & 0 & d_{34} & 0 & d_{36} \end{bmatrix}$	$\begin{cases} d_{14} = d_{25} = d_{36} \\ d_{16} = d_{21} \\ d_{23} = d_{34} \end{cases}$
	Class m	$\begin{bmatrix} d_{11} & d_{12} & d_{13} & 0 & d_{15} & 0 \\ 0 & 0 & 0 & d_{24} & 0 & d_{26} \\ d_{31} & d_{32} & d_{33} & 0 & d_{35} & 0 \end{bmatrix}$	$\begin{cases} d_{12} = d_{26} \\ d_{13} = d_{35} \\ d_{15} = d_{31} \\ d_{24} = d_{32} \end{cases}$
Orthorhombic	Class 222	$\begin{bmatrix} 0 & 0 & 0 & d_{14} & 0 & 0 \\ 0 & 0 & 0 & 0 & d_{25} & 0 \\ 0 & 0 & 0 & 0 & 0 & d_{36} \end{bmatrix}$	$\{\, d_{14} = d_{25} = d_{36}$

Table A4.1 (continued)

	Class mm2	$\begin{bmatrix} 0 & 0 & 0 & 0 & d_{15} & 0 \\ 0 & 0 & 0 & d_{24} & 0 & 0 \\ d_{31} & d_{32} & d_{33} & 0 & 0 & 0 \end{bmatrix}$	$\begin{cases} d_{15} = d_{31} \\ d_{24} = d_{32} \end{cases}$
Tetragonal	Class 4	$\begin{bmatrix} 0 & 0 & 0 & d_{14} & d_{15} & 0 \\ 0 & 0 & 0 & d_{15} & \bar{d}_{14} & 0 \\ d_{31} & d_{31} & d_{33} & 0 & 0 & 0 \end{bmatrix}$	$\begin{cases} d_{15} = d_{31} \\ d_{14} = 0 \end{cases}$
	Class $\bar{4}$	$\begin{bmatrix} 0 & 0 & 0 & d_{14} & d_{15} & 0 \\ 0 & 0 & 0 & \bar{d}_{15} & d_{14} & 0 \\ d_{31} & \bar{d}_{31} & 0 & 0 & 0 & d_{36} \end{bmatrix}$	$\begin{cases} d_{14} = d_{36} \\ d_{15} = d_{31} \end{cases}$
	Class 422	$\begin{bmatrix} 0 & 0 & 0 & d_{14} & 0 & 0 \\ 0 & 0 & 0 & 0 & \bar{d}_{14} & 0 \\ 0 & 0 & 0 & 0 & 0 & 0 \end{bmatrix}$	$\{ d_{14} = 0$
	Class 4mm	$\begin{bmatrix} 0 & 0 & 0 & 0 & d_{15} & 0 \\ 0 & 0 & 0 & d_{15} & 0 & 0 \\ d_{31} & d_{31} & d_{33} & 0 & 0 & 0 \end{bmatrix}$	$\{ d_{15} = d_{31}$
	Class $\bar{4}$2m	$\begin{bmatrix} 0 & 0 & 0 & d_{14} & 0 & 0 \\ 0 & 0 & 0 & 0 & d_{14} & 0 \\ 0 & 0 & 0 & 0 & 0 & d_{36} \end{bmatrix}$	$\{ d_{14} = d_{36}$
Cubic	Class 432	$\begin{bmatrix} 0 & 0 & 0 & 0 & 0 & 0 \\ 0 & 0 & 0 & 0 & 0 & 0 \\ 0 & 0 & 0 & 0 & 0 & 0 \end{bmatrix}$	
	Class $\bar{4}$3m	$\begin{bmatrix} 0 & 0 & 0 & d_{14} & 0 & 0 \\ 0 & 0 & 0 & 0 & d_{14} & 0 \\ 0 & 0 & 0 & 0 & 0 & d_{14} \end{bmatrix}$	
	Class 23	$\begin{bmatrix} 0 & 0 & 0 & d_{14} & 0 & 0 \\ 0 & 0 & 0 & 0 & d_{14} & 0 \\ 0 & 0 & 0 & 0 & 0 & d_{14} \end{bmatrix}$	
Trigonal	Class 3	$\begin{bmatrix} d_{11} & \bar{d}_{11} & 0 & d_{14} & d_{15} & \bar{d}_{22} \\ \bar{d}_{22} & d_{22} & 0 & d_{15} & \bar{d}_{14} & \bar{d}_{11} \\ d_{31} & d_{31} & d_{33} & 0 & 0 & 0 \end{bmatrix}$	$\begin{cases} d_{15} = d_{31} \\ d_{14} = 0 \end{cases}$
	Class 32	$\begin{bmatrix} d_{11} & \bar{d}_{11} & 0 & d_{14} & 0 & 0 \\ 0 & 0 & 0 & 0 & \bar{d}_{14} & \bar{d}_{11} \\ 0 & 0 & 0 & 0 & 0 & 0 \end{bmatrix}$	$\{ d_{14} = 0$
	Class 3m	$\begin{bmatrix} 0 & 0 & 0 & 0 & d_{15} & \bar{d}_{22} \\ \bar{d}_{22} & d_{22} & 0 & d_{15} & 0 & 0 \\ d_{31} & d_{31} & d_{33} & 0 & 0 & 0 \end{bmatrix}$	$\{ d_{15} = d_{31}$

Hexagonal Class 6

$$\begin{bmatrix} 0 & 0 & 0 & d_{14} & d_{15} & 0 \\ 0 & 0 & 0 & d_{15} & \overline{d_{14}} & 0 \\ d_{31} & d_{31} & d_{33} & 0 & 0 & 0 \end{bmatrix} \qquad \begin{cases} d_{15} = d_{31} \\ d_{14} = 0 \end{cases}$$

Class $\bar{6}$

$$\begin{bmatrix} d_{11} & \overline{d_{11}} & 0 & 0 & 0 & \overline{d_{22}} \\ \overline{d_{22}} & d_{22} & 0 & 0 & 0 & \overline{d_{11}} \\ 0 & 0 & 0 & 0 & 0 & 0 \end{bmatrix}$$

Class 622

$$\begin{bmatrix} 0 & 0 & 0 & d_{14} & 0 & 0 \\ 0 & 0 & 0 & 0 & \overline{d_{14}} & 0 \\ 0 & 0 & 0 & 0 & 0 & 0 \end{bmatrix} \qquad \{\, d_{14} = 0$$

Class 6mm

$$\begin{bmatrix} 0 & 0 & 0 & 0 & d_{15} & 0 \\ 0 & 0 & 0 & d_{15} & 0 & 0 \\ d_{31} & d_{31} & d_{33} & 0 & 0 & 0 \end{bmatrix} \qquad \{\, d_{15} = d_{31}$$

Class $\bar{6}$m2

$$\begin{bmatrix} 0 & 0 & 0 & 0 & 0 & \overline{d_{22}} \\ \overline{d_{22}} & d_{22} & 0 & 0 & 0 & 0 \\ 0 & 0 & 0 & 0 & 0 & 0 \end{bmatrix}$$

There follows an illustrative example of the derivation of the supplementary relations which apply when Kleinman symmetry holds:

Consider the tetragonal crystal class 4. The entry in the above table indicates that $d_{25} = -d_{14}$. If Kleinman symmetry holds then $\chi^{(2)}_{yzx} = \chi^{(2)}_{yxz} = \chi^{(2)}_{xyz} = \chi^{(2)}_{xzy}$, which may be expressed as $d_{25} = d_{14}$. By combining these relations, we obtain the result $d_{14} = -d_{14} = 0$.

TABLE A4.2 Formulae for d_{eff} (defined by (5.28)) for the uniaxial crystal classes assuming Kleinman symmetry is valid. The waves propagate at an angle θ to the optic axis (**z**) in a plane orientated at an angle ϕ to the crystal (**x**) axis. Extraordinary and ordinary rays are denoted e and o respectively. (From Zernike and Midwinter (1973), by permission.)

Crystal class	Two e rays and one o ray	Two o rays and one e ray
6 and 4	0	$d_{15}\sin\theta$
622 and 422	0	0
6mm and 4mm	0	$d_{15}\sin\theta$
$\bar{6}$m2	$d_{22}\cos^2\theta\cos 3\phi$	$-d_{22}\cos\theta\sin 3\phi$
3m	$d_{22}\cos^2\theta\cos 3\phi$	$d_{15}\sin\theta - d_{22}\cos\theta\sin 3\phi$
$\bar{6}$	$\cos^2\theta(d_{11}\sin 3\phi + d_{22}\cos 3\phi)$	$\cos\theta(d_{11}\cos 3\phi - d_{22}\sin 3\phi)$
3	$\cos^2\theta(d_{11}\sin 3\phi + d_{22}\cos 3\phi)$	$d_{15}\sin\theta + \cos\theta(d_{11}\cos 3\phi - d_{22}\sin 3\phi)$
32	$d_{11}\cos^2\theta\sin 3\phi$	$d_{11}\cos\theta\cos 3\phi$
$\bar{4}$	$\sin 2\theta(d_{14}\cos 2\phi - d_{15}\sin 2\phi)$	$-\sin\theta(d_{14}\sin 2\phi + d_{15}\cos 2\phi)$
$\bar{4}$2m	$d_{14}(\sin 2\theta\cos 2\phi)$	$-d_{14}\sin\theta\sin 2\phi$

APPENDIX 5

Relations between n_2 and $\chi^{(3)}$

The numerical relationship between the third-order susceptibility $\chi^{(3)}$ and the nonlinear refraction coefficient n_2 depends on several factors, which include: whether n_2 is defined in terms of the field-amplitude squared or intensity; whether or not a factor of $\frac{1}{2}$ is included in the definition of field amplitudes and yet another factor of $\frac{1}{2}$ in the definition of n_2; whether or not the K factor is incorporated into the definition of the susceptibility; and, not least, which system of units is being used. Thus, there are at least 2^5 distinct relationships which can be concocted, and it is hardly surprising, therefore, that this is a continuing source of discrepancy and some confusion. Here we set out the definitions followed in this book, and derive corresponding relations for the principal variations encountered in the literature. Some similar relations are given by Gibbs (1985).

We begin by defining the nonlinear refraction coefficient (SI units) as in (6.62):

$$n = n_0 + \delta n = n_0 + n_2|E|^2 , \tag{A5.1}$$

where (adopting the scalar notation of §2.3.4) E is the slowly-varying envelope of a quasi-monochromatic field:

$$\mathbf{E} = \tfrac{1}{2}\Big[E\,\mathbf{e}\exp(-i\omega t) + E^*\mathbf{e}^*\exp(i\omega t)\Big] . \tag{A5.2}$$

The total polarisation at ω is of amplitude P, given by (6.59):

$$\begin{aligned} P &\simeq P^{(1)} + P^{(3)} \\ &= \varepsilon_0\Big\{\chi^{(1)}(-\omega;\omega) + \tfrac{3}{4}\chi^{(3)}(-\omega;\omega,-\omega,\omega)|E|^2\Big\}E , \end{aligned} \tag{A5.3}$$

where the factor $K(-\omega;\omega,-\omega,\omega) = \frac{3}{4}$ is calculated from (2.56) and is listed in Table 2.1. In §6.3 it is shown that the corresponding expression for n_2 is

$$n_2 = \frac{3\mathrm{Re}\,\chi^{(3)}(-\omega;\omega,-\omega,\omega)}{8n_0} , \tag{A5.4}$$

and we note that n_2 has the same units as $\chi^{(3)}$, namely $\mathrm{m}^2\ \mathrm{V}^{-2}$.

A common alternative definition for the nonlinear refraction coefficient is in terms of the intensity I_ω:

$$n = n_0 + n_2^I I_\omega , \tag{A5.5}$$

where the cycle-averaged intensity is

$$I_\omega = \tfrac{1}{2}\varepsilon_0 c\, n_0 |E|^2 . \tag{A5.6}$$

It follows that

$$n_2^I = 2n_2/\varepsilon_0 c\, n_0 \tag{A5.7}$$

and

$$n_2^I = \frac{3\mathrm{Re}\,\chi^{(3)}(-\omega;\omega,-\omega,\omega)}{4\,\varepsilon_0 c\, n_0^2} , \tag{A5.8}$$

where n_2^I takes the units $\mathrm{m^2\,W^{-1}}$.

We now give the corresponding set of definitions when the esu system is used. In place of (A5.1) is the definition

$$n = n_0 + n_2(\mathrm{esu})\,|E|^2 , \tag{A5.9}$$

where E is expressed in statvolt cm^{-1}, and 1 statvolt$=(c/10^4)$ V m^{-1} with $c=3\times10^8$. The dimensions of n_2(esu) are therefore (statvolt $\mathrm{cm}^{-1})^{-2}$. The polarisation is given by $P=(\varepsilon-1)E/4\pi$, the linear refractive index is $n_0=\sqrt{1+4\pi\,\mathrm{Re}\,\chi^{(1)}_{\mathrm{esu}}(-\omega;\omega)}$ and, corresponding to (6.60), the dielectric constant ε is expressed in esu as

$$\varepsilon = 1 + 4\pi\left[\chi^{(1)}_{\mathrm{esu}}(-\omega;\omega) + \tfrac{3}{4}\,\chi^{(3)}_{\mathrm{esu}}(-\omega;\omega,-\omega,\omega)\,|E|^2\right]. \tag{A5.10}$$

Thus, corresponding to (A5.4), the nonlinear refraction coefficient is given by

$$n_2(\mathrm{esu}) = \frac{3\pi}{2n_0}\,\mathrm{Re}\,\chi^{(3)}_{\mathrm{esu}}(-\omega;\omega,-\omega,\omega) . \tag{A5.11}$$

When measured values are quoted in the current literature, the most commonly-used coefficients are n_2 ($\mathrm{m^2\,V^{-2}}$), n_2^I ($\mathrm{m^2\,W^{-1}}$), n_2(esu) and $\chi^{(3)}_{\mathrm{esu}}$. The relations between the first two and last two are (A5.7) and (A5.11) respectively. The relations between the coefficients in different systems of units (with $c=3\times10^8$ throughout) are

$$n_2^I(\mathrm{m^2W^{-1}}) = \frac{120\pi^2}{c\,n_0^2}\,\mathrm{Re}\,\chi^{(3)}_{\mathrm{esu}}(-\omega;\omega,-\omega,\omega)$$

$$= 3.9\times10^{-6}\,\mathrm{Re}\,\chi^{(3)}_{\mathrm{esu}}(-\omega;\omega,-\omega,\omega)\,/n_0^2 \tag{A5.12a}$$

and

$$n_2^I(\mathrm{m^2W^{-1}}) = \frac{80\pi}{c\,n_0}\,n_2(\mathrm{esu})$$

$$= 8.4\times10^{-7}\,n_2(\mathrm{esu})\,/n_0\,. \qquad \text{(A5.13a)}$$

If n_2^I is expressed in the practical units of $\mathrm{cm^2\ kW^{-1}}$, then

$$n_2^I(\mathrm{cm^2\ kW^{-1}}) = 39\ \mathrm{Re}\,\chi^{(3)}_{\mathrm{esu}}(-\omega;\omega,-\omega,\omega)\,/n_0^2 \qquad \text{(A5.12b)}$$

and

$$n_2^I(\mathrm{cm^2\ kW^{-1}}) = 8.4\,n_2(\mathrm{esu})\,/n_0\,. \qquad \text{(A5.13b)}$$

For a typical Group III – V semiconductor with refractive index $n_0 \simeq 3.5$, these provide the easily-remembered relations $n_2^I(\mathrm{cm^2\ kW^{-1}}) \simeq 3\,\mathrm{Re}\,\chi^{(3)}_{\mathrm{esu}}$ and $n_2^I(\mathrm{cm^2\ kW^{-1}}) \simeq 2n_2(\mathrm{esu})$.

In §2.3, an explicit numerical factor $K(-\omega_\sigma;\omega_1,\dots,\omega_n)$ is included in the general expression (2.59) for the nonlinear polarisation. The advantages of doing so (and of *not* absorbing this K factor into the definitions of the susceptibilities themselves) are discussed in §2.3, and this is the procedure followed in this book. However, this is not followed universally, and it is quite common to see the factor $K=\frac{3}{4}$ incorporated into the definition of the susceptibility $\chi^{(3)}(-\omega;\omega,-\omega,\omega)$, so that the expression (A5.3) is replaced by

$$P = \varepsilon_0\left[\chi^{(1)}(-\omega;\omega) + \chi^{(3)\prime}(-\omega;\omega,-\omega,\omega)|E|^2\right]E\,. \qquad \text{(A5.14)}$$

Similarly, the esu expression (A5.10) is replaced by

$$\varepsilon = 1 + 4\pi\left[\chi^{(1)}_{\mathrm{esu}}(-\omega;\omega) + \chi^{(3)\prime}_{\mathrm{esu}}(-\omega;\omega,-\omega,\omega)|E|^2\right]. \qquad \text{(A5.15)}$$

The relations (A5.4), (A5.8) and (A5.11) then still apply, but with the substitutions $\chi^{(3)} \rightarrow \frac{4}{3}\chi^{(3)\prime}$ (in SI or esu as appropriate) made throughout. If, in *addition* to this, the field is defined as

$$\mathbf{E} = E\,\mathbf{e}\exp(-i\omega t) + E^*\mathbf{e}^*\exp(i\omega t)\,, \qquad \text{(A5.16)}$$

(*i.e.*, the factor $\frac{1}{2}$ is omitted from (A5.2)), then the formula (A5.6) for the intensity becomes $I_\omega = 2\varepsilon_0 c\,n_0\,|E|^2$, and the resulting expression for n_2^I, in place of (A5.8), is

$$n_2^I(\mathrm{m^2W^{-1}}) = \frac{\mathrm{Re}\,\chi^{(3)\prime}(-\omega;\omega,-\omega,\omega)}{4\,\varepsilon_0\,c\,n_0^2}\,. \qquad \text{(A5.17)}$$

Yet another starting point sometimes encountered in the literature is the definition $n = n_0 + n_2 \langle\mathbf{E}\cdot\mathbf{E}\rangle = n_0 + \frac{1}{2}n_2|E|^2$ in place of (A5.1) and (A5.9), and this has the obvious effect of introducing an additional factor of 2 into the various relations for n_2 in either system of units.

APPENDIX 6

The envelope-function approximation

A problem which often arises in semiconductor physics is that of calculating the effect of a slowly-varying applied potential on the one-electron energy eigenfunctions in a crystal. The envelope-function approximation provides a powerful way of doing this. It has reached a high degree of sophistication. We give here the most elementary version of the theory, in which it is supposed that the perturbed wave function $\psi(\mathbf{r})$ may be expressed in terms of the Bloch functions $\psi_{\mathbf{k}}(\mathbf{r})$ associated with a single spherically-symmetric energy valley centred on $\mathbf{k}=0$:

$$\psi(\mathbf{r}) = \sum_{\mathbf{k}} a_{\mathbf{k}}\, \psi_{\mathbf{k}}(\mathbf{r}) . \tag{A6.1}$$

Then the envelope function is defined by

$$\phi(\mathbf{r}) = \sum_{\mathbf{k}} a_{\mathbf{k}} \exp(i\mathbf{k}\cdot\mathbf{r}) . \tag{A6.2}$$

We seek an equation for $\phi(\mathbf{r})$. Once it has been determined we see from (A6.1) that the coefficient $a_{\mathbf{k}}$ in $\psi(\mathbf{r})$ is just V^{-1} times the Fourier transform of $\phi(\mathbf{r})$:

$$a_{\mathbf{k}} = V^{-1} \int \mathrm{d}\mathbf{r}\, \phi(\mathbf{r}) \exp(-i\mathbf{k}\cdot\mathbf{r}) , \tag{A6.3}$$

where V is the macroscopic volume of the crystal. Kriechbaum (1986) has recently given a succinct treatment of the general case in which the summand (A6.1) is extended over several energy bands. The general case is also treated in the pioneering papers of Kohn and Luttinger (1955).

Let $V(\mathbf{r})$ denote the periodic potential and $V_{\text{ext}}(\mathbf{r})$ the applied potential. Then Schrödinger's equation for $\psi(\mathbf{r})$ is:

$$\left[-\frac{\hbar^2}{2m}\nabla^2 + V(\mathbf{r}) + V_{\text{ext}}(\mathbf{r}) \right] \psi(\mathbf{r}) = \mathbb{E}\, \psi(\mathbf{r}) , \tag{A6.4}$$

where $\mathbb{E}$ denotes the energy. By making the substitution (A6.1), multiplying across by $\psi_{\mathbf{k}'}(\mathbf{r})$ and integrating over V, we obtain:

$$[\mathbb{E}(\mathbf{k}')-\mathbb{E}]\,a_{\mathbf{k}'} + \sum_{\mathbf{k}} \langle\mathbf{k}'|V_{\text{ext}}|\mathbf{k}\rangle\, a_{\mathbf{k}} = 0\,, \tag{A6.5}$$

where $\langle\mathbf{k}'|V_{\text{ext}}|\mathbf{k}\rangle$ is the matrix element of $V_{\text{ext}}(\mathbf{r})$ in the Dirac bra–ket form (3.7a), and we have used the equation satisfied by $\psi_{\mathbf{k}'}(\mathbf{r})$:

$$\left[-\frac{\hbar^2}{2m}\nabla^2 + V(\mathbf{r})\right]\psi_{\mathbf{k}'}(\mathbf{r}) = \mathbb{E}(\mathbf{k}')\,\psi_{\mathbf{k}'}(\mathbf{r})\,. \tag{A6.6}$$

Now we may approximate $\langle\mathbf{k}'|V_{\text{ext}}|\mathbf{k}\rangle$ as follows:

$$\begin{aligned}\langle\mathbf{k}'|V_{\text{ext}}|\mathbf{k}\rangle &= \int_V \mathrm{d}\mathbf{r}\,\exp[i(\mathbf{k}-\mathbf{k}')\cdot\mathbf{r}]\,u_{\mathbf{k}'}{}^*(\mathbf{r})\,u_{\mathbf{k}}(\mathbf{r})\,V_{\text{ext}}(\mathbf{r}) \\ &\simeq \sum_{\mathbf{R}} \exp[i(\mathbf{k}-\mathbf{k}')\cdot\mathbf{R}]\,V_{\text{ext}}(\mathbf{R})\,V_{\text{c}}/V \\ &\simeq V^{-1}\int\mathrm{d}\mathbf{r}\,V_{\text{ext}}(\mathbf{r})\,\exp[i(\mathbf{k}-\mathbf{k}')\cdot\mathbf{r}] \\ &= \widetilde{V}_{\text{ext}}(\mathbf{k}'-\mathbf{k})/V\,. \end{aligned} \tag{A6.7}$$

In the second line we suppose that the values of $\mathbf{k}$ are all close to an extremum of the energy band at $\mathbf{k}=0$ so that the exponential factor varies slowly over a unit cell. We suppose that the same is true of $V(\mathbf{r})$. Then we may set $\mathbf{r}\simeq\mathbf{R}$ in both these functions throughout the unit cell with one corner located at $\mathbf{R}$. Since the Bloch functions are normalised in the macroscopic volume V, it follows that the integral of the product $u_{\mathbf{k}'}{}^*(\mathbf{r})\,u_{\mathbf{k}}(\mathbf{r})$, which involves the periodic parts of the Bloch functions, over the unit cell volume V_{c} is V_{c}/V when $\mathbf{k}'=\mathbf{k}$ and may be approximated by V_{c}/V when $\mathbf{k}'\simeq\mathbf{k}$. In the third line of (A6.7) we have taken V^{-1} outside the summation which is then recognisable as an approximation to the integral which is written there. Thus we see that $\langle\mathbf{k}'|V_{\text{ext}}|\mathbf{k}\rangle$ is approximately equal to V^{-1} times the Fourier transform of $V_{\text{ext}}(\mathbf{r})$; the Fourier transform is defined by the last equality in (A6.7) and we denote it by $\widetilde{V}_{\text{ext}}(\mathbf{k}'-\mathbf{k})$.

To complete the calculation we write $\mathbb{E}(\mathbf{k}')\simeq\hbar^2(\mathbf{k}')^2/2m^*$, where m^* is the effective mass, in (A6.5) and use the approximation (A6.7). Then we have:

$$\left[\frac{\hbar^2(\mathbf{k}')^2}{2m^*} - \mathbb{E}\right]a_{\mathbf{k}'} + V^{-1}\sum_{\mathbf{k}}\widetilde{V}_{\text{ext}}(\mathbf{k}'-\mathbf{k})\,a_{\mathbf{k}} = 0\,. \tag{A6.8}$$

When this equation is multiplied by $\exp(i\mathbf{k}'\cdot\mathbf{r})$ and summed over $\mathbf{k}'$ we obtain the desired equation for the envelope function $\phi(\mathbf{r})$:

$$\left[-\frac{\hbar^2}{2m^*}\nabla^2 + V_{\text{ext}}(\mathbf{r})\right]\phi(\mathbf{r}) = \mathbb{E}\,\phi(\mathbf{r})\,. \tag{A6.9}$$

We see that the effect of the periodic potential appears only through the effective mass m^*. That is what makes the envelope function formalism so useful.

We note here that $\phi(\mathbf{r})$ is normalised to V when $\psi(\mathbf{r})$ is normalised to unity in V. Thus we see from (A6.1), (A6.2) and the orthonormality properties of the Bloch functions and the plane wave functions (see §3.2.1) that

$$\begin{aligned}\int_V \mathrm{d}\mathbf{r}\, |\phi(\mathbf{r})|^2 &= V \sum_{\mathbf{k}} |a_{\mathbf{k}}|^2 \\ &= V \int \mathrm{d}\mathbf{r}\, |\psi(\mathbf{r})|^2 \\ &= V. \end{aligned} \tag{A6.10}$$

APPENDIX 7

$\mathbf{k}\cdot\mathbf{p}$ theory of a standard semiconductor with a small energy gap

When the energy gap E_g in Fig. 8.2(a) is small compared to the separation between the valence band and other energy bands, we may make an elementary application of $\mathbf{k}\cdot\mathbf{p}$ theory (Kane, 1982) to find the form of the conduction and valence bands near $\mathbf{k}=0$. The results apply to the conduction and light-hole bands in the Group III–V semiconductors GaAs and InSb.

The Schrödinger equation (8.1) may be written in the form:

$$\frac{\mathbf{p}^2}{2m}\psi(\mathbf{r}) + V(\mathbf{r})\,\psi(\mathbf{r}) = E\,\psi(\mathbf{r})\,, \tag{A7.1}$$

where

$$\mathbf{p} = -i\hbar\nabla \tag{A7.2}$$

and $V(\mathbf{r})$ has the crystal periodicity. We may rewrite (A7.1) in terms of the periodic part $u_{\mathbf{k}}(\mathbf{r})$ of the Bloch function

$$\psi_{\mathbf{k}}(\mathbf{r}) = \exp(i\mathbf{k}\cdot\mathbf{r})\,u_{\mathbf{k}}(\mathbf{r}) \tag{A7.3}$$

with energy

$$E = E(\mathbf{k})\,, \tag{A7.4}$$

by substituting these formulae into the equation. Thus we find that

$$\left[\frac{\mathbf{p}^2}{2m} + V(\mathbf{r})\right] u_{\mathbf{k}}(\mathbf{r}) + \frac{\hbar}{m}\,\mathbf{k}\cdot\mathbf{p}\,u_{\mathbf{k}}(\mathbf{r}) = \left[E(\mathbf{k}) - \frac{\hbar^2}{2m}\mathbf{k}^2\right] u_{\mathbf{k}}(\mathbf{r})\,. \tag{A7.5}$$

When $\mathbf{k}=0$ we write $u_c(\mathbf{r})$ and $u_v(\mathbf{r})$ for the real solutions of this equation at the energies 0 and $-E_g$ corresponding to the edges of the conduction and valence bands. When $\mathbf{k}$ is small, we treat the term involving $\mathbf{k}\cdot\mathbf{p}$ in (A7.5) as a perturbation and write

$$u_{\mathbf{k}}(\mathbf{r}) = a\,u_c(\mathbf{r}) + b\,u_v(\mathbf{r})\,. \tag{A7.6}$$

Then, in the basis provided by $u_c(\mathbf{r})$ and $u_v(\mathbf{r})$, we obtain the matrix equations:

$$0\,a + \left[\frac{\hbar}{m}\mathbf{k}\cdot\mathbf{p}_{\mathrm{cv}}\right] b = \left[\mathrm{E}(\mathbf{k}) - \frac{\hbar^2}{2m}\mathbf{k}^2\right] a \tag{A7.7}$$

$$-\mathrm{E}_{\mathrm{g}}b + \left[\frac{\hbar}{m}\mathbf{k}\cdot\mathbf{p}_{\mathrm{cv}}^*\right] a = \left[\mathrm{E}(\mathbf{k}) - \frac{\hbar^2}{2m}\mathbf{k}^2\right] b\,. \tag{A7.8}$$

Here we have made use of both the Hermitian character of **p** (see §3.2.3) and the fact that it is an imaginary operator. These properties guarantee that **p** is diagonal in the real basis provided by $u_{\mathrm{c}}(\mathbf{r})$ and $u_{\mathrm{v}}(\mathbf{r})$ and that

$$\begin{aligned}\mathbf{p}_{\mathrm{cv}} &= \int \mathrm{d}\mathbf{r}\, u_{\mathrm{c}}(\mathbf{r})\, \mathbf{p}\, u_{\mathrm{v}}(\mathbf{r}) \\ &= \left[\int \mathrm{d}\mathbf{r}\, u_{\mathrm{v}}(\mathbf{r})\, \mathbf{p}\, u_{\mathrm{c}}(\mathbf{r})\right]^*.\end{aligned} \tag{A7.9}$$

The determinantal equation for the consistency of (A7.7) and (A7.8) is

$$\left[\mathrm{E}(\mathbf{k}) - \frac{\hbar^2}{2m}\mathbf{k}^2\right]\left[\mathrm{E}(\mathbf{k}) - \frac{\hbar^2}{2m}\mathbf{k}^2 + \mathrm{E}_{\mathrm{g}}\right] = \frac{\hbar^2}{m^2}\,|\mathbf{p}_{\mathrm{cv}}\cdot\mathbf{k}|^2\,. \tag{A7.10}$$

When $\mathbf{k}=(0,0,k_z)$ the right-hand side of this equation reduces to $\hbar^2 p_{\mathrm{cv}}^2 k_z^2/m^2$, where p_{cv} is the matrix element of p_z. The standard band structure is spherically symmetrical and we may replace k_z^2 by k^2 (strictly speaking, we should take proper account of the threefold valence-band degeneracy at this point – see Kane (1982)). Thus we obtain, when $k \to 0$, spherically-symmetric conduction and valence bands with electron and hole masses given by

$$m_{\mathrm{e,h}}^{-1} = m^{-1}\left(\frac{2|p_{\mathrm{cv}}|^2}{m\mathrm{E}_{\mathrm{g}}} \pm 1\right). \tag{A7.11}$$

In typical Group III – V semiconductors $m_{\mathrm{e}} \ll m$ and the second term in (A7.11) for m_{e}^{-1} is negligible. Hence we may express $|p_{\mathrm{cv}}|^2$ in the useful form:

$$|p_{\mathrm{cv}}|^2/m = (\mathrm{E}_{\mathrm{g}}/2)(m/m_{\mathrm{e}})\,. \tag{A7.12}$$

The quantity $P = \hbar|p_{\mathrm{cv}}|/m$ is known as the Kane momentum parameter and has a value close to 10^{-28} J m (10^{-19} esu) for most of the semiconductors to which Kane's **k**·**p** theory applies.

APPENDIX 8

The Kramers-Kronig dispersion relations

Consider an analytic function $f(\omega)$ of a complex frequency ω such that $f(\omega)\to 0$ when $|\omega|\to\infty$ and all poles of $f(\omega)$ lie in the lower half-plane. Then the following relations exist between the real and imaginary parts of $f(\omega)$, where ω and ω' are real:

$$\operatorname{Re} f(\omega') = \frac{1}{\pi} \mathrm{P} \int_{-\infty}^{\infty} \mathrm{d}\omega \, \frac{\operatorname{Im} f(\omega)}{\omega-\omega'} \tag{A8.1}$$

$$\operatorname{Im} f(\omega') = -\frac{1}{\pi} \mathrm{P} \int_{-\infty}^{\infty} \mathrm{d}\omega \, \frac{\operatorname{Re} f(\omega)}{\omega-\omega'} . \tag{A8.2}$$

In these equations P denotes the principal part of the following integrals, which are taken to lie along the real axis.

To prove these relations we take the contour integral of $f(\omega)/(\omega-\omega')$ round a closed contour consisting of the real axis indented by a small semicircle in the upper half-plane, which is centred on ω', and is closed by a large semicircle in the upper half-plane which is centred on the origin. Since the contour encloses no singularities, the contour integral vanishes. Moreover, since $f(\omega)\to 0$ when $|\omega|\to\infty$, the contribution from the large semicircle vanishes in the limit as the radius of the semicircle tends to ∞. Finally, the contribution from the small semicircle is $-i\pi f(\omega')$ in the limit as the radius of the small semicircle approaches zero. Taking the real and imaginary parts of the resulting equation gives (A8.1) and (A8.2).

Appropriate response functions satisfy the conditions required of $f(\omega)$. In particular, the relations (A8.1) and (A8.2) apply to the susceptibility tensors $\boldsymbol{\chi}^{(n)}$ when they are correctly formulated, as described in §§2.2.1 and 4.5.1. Certain other optical-response functions also satisfy the necessary conditions: for example, in the linear regime, when the absorption coefficient $\alpha(\omega)$ due to a particular mechanism (*e.g.*, interband absorption in a semiconductor, *cf.* §8.3) has been measured or calculated, the relation (A8.2) can be used to determine the associated component of the total refractive index $n(\omega)$; here we denote this component by $n'(\omega)$. This is because, as we see from the discussion of these quantities given in §8.3 and physical intuition, $\alpha(\omega)-i2\omega n'(\omega)/c$ satisfies the conditions of the theorem. More generally, so does the linear conductivity $\sigma^{(1)}_{\mu\alpha}(-\omega;\omega)$ for all μ and α. Moreover, in each of these cases, the real and imaginary parts of the function are respectively even and odd functions of ω on the real axis. When this is true, the relations (A8.1) and (A8.2) may readily be put in the more familiar form:

$$\operatorname{Re} f(\omega') = \frac{1}{\pi} \mathrm{P} \int_0^\infty \mathrm{d}\omega \, \frac{2\omega \operatorname{Im} f(\omega)}{\omega^2 - (\omega')^2} \tag{A8.3}$$

$$\operatorname{Im} f(\omega') = -\frac{2\omega'}{\pi} \mathrm{P} \int_0^\infty \mathrm{d}\omega \, \frac{\operatorname{Re} f(\omega)}{\omega^2 - (\omega')^2} \,. \tag{A8.4}$$

It is a quite common practice for these Kramers-Kronig relations to be applied in a regime in which $f(\omega)$ is itself a function of the optical intensity. Thus, for example, an optical field of given intensity I may cause the absorption coefficient $\alpha(\omega,I)$ to differ from its linear value $\alpha(\omega,0)$ by an amount $\Delta\alpha(\omega,I)$. (An example of such a process is the bleaching of interband absorption in a semiconductor, as described in §8.4.) As a result of this induced change in the absorption spectrum, the refractive index $n(\omega,I)$ will undergo a modification $\Delta n(\omega,I)$ from its linear value. By substituting $f(\omega)=\alpha(\omega,0)+\Delta\alpha(\omega,I)-i\{2\omega[n'(\omega,0)+\Delta n(\omega,I)]/c\}$ in (A8.4), we find

$$\Delta n(\omega',I) = \frac{c}{\pi} \mathrm{P} \int_0^\infty \mathrm{d}\omega \, \frac{\Delta\alpha(\omega,I)}{\omega^2 - (\omega')^2} \,. \tag{A8.5}$$

This relation is often used to deduce $\Delta n(\omega,I)$ from measured values of the absorption change $\Delta\alpha(\omega,I)$. Strictly speaking, as emphasised by Miller *et al* (1981a), the Kramers-Kronig relations are valid only in the regime in which all external forces acting on the system – such as the optical intensity I – remain constant. Thus, strictly, a measurement of $\Delta\alpha$ made by varying I invalidates the relations (A8.1) – (A8.4). More specifically, (A8.1) – (A8.4) are valid for semiconductors only when the carrier density is fixed. In practice, however, the procedure using I as a parameter, as described by (A8.5), can give good results provided the absorption change $\Delta\alpha(\omega,I)$ is significant only within some finite spectral region $\omega_1<\omega<\omega_2$, so that (A8.5) can be approximated by replacing the lower and upper limits of integration by ω_1 and ω_2, respectively (Weiner *et al,* 1987). Results calculated in this way from measured values of $\Delta\alpha$ have been found to be in good agreement with values of Δn obtained by direct measurement – as discussed for semiconductors in §8.4 (see also Weiner *et al,* 1987; Chemla *et al,* 1984).

APPENDIX 9

Electric-multipole and magnetic interactions

Throughout this book we consider the optical response of materials in the electric-dipole approximation. As mentioned in §2.5, this is a perfectly satisfactory approximation for the majority of practical cases in nonlinear optics because other interactions, such as electric quadrupole and magnetic dipole, are almost always very weak in comparison. However, there are a few cases in which these electric-multipole and magnetic effects need to be considered. For example, second-harmonic generation *via* electric-dipole interaction is forbidden in centrosymmetric media from symmetry considerations (see §5.3). Yet a weak effect is sometimes observed in centrosymmetric solids – such as the crystal calcite (Terhune *et al,* 1962) – which can be ascribed to an electric-quadrupole interaction. This is one of the contributory mechanisms being considered currently in an attempt to explain the observation that glass optical fibres can, in certain circumstances, perform efficient second-harmonic generation (Terhune and Weinberger, 1987). Also it is found that certain atomic gases are good systems for the observation of multipole and magnetic nonlinear-optical effects (Hanna *et al,* 1979). It is possible to tune the optical frequencies in the vicinity of selected electronic transitions that are forbidden in the electric-dipole approximation, but which are allowed *via* electric-quadrupole and magnetic-dipole interactions; with resonance enhancement, these nonlinear effects can therefore be significant. It is sometimes necessary to take account of such effects in the analysis of very sensitive spectroscopic measurements. As a further example, the electric-dipole approximation may be invalid for highly-extended charge distributions, such as long conjugated-chain molecules.

Here we outline how our explicit formulae for the polarisation $\mathbf{P}$ in terms of the susceptibilities $\boldsymbol{\chi}^{(n)}$ and field $\mathbf{E}$ can be modified in a simple way to take account of electric-multipole and magnetic interactions. In a magnetic dielectric both the polarisation $\mathbf{P}$ and magnetisation $\mathbf{M}$ are source terms in Maxwell's equations. In the main text we neglect the magnetisation ($\mathbf{M}=0$) and we take $\mathbf{P}=\mathbf{P}^{\mathrm{D}}$, where $\mathbf{P}^{\mathrm{D}}$ denotes the electric-dipole polarisation. Here we shall concern ourselves only with the lowest-order additional contributions to $\mathbf{P}$ and $\mathbf{M}$:

$$\mathbf{P} = \mathbf{P}^{\mathrm{D}} + \mathbf{P}^{\mathrm{Q}}\,, \qquad \mathbf{M} = \mathbf{M}^{\mathrm{D}}\,, \tag{A9.1}$$

where $\mathbf{P}^{\mathrm{Q}}$ is the quadrupolar polarisation and $\mathbf{M}^{\mathrm{D}}$ is the dipolar magnetisation; diamagnetic contributions are neglected for simplicity. The corresponding constitutive relations for the electric displacement **D** and magnetic induction **B** are

$$\mathbf{D} = \varepsilon_0\,\mathbf{E} + \mathbf{P}\,, \qquad \mathbf{B} = \mu_0\,(\mathbf{H} + \mathbf{M})\,. \tag{A9.2}$$

For simplicity we assume here, as in §4.2, that the nonlinear medium consists of an assembly of identical and noninteracting molecules. In §4.2, the electric-dipole polarisation $\mathbf{P}^{\mathrm{D}}$ in a small volume V is calculated as the expectation value of the electric-dipole operator $\mathbf{Q}$:

$$\mathbf{P}^{\mathrm{D}} = Ne\mathbf{r} \equiv V^{-1}\mathbf{Q}\,, \tag{A9.3}$$

where there are N molecules per unit volume, and the energy of the interaction between the molecular system and the radiation field in the electric-dipole (ED) approximation is given by (4.4):

$$H_{\mathrm{ED}} = -\mathbf{Q}\cdot\mathbf{E} = -V\mathbf{P}^{\mathrm{D}}\cdot\mathbf{E}\,. \tag{A9.4}$$

We now incorporate the other interactions by forming the multipole expansion of the total interaction Hamiltonian H_{I}, taking the terms corresponding to (A9.1) (Bottcher, 1952; Loudon, 1983):

$$H_{\mathrm{I}} = H_{\mathrm{ED}} + H_{\mathrm{EQ}} + H_{\mathrm{MD}}\,. \tag{A9.5}$$

The last two terms are the electric-quadrupolar and magnetic-dipolar interactions, respectively, given by

$$H_{\mathrm{EQ}} = -V\mathbf{P}^{\mathrm{Q}}\cdot\mathbf{E}\,, \qquad H_{\mathrm{MD}} = -V\mu_0\,\mathbf{M}^{\mathrm{D}}\cdot\mathbf{H}\,, \tag{A9.6}$$

and the new operators are

$$\mathbf{P}^{\mathrm{Q}} = N\,\mathbf{q}\cdot\boldsymbol{\nabla}\,, \qquad \mathbf{M}^{\mathrm{D}} = -N\,\mathbf{m}\,. \tag{A9.7}$$

Here $\mathbf{q}$ is the molecular electric-quadrupole tensor:

$$\mathbf{q} = \frac{-e}{2}\sum_j \boldsymbol{\mu}_j\boldsymbol{\mu}_j + \frac{e}{2}\sum_k Z_k\,\boldsymbol{\mu}_k\boldsymbol{\mu}_k\,, \tag{A9.8}$$

where the summation is over all the charged particles (electrons and ion cores) in the molecule, and the symbols have the meanings defined previously in (4.1), *i.e.*, the position vector of the jth electron (charge $-e$) is denoted by $\boldsymbol{\mu}_j$, and the position vector of the kth nucleus of charge $Z_k e$ is denoted by $\boldsymbol{\mu}_k$. Similarly, $\mathbf{m}$ is the total electronic angular-momentum operator for the molecule:

$$\mathbf{m} = \frac{\mu_{\mathrm{B}}}{\hbar}\sum_j \boldsymbol{\mu}_j\times\mathbf{p}_j\,, \tag{A9.9}$$

where $\mathbf{p}_j$ is the momentum conjugate to $\boldsymbol{\mu}_j$, and $\mu_{\mathrm{B}} = e\hbar/2m$ is the Bohr magneton. The operator $\mathbf{m}$ may also be written $\mathbf{m}=\mu_B(\mathbf{L}+2\mathbf{S})$, where $\hbar\mathbf{L}$ and $\hbar\mathbf{S}$ are the electronic orbital- and spin-angular momentum operators, respectively.

It is now straightforward to calculate the interaction energies H_{EQ} and H_{MD} in the case of a plane-wave field:

$$\mathbf{E}(\mathbf{r},t) = \tfrac{1}{2}\{\mathbf{E}_\omega \exp[i(\mathbf{k}\cdot\mathbf{r}-\omega t)] + \mathbf{E}_\omega^* \exp[-i(\mathbf{k}\cdot\mathbf{r}-\omega t)]\}. \quad \text{(A9.10)}$$

By substituting (A9.10) in (A9.6) and evaluating $\nabla\cdot\mathbf{E}$, we obtain the results:

$$H_{EQ} = -\frac{VN}{2} i\mathbf{k}\cdot\mathbf{q}\cdot\mathbf{E}_\omega \exp[i(\mathbf{k}\cdot\mathbf{r}-\omega t)] + \text{h.c.}, \quad \text{(A9.11)}$$

$$H_{MD} = \frac{VN}{2\omega} \mathbf{m}\cdot(\mathbf{k}\times\mathbf{E}_\omega) \exp[i(\mathbf{k}\cdot\mathbf{r}-\omega t)] + \text{h.c.}, \quad \text{(A9.12)}$$

where h.c. denotes the Hermitian conjugate [defined by (3.14)]. We have assumed $\mathbf{M} \ll \mathbf{H}$ and used the property $\mathbf{B}=(\mathbf{k}\times\mathbf{E})/\omega$ for a plane wave.

The important point here is that the interactions H_{ED}, H_{EQ} and H_{MD} are each *linear* in the field. Consequently, the power-law dependences of nonlinear-optical processes are identical to those in the electric-dipole approximation: for example, the expression for $\mathbf{P}^{(2)}$ contains products of two fields; the second-harmonic intensity is proportional to the square of the fundamental intensity; *etc.* A further consequence is that very straightforward modifications can be made to the explicit formulae for the susceptibilities to take account of these other interactions (Hanna *et al,* 1979). For example, each of the operators $\mathbf{e}_j\cdot\mathbf{r}$ $(j=1,\ldots,n)$ that appear in the general formula (4.76) for $\chi^{(n)}(-\omega_\sigma;\omega_1,\ldots,\omega_n)$ must be replaced by $\mathbf{e}_j\cdot e\mathbf{r} + i\mathbf{k}\cdot\mathbf{q}\cdot\mathbf{e}_j + \mathbf{m}\cdot(\mathbf{k}\times\mathbf{e}_j)/\omega$. Similarly, the operator $\mathbf{e}_\sigma^*\cdot\mathbf{P}^D$ is replaced by $\mathbf{e}_\sigma^*\cdot\mathbf{P}^Q$ or $\mathbf{e}_\sigma^*\cdot\mathbf{M}^D$ if an electric-quadrupole polarisation or magnetic-dipole magnetisation associated with the frequency ω_σ is to be calculated.

With this prescription, expressions for the nonlinear-optical response involving these other interactions can be written down directly using the formulae derived in Chapter 4. A specific example (the susceptibility describing electric-quadrupole-resonant sum-frequency generation) is discussed in detail by Hanna *et al* (1979).

APPENDIX 10

Gaussian beam optics

Many lasers are designed so that the output beam is the lowest-order spatial mode (TEM_{00}), which has a Gaussian intensity distribution in the transverse direction (Kogelnik and Li, 1966). The field is given by

$$\mathbf{E}(\mathbf{r},t) = \tfrac{1}{2}[\mathbf{E}(\mathbf{r})\exp(-i\omega t) + \mathbf{E}^*(\mathbf{r})\exp(i\omega t)] \qquad \text{(A10.1)}$$

where

$$\mathbf{E}(\mathbf{r}) = \mathbf{E}_0 \exp(ikz)(1+i\xi)^{-1}\exp\{-kr^2/[b(1+i\xi)]\}, \qquad \text{(A10.2)}$$

$\mathbf{r}$ is the radial coordinate, $k=\omega n/c$ is the magnitude of the wave vector in the medium, and $\xi=2(z-z_0)/b$ is a normalised z coordinate which takes the position of the beam waist ($z=z_0$) as its origin.

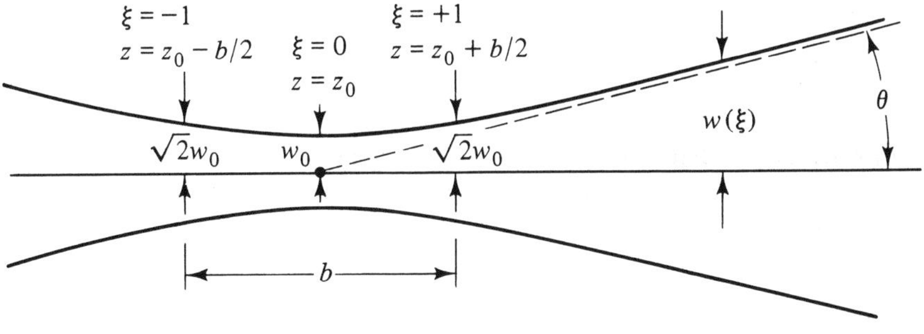

Fig. A10.1 Gaussian mode in the region of the beam waist

The distance b is termed the 'confocal parameter' and, as shown in Fig. A10.1, it is the distance over which the spot size w (e^{-1} field radius) does not exceed $\sqrt{2}$ times the value w_0 which it has at the waist position $z=z_0$ ($\xi=0$). The spot size varies as

$$w(\xi) = w_0\sqrt{1+\xi^2}, \qquad \text{(A10.3)}$$

and therefore

$$b = kw_0^2. \qquad \text{(A10.4)}$$

Also $w=0.849D$, where D is the beam diameter defined at half-maximum intensity. The intensity I_0 on the axis ($r=0$) is related to the total power

P in the beam by:

$$I_0 = 2P/\pi w^2 . \tag{A10.5}$$

The domain $|z-z_0|<b/2$ is known as the 'near-field' region. The 'far-field' region, on the other hand, is $|z-z_0|>>b/2$. In that region the Gaussian beam behaves as though it emanates from a point source of geometrical optics; the beam diverges with a half-angle $\theta=2w_0/b=2/kw_0$, and $w(\xi)\rightarrow\xi w_0=2w_0(z-z_0)/b$.

If we consider the phase variation along the beam axis ($r=0$), we see from (A10.2) that in addition to the exp(ikz) term there is a contribution from the factor $(1+i\xi)^{-1}=\exp(-i\tan^{-1}\xi)/\sqrt{1+\xi^2}$. This additional contribution to the phase varies from $+\pi/2$ at $\xi=-\infty$, to 0 at $\xi=0$, to $-\pi/2$ at $\xi=+\infty$; there is therefore a total phase lag of π in going through the focus. This phase lag is found to be important in some phase-matched parametric interactions with focused beams (see §7.2.1).

The importance of the Gaussian mode, as opposed to any other transverse spatial distribution of the field in an optical beam, stems from two related facts: first, the transformation of a Gaussian beam by a spherical lens gives another Gaussian beam; second, the Gaussian mode (as opposed to higher-order modes) allows for production of the minimum spot size. As depicted in Fig. A10.2, a thin spherical lens of focal length f positioned a distance d_1 from a beam waist of size w_1 will transform the beam to one having a waist w_2 at a further distance d_2, given by:

$$(w_2)^{-2} = (w_1)^{-2}(1-d_1/f)^2 + (kw_1/2f)^2 , \tag{A10.6}$$

$$d_2 = f + \frac{(d_1-f)f^2}{(d_1-f)^2+(kw_1^2/2)^2} . \tag{A10.7}$$

An excellent tutorial review of Gaussian beam optics is by Kogelnik and Li (1966).

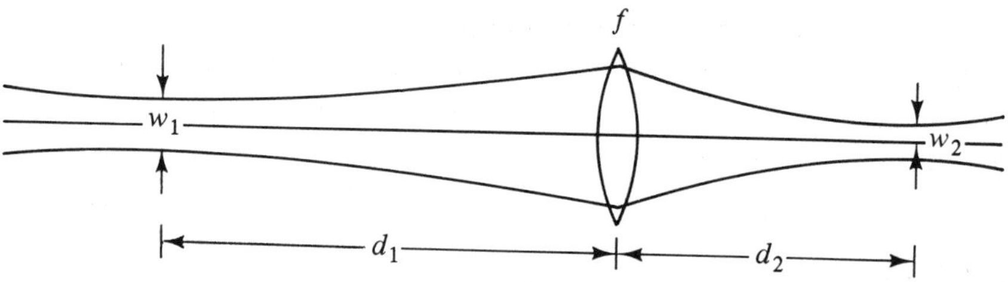

Fig. A10.2 Gaussian beam transformation by a spherical lens

Glossary of mathematical symbols

Listed are the major symbols that appear in the text (symbols used only locally are not listed). The *italic figure in brackets* indicates the page on which a symbol makes its first appearance or has its main definition.

Notation

x	scalar
$\mathbf{v}$	vector
v_α	component of $\mathbf{v}$ in the direction of the cartesian coordinate axis labelled α
$\mathbf{T}^{(n)}$	$(n+1)$-rank tensor
$T^{(n)}_{\mu\alpha_1\cdots\alpha_n}$	component of $\mathbf{T}^{(n)}$ where $\mu,\alpha_1,\ldots,\alpha_n$ are labels for the cartesian coordinate axes (*i.e.*, each label takes the value x, y or z ; see §2.1.1).

a	label for lower state of two-level atom [*157*]
$a_{\mathbf{kk}'}$	coefficient in envelope-function approximation [*258, 309*]
a_0	classical Bohr radius [*295*]
a_0^*	exciton Bohr radius [*259*]
$\mathbf{A}(t)$	vector potential of electromagnetic field [*110*]
b	label for upper state of two-level atom [*157*]
b	Gaussian-beam confocal parameter [*224, 319*]
$\mathbf{B}(t)$	magnetic induction vector [*211, 317*]
c	velocity of light in vacuum [*295*]
c.c.	complex conjugate [*3*]
d_{eff}	effective value of second-order susceptibility [*131*]
$d\sigma_R/d\Omega$	differential Raman cross-section [*231*]
$\mathbf{d}(-\omega_\sigma;\omega_1,\omega_2)$	second-order susceptibility tensor [*130*]
D	Debye unit of electric-dipole moment [*295*]
$\mathbf{D}$	electric displacement vector [*317*]
e	modulus of electronic charge [*295*]
$\mathbf{e}$, $\mathbf{e}_j$	unit vector in polarisation direction of the field $\mathbf{E}_\omega$, $\mathbf{E}_{\omega_j}$ [*28*]
er^α_{ab}	(ab)th matrix element of er_α [*67*]
$e\mathbf{r}$	molecular electric-dipole moment [*62, 65*]
$e\hat{\mathbf{r}}$	molecular dipole moment in centred coordinates [*114*]
$e\mathbf{r}(t)$	electric-dipole moment operator in interaction picture [*63, 65*]

$e\mathbf{r}_{ab}$	(ab)th matrix element of $e\mathbf{r}$ [*102*]
$e\mathbf{r}^{(n)}$	nth-order element of perturbation series for $e\mathbf{r}$ [*77*]
$e\mathbf{r}_{\omega}^{(n)}$	amplitude of monochromatic component of $e\mathbf{r}$ at frequency ω [*78*]
$\exp(x)$	exponential (e^x)
$E(t)$, $E_j(t)$	scalar amplitude of $\mathbf{E}_{\omega}(t)$, $\mathbf{E}_{\omega_j}(t)$ [*159*]
$\hat{E}(\omega)$	scalar amplitude of $\hat{\mathbf{E}}(\omega)$ [*212*]
E_j	scalar amplitude of $\mathbf{E}_{\omega_j}$ [*28*]
$\hat{E}_j$	scalar amplitude of $\hat{\mathbf{E}}_{\omega_j}$ [*218*]
$\mathbf{E}(t)$	total electric field [*12*]
$\mathbf{E}(\omega)$	Fourier transform of $\mathbf{E}(t)$ [*17*]
$\hat{\mathbf{E}}(\omega)$	travelling-wave amplitude of $\mathbf{E}(\omega)$ [*212*]
$\mathbf{E}_{\omega}$	amplitude of monochromatic electric-field component at frequency ω [*23*]
$\hat{\mathbf{E}}_{\omega}$	electric-field amplitude of monochromatic travelling wave at frequency ω [*217*]
$\mathbf{E}_{\omega}^{m}$	amplitude of monochromatic component of $\mathbf{E}_m(t)$ at frequency ω [*81*]
$\mathbf{E}_m(t)$	total electric field acting locally at site of mth molecule [*77*]
$\mathbf{E}_{\omega}(t)$	slowly-varying amplitude of quasi-monochromatic electric-field component at frequency ω [*29*]
$\mathbf{E}_{\omega}(\omega)$	Fourier transform of $\mathbf{E}_{\omega}(t)$ [*151*]
$\mathbb{E}$	energy [*46*]
$\mathbb{E}_{\mathrm{F}}$	chemical potential, Fermi energy [*108, 251*]
$\mathbb{E}_{\mathrm{Fc}}$, $\mathbb{E}_{\mathrm{Fv}}$	quasi-Fermi levels [*263*]
$\mathbb{E}_{\mathrm{g}}$	semiconductor forbidden energy gap [*248*]
$\mathbb{E}_n$	energy eigenvalue [*46*]
$\mathbb{E}_n(\mathbf{k})$	energy of one-electron eigenstate with wave vector $\mathbf{k}$, band index n [*247*]
$\mathbb{E}_{n'n}(\mathbf{k})$	energy of interband transition [*253*]
$\mathbb{E}_{\mathrm{Ry}}$	classical Rydberg energy [*295*]
$\mathbb{E}_{\mathrm{Ry}}^{*}$	exciton Rydberg energy [*259*]
$f(\omega)$	scalar local-field factor at frequency ω in medium with isotropic or cubic symmetry [*83*]
$f_{\mathrm{c}}(\mathbb{E})$, $f_{\mathrm{v}}(\mathbb{E})$	Fermi-Dirac function for conduction, valence band [*265*]
$f_0(a)$	thermal average of occupation number of one-electron state a [*108*]
$\mathbf{f}(\omega)$	second-rank tensor whose elements are local-field factors at frequency ω [*82*]
g_{R}	gain per unit pump intensity for stimulated Raman scattering [*206, 229*]
G	parametric gain coefficient [*222, 227*]
G_{R}	gain coefficient for stimulated Raman scattering [*229*]

$\hbar$	Dirac constant [*295*]
H	Hamiltonian operator [*38*]
$H_{\mathrm{I}}(t)$	perturbation Hamiltonian operator [*49*]
$H'_{\mathrm{I}}(t)$	perturbation Hamiltonian operator in interaction picture [*53*]
$\widetilde{H}_{\mathrm{I}}(t)$	effective two-level perturbation Hamiltonian for multi-photon processes [*203*]
H_{R}	Hamiltonian operator representing relaxation processes [*159, 164*]
H_0	equilibrium Hamiltonian operator [*46*]
$\mathbf{H}(t)$	magnetic field vector [*212, 317*]
i	imaginary operator ($i^2 = -1$) [*3*]
I_{N}	equivalent noise intensity [*227*]
I_{sat}	saturation intensity [*179*]
I_{ω}	intensity of monochromatic travelling wave at frequency ω [*23*]
$\mathbf{I}$	second-rank unit tensor [*106*]
$\mathbf{J}(t)$	volume current density [*110, 211*]
$\mathbf{J}^{(n)}(t)$	nth-order component of volume current density [*110*]
k	Boltzmann constant [*295*]
k	magnitude of $\mathbf{k}$ [*256*]
k_j	magnitude of wave vector (propagation constant) of optical wave with frequency ω_j [*217*]
k_{P}	sum of propagation constants $k_1 + k_2 + \cdots + k_n$ when waves are collinear [*217*]
$\mathbf{k}$	wave vector of optical wave [*33, 212*]
$\mathbf{k}$	wave vector of wave function in reciprocal space [*247*]
$\mathbf{k}_j$	wave vector of optical wave with frequency ω_j [*228*]
$\mathbf{k}_{\mathrm{P}}$	vector sum of optical wave vectors $\mathbf{k}_1 + \mathbf{k}_2 + \cdots + \mathbf{k}_n$ [*33*]
$K(-\omega_\sigma;\omega_1,\ldots,\omega_n)$	conventional numerical factor in expression for nth-order polarisation at frequency ω_σ [*24*]
$\ln(x)$	natural logarithm (base e)
L_{a}	aperture length [*224*]
L_{c}	coherence length [*221*]
$\mathbf{L}_{\omega}$	Lorentz tensor [*81*]
m	electron rest mass [*295*]
$m_{\mathrm{e}}, m_{\mathrm{h}}$	effective mass of electron, hole [*248*]
m_{r}	reduced mass of electrons and holes [*256*]
m_{t}	sum of electron and hole effective masses [*259*]
$\mathbf{m}$	molecular angular-momentum operator [*317*]
M_k	mass of kth ion core [*106*]
$\mathbf{M}$	magnetisation vector [*316*]
n	order of nonlinearity [*12, 16*]
n	energy band index in semiconductor [*247*]

n	principal quantum number of bound exciton state [*259*]
$n(\omega)$	refractive index at frequency ω [*23, 175*]
n_E	extraordinary refractive index of birefringent medium [*213*]
n_j	refractive index at frequency ω_j [*220*]
n_O	ordinary refractive index of birefringent medium [*213*]
n_{ph}	phonon number (Planck distribution function) [*252*]
n_0	refractive index of semiconductor in absence of free carriers (in Chapter 8) [*254*]
n_2^I	nonlinear refraction coefficient (defined with respect to optical intensity I) [*307*]
$n_e(\theta)$	refractive index for an extraordinary ray propagating at an angle θ to the optic axis in a birefringent medium [*214*]
$n_0(\omega)$	linear refractive index at frequency ω [*174*]
$n_2(\omega)$	nonlinear refraction coefficient at frequency ω [*175*]
N	number density (of atoms, molecules, *etc.*) [*3, 65*]
$N^{(e)}$	number density of electrons [*108*]
$N_0^{(e)}$	total electron density in conduction and valence bands in thermal equilibrium [*252*]
$N_n(\mathbb{E})$	density of states (number of states per unit energy range per unit volume of crystal) in energy band n of a semiconductor [*248*]
$N_{cv}(\mathbb{E})$	joint density of states for the valence and conduction bands [*256*]
$\mathbf{N}$	Poynting vector [*213*]
O_R	rotation operator [*145*]
p_{ab}^{α}	(ab)th matrix element of momentum operator p_α [*108*]
$p_{nn'}(\mathbf{k})$	cartesian μ-component of $\mathbf{p}_{nn'}(\mathbf{k})$ [*253, 256*]
$\mathbf{p}$	one-electron momentum operator [*106, 252*]
$\mathbf{p}_j$	momentum of jth particle [*106*]
$\mathbf{p}_{ab}$	(ab)th matrix element of momentum operator $\mathbf{p}$ [*109*]
$\mathbf{p}_{nn'}(\mathbf{k})$	matrix element of one-electron momentum operator [*252*]
$P(t)$	scalar amplitude of $\mathbf{P}_\omega(t)$ [*163*]
$P(\psi)$	projection operator for state with wave vector ψ [*43*]
$P_j^{(n)}$	scalar amplitude of $\mathbf{P}_{\omega_j}^{(n)}$ [*28*]
$\mathbf{P}(t)$	total polarisation [*12, 57*]
$\mathbf{P}_{\omega_j}^{(n)}$	amplitude of monochromatic component at frequency ω_j of nth-order polarisation [*23*]
$\mathbf{P}_{\omega_j}^{NL}$	amplitude of monochromatic component at frequency ω_j of nonlinear polarisation [*217*]
$\mathbf{P}^{(n)}(t)$	nth-order component of total polarisation [*12*]
$\mathbf{P}^{(n)}(\omega)$	Fourier transform of $\mathbf{P}^{(n)}(t)$ [*22*]

$\mathbf{P}^{\mathrm{NL}}(t)$	total nonlinear polarisation [*215*]
$\mathbf{P}^{\mathrm{NL}}(\omega)$	Fourier transform of $\mathbf{P}^{\mathrm{NL}}(t)$ [*215*]
$\mathbf{P}^{(n)}_{\omega_j}(t)$	slowly-varying amplitude of quasi-monochromatic component at frequency ω_j of nth-order polarisation [*29*]
$\mathbb{P}_{\omega}$, $\mathbb{P}_{3\omega}$	power of optical beam at frequency ω, 3ω [*225*]
$\mathbf{q}$	molecular electric-quadrupole tensor [*317*]
$\mathbf{Q}$	dipole moment of charged particle system [*57*]
$\mathbf{Q}(t)$	dipole moment operator in interaction picture [*58*]
$\mathbf{r}$	*see* $e\mathbf{r}$
$\mathbf{r}$	electron position vector (in Chapters 8 and 9) [*247*]
$\mathbf{r}_{\mathrm{e}}$, $\mathbf{r}_{\mathrm{h}}$	electron, hole position vector (in Chapters 8 and 9) [*258*]
$\mathbf{R}$	coordinate-axis transformation matrix [*135–6*]
$\mathbf{R}$	rotating-frame Bloch-Feynman vector (Bloch vector) [*167*]
$\mathbf{R}^{(n)}(t_1, ..., t_n)$	nth-order polarisation response function [*17*]
$\mathrm{sgn}(x)$	sign of x
$\mathrm{sinc}(x)$	$\sin(x)/x$
$\mathbf{s}$	unit vector in the direction of optical wave propagation [*212*]
$\mathbf{S}$	intrinsic permutation operator [*16, 61*]
$\mathbf{S}_{\mathrm{T}}$	overall permutation operator [*71–3*]
t	time variable [*12*]
T	absolute temperature [*47*]
$\mathrm{Tr}(O)$	trace of matrix O (sum of its diagonal elements) [*41*]
T_1	longitudinal relaxation time ('recirculation' time of transition) [*164*]
T_2	transverse relaxation time for homogeneous transition (dephasing time, equivalent to $1/\Gamma$) [*164*]
T_2^*	transverse relaxation time for inhomogeneous transition [*165*]
u	dispersive (in-phase) component of Bloch vector [*162*]
$u_i(\mathbf{\Theta})$	monomolecular wave function [*64, 65*]
U	energy density of optical field [*182*]
$U_0(t)$	unperturbed time-development operator [*52*]
v	absorptive (quadrature) component of Bloch vector [*162*]
v_{g}	group velocity [*237*]
V	volume element, volume of crystal [*56, 105*]
w	fractional population inversion, third component of Bloch vector [*162*]
w	Gaussian beam radius [*319*]
w_0	thermal-equilibrium fractional population inversion [*165*]
w_0	Gaussian beam radius at waist position [*319*]
W_i	power input to unit volume of the medium from optical wave at frequency ω_i [*124*]

X	dimensionless vibrational coordinate of molecule [*197–8*]
X_Ω	slowly-varying envelope of vibrational coordinate X [*198*]
$\mathbf{X}$	nuclear conformation [*196*]
Z_k	charge of kth ion in units of e [*57*]
α, $\alpha(\omega)$	intensity absorption coefficient at frequency ω [*175, 214*]
α_R	Raman polarisability [*204*]
$\alpha_{aa}(-\omega;\omega)$	ground state polarisability [*182*]
$\alpha_{fg}(\omega_1;\omega_2)$	first-order transition hyperpolarisability [*96*]
$\boldsymbol{\alpha}$	linear polarisability tensor for d.c. applied field [*78–9*]
$\boldsymbol{\alpha}(-\omega;\omega)$	linear polarisability tensor [*79*]
β	Rabi frequency [*160*]
$\boldsymbol{\beta}$	second-order hyperpolarisability tensor for d.c. applied field [*78–9*]
$\boldsymbol{\beta}(-\omega_\sigma;\omega_1,\omega_2)$	second-order hyperpolarisability tensor [*79*]
$\boldsymbol{\gamma}$	third-order hyperpolarisability tensor for d.c. applied field [*78–9*]
$\boldsymbol{\gamma}(-\omega_\sigma;\omega_1,\omega_2,\omega_3)$	third-order hyperpolarisability tensor [*79*]
$\boldsymbol{\gamma}^{(n)}(-\omega_\sigma;\omega_1,\ldots,\omega_n)$	nth-order hyperpolarisability tensor [*77, 87*]
Γ	damping parameter [*3*]
Γ_{ab}	transition dephasing parameter (equivalent to $1/T_2$) [*92*]
$\Gamma_{ji}(R)$	matrix representation of rotation R [*146*]
$\Gamma_{nn'}(\mathbf{k})$	damping parameter for optical transition in a semiconductor [*253, 256*]
$\delta(x)$	Dirac delta-function [*18*]
δ_{ij}	Kronecker delta [*39*]
$\delta(i,j)$	Kronecker delta (alternative notation) [*124*]
$\delta\mathbb{E}_a$	optical Stark shift of energy level a [*160*]
$\delta\Omega_{ba}$	width of homogeneously-broadened line [*179*]
$\delta\Omega_{ba}^\mathrm{I}$	width of inhomogeneously-broadened line [*185*]
Δ	optical frequency detuning from resonance [*153, 202*]
Δk	magnitude of phase mismatch (in a collinear-wave interaction) [*218*]
$\Delta\mathbf{k}$	vector phase mismatch [*228*]
Δn	induced change in refractive index [*274, 315*]
$\Delta\alpha$	induced change in absorption [*266, 315*]
$\varepsilon(\omega)$	dielectric constant [*83, 174*]
ε_0	permittivity of free space [*295*]
$\boldsymbol{\varepsilon}(\omega)$	dielectric tensor [*174, 212*]
ς	symbol representing all position and spin coordinates of a physical system [*37*]
η	normalisation constant of Boltzmann distribution [*46, 48*]
$\theta(t)$	field area [*171*]
θ_m	phase-matching angle [*222*]

$\boldsymbol{\Theta}$	vector of molecular configuration coordinates [*65*]
$\kappa(\omega)$	imaginary component of refractive index at frequency ω [*175, 214*]
λ	wavelength [*6*]
μ	conventional label for cartesian component of induced polarisation [*21*]
μ_{B}	Bohr magneton [*293, 317*]
μ_0	permeability of free space [*295*]
$\boldsymbol{\mu}_j$	position vector of jth particle [*57*]
$\xi_{nlm_l}(\mathbf{r})$	hydrogenic wave function [*259*]
π_{ab}^{α}	(ab)th matrix element of π_α [*108*]
$\boldsymbol{\pi}$	monomolecular current operator [*107*]
$\boldsymbol{\Pi}$	current operator [*106*]
$\boldsymbol{\Pi}(t)$	current operator in interaction picture [*106*]
ρ, $\rho(t)$	density operator [*45*]
ρ	walk-off angle [*214*]
$\rho_n(t)$	nth-order density operator [*49*]
ρ_0	thermal-equilibrium density operator [*47, 65*]
$\rho_0(a)$	density-matrix element for state labelled a [*66*]
$\sigma_{\mathrm{a}}(\omega)$	absorption cross-section at frequency ω [*181*]
σ_0	d.c. conductivity of free electron gas [*254*]
$\mathrm{d}\sigma_{\mathrm{R}}/\mathrm{d}\Omega$	differential Raman cross-section [*231*]
$\boldsymbol{\sigma}^{(n)}(-\omega_\sigma;\omega_1,\ldots,\omega_n)$	nth-order conductivity tensor [*110*]
τ	time variable [*13*]
τ_{c}	characteristic time scale for fluctuations of the optical field [*153*]
τ_{p}	pulse length [*155*]
τ_{R}	carrier recombination time [*264*]
$\phi(\mathbf{r})$	envelope function [*258, 309*]
$\boldsymbol{\Phi}^{(n)}(-\omega_\sigma;\omega_1,\ldots,\omega_n)$	nth-order envelope response function [*30*]
χ_{R}	Raman susceptibility [*229*]
$\chi^{(n)}(-\omega_\sigma;\omega_1,\ldots,\omega_n)$	nth-order scalar susceptibility [*28*]
$\boldsymbol{\chi}_{\mathrm{mic}}^{(n)}(-\omega_\sigma;\omega_1,\ldots,\omega_n)$	nth-order microscopic susceptibility tensor [*79, 87*]
$\boldsymbol{\chi}^{(n)}(-\omega_\sigma;\omega_1,\ldots,\omega_n)$	nth-order susceptibility tensor [*21*]
$\overline{\boldsymbol{\chi}}(\omega;\mathbf{E}_\omega)$	overall susceptibility at frequency ω of medium under the influence of an optical field $\mathbf{E}_\omega$ [*175*]
ψ, ψ_n	wave function [*37*]
ω	optical frequency [*3*]
ω_{p}, ω_{pe}	plasma frequency [*253, 254*]
ω_{ph}	phonon frequency [*257*]
ω_σ	sum of frequencies $\omega_1+\omega_2+\cdots+\omega_n$ [*21*]
Ω_{ab}	energy of transition between states a and b in units of $\hbar$ [*67*]
$\boldsymbol{\Omega}$	pseudo-field vector, torque vector [*167*]

$< \cdots >$	cycle average [*23*]
$< \cdots >$	orientation average [*85–6*]
$<O>$	expectation value of operator O [*38*]
$<a\|O\|b>$	Dirac bra–ket notation; (ab)th matrix element of operator O [*39*]
∇	vector del operator [*211*]

Bibliography

An extensive list of books on nonlinear optics is highlighted by the letter **B** in the left-hand margin.
The *italic figures in brackets* indicate the pages of text in which references are cited.

Abramowitz, M. and Stegun, I.A. eds. (1965) *Handbook of Mathematical Functions,* Chap. 7. New York: Dover. [*185*]

Abrams, R.L., Lam, J.F., Lind, R.C., Steel, D.G. and Liao, P.F. (1983) *Optical Phase Conjugation,* Chap. 8, ed. R.A. Fisher. New York: Academic Press. [*191*]

Adams, A.R. (1986) *Electron. Lett.,* **22**, 249–50. [*280*]

Adams, M.J. (1981) *An Introduction to Optical Waveguides.* Chichester: John Wiley. [*239*]

B Agrawal, G.P (1989) *Nonlinear Fiber Optics.* New York: Academic Press. [*239, 241, 245*]

B Akhmanov, S.A. and Khokhlov, R.V. (1964) *Problems of Nonlinear Optics.* Moscow: Akad. Nauk. SSR.; English edn. (1972) New York: Gordon & Breach.

B Alfano, R.R. ed. (1989) *The Supercontinuum Laser Source.* New York: Springer-Verlag.

B Allen, L. and Eberly, J.H. (1975) *Optical Resonance and Two-Level Atoms.* New York: Wiley-Interscience. [*156, 165, 167, 171*]

Anderson, J.C. (1964) *Dielectrics.* London: Chapman & Hall. [*85*]

Andrews, D.L. and Thirunamachandran, T. (1977) *J. Chem. Phys.,* **67**, 5026–33. [*86*]

Armstrong. J.A., Bloembergen, N., Ducuing, J. and Pershan, P.S. (1962) *Phys. Rev.,* **127**, 1918–39. [*80, 222*]

B Baldwin, G.C. (1969) *An Introduction to Nonlinear Optics.* New York: Plenum.

Banyai, L. and Koch, S.W. (1986) *Z. Phys. B: Condensed Matter,* **63**, 283–91. [*263, 271–4*]

Bastard, G., Mendez, E.E., Chang, L.L. and Esaki, L. (1982) *Phys. Rev. B,* **26**, 1974–9. [*284*]

Bey, P.P. and Rabin, H. (1967) *Phys. Rev.,* **162**, 794–800. [*144*]

Bjorklund, G.C. (1975) *IEEE J. Quant. Electron.,* **QE-11**, 287–96. [*225, 228*]

Bloch, F. (1946) *Phys. Rev.,* **70**, 460–74. [*164, 167*]

B Bloembergen, N. (1965) *Nonlinear Optics.* New York: Benjamin. [*6*]

Bloembergen, N. (1982) *Rev. Mod. Phys.,* **54**, 685–95.

Bloembergen, N. and Shen, Y.R. (1964) *Phys. Rev. Letters,* **12**, 504–7. [*232*]

Bloembergen, N., Logan, H. and Lynch, R.T. (1978) *Indian J. Pure Appl. Phys.,* **16**, 151–8. [*70, 207*]

Blow, K.J. and Doran, N.J. (1985) *Opt. Commun.,* **52**, 367–70. [*243*]

Blow, K.J. and Doran, N.J. (1987) *IEE Proc. Pt. J (Optoelectronics),* **134**, 138–44. [*241, 242*]

Bottcher, C.J. (1952) *Theory of Electric Polarization.* Amsterdam: Elsevier. [*83, 317*]

Boyd, G.D. and Kleinman, D.A. (1968) *J. Appl. Phys.,* **39**, 3597–639. [*224*]

Brewer, R.G. (1977) *Nonlinear Spectroscopy,* pp. 87–137, Proc. Int. School Phys. 'Enrico Fermi' Course 64, ed. N. Bloembergen. Amsterdam: North-Holland. [*158, 165, 171*]

Brewer, R.G. and Hahn, E.L. (1975) *Phys. Rev. A,* **11**, 1641–9. [*207*]

Brink, D.M. and Satchler, G.R. (1971) *Angular Momentum,* 2nd edn. Oxford: Clarendon Press. [*149*]

Burns, G. (1977) *Introduction to Group Theory with Applications.* New York: Academic Press. [*134, 146*]

Butcher, P.N. (1965) *Nonlinear Optical Phenomena,* Bulletin 200, Engineering Experiment Station. Ohio State Univ. [*xii, 143*]

Butcher, P.N. (1986) *Crystalline Semiconductor Materials and Devices,* pp. 131–94, eds. P.N. Butcher, N.H. March and M.P. Tosi. New York: Plenum Press. [*251, 263*]

Butcher, P.N. and McLean, T.P. (1963) *Proc. Phys. Soc. (GB),* **81**, 219–32. [*131, 254*]

Butcher, P.N. and McLean, T.P. (1964) *Proc. Phys. Soc. (GB),* **83**, 579–89. [*131*]

Byer, R.L. (1977) *Nonlinear Optics,* Chap. 2, eds. P.G. Harper and B.S. Wherrett. London: Academic Press. [*117, 222, 224, 227*]

Byer, R.L. (1989) *Laser Focus World,* March 1989, pp. 77–86. [*222*]

Chang, L.L. and Ploog, K. eds. (1985) *Molecular Beam Epitaxy and Heterostructures.* The Netherlands: Nijhoff. [*276*]

Chemla, D.S. (1975) *Phys. Rev. B,* **12**, 3275–9. [*84, 85*]

Chemla, D.S. (1980) *Rept. Prog. Phys.,* **43**, 1191–262. [*84, 114*]

Chemla, D.S. and Bonneville, R. (1978) *J. Chem. Phys.,* **68**, 2214–20. [*86*]

Chemla, D.S., Miller, D.A.B., Smith, P.W., Gossard, A.C. and Wiegmann, W. (1984) *IEEE J. Quant. Electron.,* **QE-20**, 265–75. [*288, 315*]

Chemla, D.S. and Miller, D.A.B. (1985) *J. Opt. Soc. Am. B,* **2**, 1155–71. [*276, 283, 285, 287, 288*]

Chemla, D.S., Miller, D.A.B. and Smith, P.W. (1985) *Opt. Eng.,* **24**, 556–64. [*284*]

B Chemla, D.S. and Zyss, J. eds. (1987) *Nonlinear Optical Properties of Organic Molecules and Crystals,* **1**: *Quadratic Nonlinear Optical Effects;* **2**: *Cubic Nonlinear Optical Effects.* Orlando: Academic Press. [*113, 117, 149, 275*]

Cohen, H.D. and Roothan, C.C.J. (1965) *J. Chem. Phys.,* **43**, S34–8. [*121*]

Cohen-Tannoudji, C. (1977) *Frontiers in Laser Spectroscopy,* pp. 3–104, eds. R. Balian, S. Haroche and S. Liberman. Amsterdam: North-Holland. [*159*]

Cojan, C., Agrawal, G.P. and Flytzanis, C. (1977) *Phys. Rev. B,* **15**, 909–25. [*84*]

Condon, E.U. and Shortley, G.H. (1957) *The Theory of Atomic Spectra.* Cambridge Univ. Press. [*149, 259, 260*]

Cotter, D. (1986) *Elec. Lett.,* **22**, 693–4. [*276*]

Cotter, D. (1987) *Opt. Quant. Electron.,* **19**, 1–17. [*245*]

Cotter, D. and Hanna, D.C. (1976) *J. Phys. B.: Atom. Molec. Phys.,* **9**, 2165–71. [*149*]

Cotter, D., Ironside, C.N., Ainslie, B.J. and Girdlestone, H.P. (1988) *Ultrafast Phenomena VI,* Springer Ser. in Chem. Phys. **48**, pp. 369–71, eds. T. Yajima, K. Yoshihara, C.B. Harris and S. Shionoya. Berlin: Springer-Verlag. [*276, 293*]

Crisp, M.D. (1973) *Phys. Rev. A,* **8**, 2128–35. [*172*]

Cyvin, S.J., Rauch, J.E. and Decius, J.C. (1965) *J. Chem. Phys.,* **43**, 4083–95. [*86*]

Darwin, C.G. (1934) *Proc. Roy. Soc. (London),* **146A**, 17. [*84*]

B Delone, N.B. and Krainov, V.P. (1988) *Fundamentals of Nonlinear Optics of Atomic Gases.* New York: John Wiley.

B Demtröder, W. (1981) *Laser Spectroscopy.* Berlin: Springer-Verlag.

de Rougemont, F. and Frey, R. (1988) *Phys. Rev. B,* **37**, 1237–44. [*263, 266–70, 274*]

Dirac, P.A.M. (1958) *Principles of Quantum Mechanics,* Oxford Univ. Press. [*37*]

Docherty, V.J., Pugh, D. and Morley, J.O. (1985) *J. Chem. Soc. Faraday Trans. 2,* **81**, 1179–92. [*119, 120*]

Dodd, R.K., Eilbeck, J.C., Gibbon, J.D. and Morris, H.C. (1982) *Solitons and Nonlinear Wave Equations.* London: Academic Press. [*241*]

Dresselhaus, G. (1956) *J. Phys. Chem. Solids,* **1**, 14–22. [*262*]

Ducuing, J. (1969) *Quantum Optics,* Proc. Int. School of Physics, 'Enrico Fermi' Course XLII, pp. 421–72, ed. R.J. Glauber. New York: Academic Press. [*70*]

Ducuing, J. (1977) *Nonlinear Spectroscopy,* Proc. Int. School of Physics, 'Enrico Fermi' Course LXIV, pp. 276–95, ed. N. Bloembergen. Amsterdam: North-Holland. [*113, 116*]

Ducuing, J. and Flytzanis, C. (1968) *C. R. Acad. Sci. Paris,* **B266**, 808–10. [*117*]

B Ducuing, J. and Flytzanis, C. (1972) *Optical Properties of Solids,* pp. 859–990, ed. F. Abelès. Amsterdam: North-Holland.

B Eichler, H.J., Günter, P. and Pohl, D.W. (1986) *Laser-Induced Dynamic Gratings,* Springer Series in Optical Sciences **50.** Berlin: Springer-Verlag. [*195*]

Ekimov, A.I and Onushchenko, A.A. (1984) *JETP Lett.,* **40**, 1136–9. [*293*]

Elliott, R.J. (1957) *Phys. Rev.,* **108**, 1384–9. [*258, 260, 262*]

Erdös, P. (1964) *Helv. Phys. Acta,* **37**, 493–504. [*143*]

B Feld, M.S. and Letokhov, V.S. eds. (1980) *Coherent Nonlinear Optics.* Berlin: Springer-Verlag.

Feneuille, S. (1977) *Rep. Prog. Phys.,* **40**, 1257–1304. [*159*]

Feynman, R.P., Vernon. Jr., F.L. and Hellwarth, R.W. (1957) *J. Appl. Phys.,* **28**, 49–52. [*167*]

B Fisher, R.A. ed. (1983) *Optical Phase Conjugation.* New York: Academic Press. [*195, 235, 236*]

B Flytzanis, C. (1975) *Quantum Electronics: A Treatise,* **1A**: *Nonlinear Optics,* pp. 9–207, eds. H. Rabin and C.L. Tang. New York: Academic Press. [*33, 34, 102–4, 113*]

Flytzanis, C., Hache, F., Ricard, D. and Roussignol, P.L. (1986) *Physics and Fabrication of Microstructures and Microdevices,* pp. 331–42, eds. M.J. Kelly and C. Weisbuch. Berlin: Springer-Verlag. [*276, 293*]

Franken, P.A., Hill, A.E., Peters, C.W. and Weinreich, G. (1961) *Phys. Rev. Lett.,* **7**, 118–9. [*xi, 6*]

Frantz, L.M. and Nodvik, J.S. (1963) *J. Appl. Phys.,* **34**, 2346–9. [*188*]

Friedmann, H. and Wilson-Gordon, A.D. (1978) *Opt. Commun.,* **24**, 5–10; *ibid,* **26**, 193–198. [*196, 201–2*]

Fröhlich, H. (1958) *Theory of Dielectrics,* 2nd edn. Oxford Univ. Press. [*85*]

Fumi, F.G. (1952) *Il Nuovo Cimento,* **9**, 739–56. [*142*]

Garrett, C.G.B. and Robinson, F.N.H. (1966) *IEEE J. Quant. Electron.,* **QE-2**, 328–9. [*6, 118*]

Genzel, L. and Martin, T.P. (1973) *Surf. Sci.,* **34**, 33–49. [*293*]

B Gibbs, H.M. (1985) *Optical Bistability: Controlling Light with Light.* Orlando: Academic Press. [*9, 35, 173, 275, 306*]

Ginsburg, V.L. (1958) *Sov. Phys.-JETP,* **7**, 1096. [*33*]

Goldstone, J.A. and Garmire, E. (1984) *Phys. Rev. Lett.,* **53**, 910–3. [*34–5*]

Grischkowsky, D., Loy, M.M.T. and Liao, P.F. (1975) *Phys. Rev. A,* **12**, 2514–33. [*206*]

Grischkowsky, D. and Balant, A.C. (1982) *Appl. Phys. Lett.,* **41**, 1–3. [*244*]

Hache, F., Ricard, D. and Flytzanis, C. (1986) *J. Opt. Soc. Am. B,* **3**, 1647–55. [*276, 294*]

Halbout, J.-M. and Tang, C.L. (1987) *Nonlinear Optical Properties of Organic Molecules and Crystals,* **1**: *Quadratic Nonlinear Optical Effects,* Chap. II–6, eds. D.S. Chemla and J. Zyss. Orlando: Academic Press. [*222*]

Hall, D.W. and Borrelli, N.F. (1988) *J. Opt. Soc. Am. B,* **5**, 1650–4. [*292, 293*]

B Hann, R.A. and Bloor, D. eds. (1989) *Organic Materials for Nonlinear Optics. Special Publn.* **69.** London: Roy. Soc. Chem.

B Hanna, D.C., Yuratich, M.A. and Cotter, D. (1979) *Nonlinear Optics of Free Atoms and Molecules,* Springer Series in Optical Sciences **17.** Berlin: Springer-Verlag. [*20, 90, 97, 101, 113, 152, 206, 225, 228, 232, 316, 318*]

B Harper, P.G. and Wherrett, B.S. eds. (1977) *Nonlinear Optics.* London: Academic Press.

Hasegawa, A. and Tappert, F. (1973) *Appl. Phys. Lett.,* **23**, 142–4. [*243*]

B Haug, H. ed. (1988) *Optical Nonlinearities and Instabilities in Semiconductors.* Orlando: Academic Press.

Haus, H.A. (1958) *IRE Trans. PGMTT,* **6**, 317. [*123*]

B Haus, H.A. (1984) *Waves and Fields in Optoelectronics.* Englewood Cliffs, NJ: Prentice-Hall.

B Hayes, W. and Loudon, R. (1978) *Scattering of Light by Crystals.* New York: John Wiley. [*34*]
Heitler, W. (1954) *The Quantum Theory of Radiation,* 3rd ed. Oxford Univ. Press. [*200*]
Hellwarth, R.W. (1977) *Progr. Quant. Electron.,* **5**, 2–68. [*149*]
Henry, N.F.M. and Lonsdale, K. eds. (1952) *International Tables for X-Ray Crystallography,* **1**: *Symmetry Groups.* Birmingham, England: Kynoch Press. [*134*]
Hermann, J.P. and Ducuing, J. (1974) *J. Appl. Phys.,* **45**, 5100–2. [*116, 117*]
Hobden, M.V. (1967) *J. Appl. Phys.,* **38**, 4365–72. [*223*]
Hopfield, J.J. (1971) *The Physics of Optoelectronic Materials,* pp. 1–16, ed. W.A. Albers Jr. New York: Plenum. [*114*]
Inkson, J.C. (1983) *Many-Body Theory of Solids,* pp. 11 ff. London: Plenum. [*250*]
Jain, R.K. and Lind, R.C. (1983) *J. Opt. Soc. Am.* **73**, 647–53. [*196, 276*]
Jaynes, E.T. (1957) *Phys. Rev.,* **106**, 620–30; *ibid,* **108**, 171–90. [*47*]
Jerphagnon, J., Chemla, D.S. and Bonneville, R. (1978) *Adv. Phys.,* **27**, 609–50. [*149*]
Joffre, M., Hulin, D., Migus, A., Antonetti, A. and Benoit à la Gullaume, C. (1988) *Opt. Letts.,* **13**, 276–8. [*210*]
Kaiser, W. ed. (1988) *Ultrashort Light Pulses and Their Applications,* Topics in Appl. Phys. **60.** Berlin: Springer-Verlag. [*238*]
Kaiser, W. and Garrett, C.G.B. (1961) *Phys. Rev. Letts.,* **8**, 404–6. [*10*]
Kaiser. W. and Maier, M. (1972) *Laser Handbook* **2**, pp. 1077–150, eds. F.T. Arecchi and E.O. Schulz-DuBois. Amsterdam: North-Holland. [*231, 232*]
Kane, E.O. (1982) *Handbook of Semiconductors,* **1**, pp. 193–217, ed. W. Paul. Amsterdam: North-Holland. [*269, 312–3*]
Kittel, C. (1986) *Introduction to Solid State Physics.* New York: Wiley. [*34, 247–8, 253, 254, 255, 272*]
Kleinman, D.A. (1962) *Phys. Rev.,* **126**, 1977–9. [*123*]
Kleinman, D.A. (1968) *Phys. Rev.,* **174**, 1027–41. [*227*]
Knox, W.H., Hirlimann, C., Miller, D.A.B., Shah, J., Chemla, D.S. and Shank, C.V. (1986) *Phys. Rev. Letts.,* **56**, 1191–3. [*263*]
Koch, S.W., Peyghambarian, N. and Gibbs, H.M. (1988) *J. Appl. Phys.,* **63**, R1–11. [*263, 271, 272–3*]
Kodama, Y. and Hasegawa, A. (1982) *Opt. Lett.,* **7**, 285–7. [*243*]
Kogelnik, H. and Li, T. (1966) *Appl. Opt.* **5**, 1550–67. [*319–20*]
Kohn, W. and Luttinger, J.M. (1955) *Phys. Rev.,* **97**, 869–83; *ibid,* **98**, 915–22. [*309*]
Kriechbaum, M. (1986) *Two-Dimensional Systems: Physics and Devices,* pp. 120–9, eds. G. Bauer, F. Kuchar and H. Heinrich. Berlin: Springer-Verlag. [*309*]
Kubo, R.J. (1957) *J. Phys. Soc. Japan,* **12**, 570–86. [*55*]
Kuhn, H. (1956) *J. Chem. Phys.,* **25**, 293–6. [*116*]
Lalama, S.J. and Garito, A.F. (1979) *Phys. Rev. A,* **20**, 1179–94. [*118*]
Landau, L.D. and Lifshitz, E.M. (1960) *Electrodynamics of Continuous Media.* New York: Pergamon. [*133*]
Lax, M. (1974) *Symmetry Principles in Solid State and Molecular Physics.* New York: Wiley (Interscience). [*134, 143*]
B Letokhov, V.S. and Chebotayev, V.P. (1977) *Nonlinear Laser Spectroscopy,* Springer Series in Optical Sciences **4.** Berlin: Springer-Verlag.
B Levenson, M.D. and Kano, S.S. (1988) *Introduction to Nonlinear Laser Spectroscopy,* rev. edn. New York: Academic Press. [*144, 158, 165, 170, 171, 189, 207*]
Levine, B.F. and Bethea, C.G. (1975) *J. Chem. Phys.,* **63**, 2666–82. [*85*]
Lin, W.Z., Fujimoto, J.G., Ippen, E.P. and Logan, R.A. (1987) *Appl. Phys. Letts.,* **50**, 124–6. [*263*]
Lind, R.C., Steel, D.G. and Dunning, G.J. (1982) *Opt. Eng.,* **21**, 190–8. [*195*]
Lomont, J.S. (1959) *Applications of Finite Groups.* New York: Academic Press. [*134, 142*]
Lorentz, H.A. (1916) *The Theory of Electrons.* Leipzig: Teubner; reprinted (1951) New York: Dover. [*82*]

B Loudon, R. (1983) *The Quantum Theory of Light,* 2nd edn. Oxford: Clarendon Press. [*75, 227, 317*]

Lowitz, D.A. (1967) *J. Chem. Phys.,* **46**, 4698–717. [*118*]

Loy, M.M.T. (1974) *Phys. Rev. Lett.,* **32**, 814–7. [*179*]

Manley, J.M. and Rowe, H.E. (1956) *Proc. IRE,* **44**, 904–13. [*123*]

B Marcuse, D. (1980) *Principles of Quantum Electronics.* New York: Academic Press.

Maruani, A. (1980) *IEEE J. Quant. Electron.,* **QE-16**, 558–66. [*233*]

Merzbacher, E. (1970) *Quantum Mechanics,* 2nd edn. New York: Wiley. [*37, 46*]

Miles, R.B. and Harris, S.E. (1973) *IEEE J. Quant. Electron.,* **QE-9**, 470–84. [*91*]

Miller, D.A.B. (1984) *J. Opt. Soc. Am. B,* **1**, 857–64. [*34, 36*]

Miller, R.C. (1964) *Appl. Phys. Lett.,* **5**, 17–19. [*34, 117*]

Miller, A., Miller, D.A.B. and Smith, S.D. (1981a) *Adv. Phys.,* **30**, 697–800. [*246, 263, 264, 268, 270*]

Miller, D.A.B., Seaton, C.T., Prise, M.E. and Smith, S.D. (1981b) *Phys. Rev. Lett.,* **47**, 197–200. [*264, 268, 270*]

Miller, D.A.B., Chemla, D.S., Damen, T.C., Gossard, A.C., Wiegmann, W., Wood, T.H. and Burrus, C.A. (1985) *Phys. Rev. B,* **32**, 1043–60. [*276, 283, 286, 287*]

Miller, D.A.B., Chemla, D.S. and Schmitt-Rink, S. (1988a) *Optical Nonlinearities and Instabilities in Semiconductors,* pp. 325–59, ed. H. Haug. Orlando: Academic Press. [*208, 286*]

Miller, D.A.B., Chemla, D.S. and Schmitt-Rink, S. (1988b) *Appl. Phys. Lett.,* **52**, 2154–6. [*287*]

Miller, R.C., Kleinman, D.A., Tsang, W.T. and Gossard, A.C. (1981c) *Phys. Rev. B,* **24**, 1134–6. [*283*]

Mollenauer, L.F., Stolen, R.H. and Gordon, J.P. (1980) *Phys. Rev. Lett.,* **45**, 1095–8. [*242*]

B Mollenauer, L.F. and White, J.C. eds. (1987) *Tunable Lasers,* Topics in Applied Physics **59.** Berlin: Springer-Verlag.

Mollenauer, L.F. and Smith, K. (1988) *Opt. Lett.,* **13**, 675–7. [*243*]

Morrell, J.A. and Albrecht, A.C. (1979) *Chem. Phys. Lett.,* **64**, 46–50. [*118*]

Narasimhamurty, T.S. (1981) *Photoelastic and Electro-Optic Properties of Crystals.* New York: Plenum. [*134, 141*]

Nelson, B.P., Cotter, D., Blow, K.J. and Doran, N.J. (1983) *Opt. Commun.,* **48**, 292–4. [*244*]

Neumann, E.-G. (1988) *Single-mode Fibers–Fundamentals,* Springer Series in Optical Sciences **57.** Berlin: Springer-Verlag. [*239*]

Nye, J.F. (1959) *Physical Properties of Crystals.* Oxford: Clarendon Press. [*129, 133, 134, 137, 138*]

Omont, A. (1977) *Prog. Quant. Electron.* **5**, 70–120. [*207*]

Onsanger, L. (1936) *J. Am. Chem. Soc.,* **58**, 1456. [*84*]

Orr, B.J. and Ward, J.F. (1971) *Mol. Phys.,* **20**, 513–526. [*102*]

Oudar, J.-L. (1977) *J. Chem. Phys.,* **67**, 446–57. [*113*]

Oudar, J.-L. (1985) *Nonlinear Optics: Materials and Devices,* pp. 91–103, eds. C. Flytzanis and J.-L. Oudar. Berlin: Springer-Verlag. [*246, 263*]

Pan, C.L., She, C.Y., Fairbank, W.M. and Billman, K.W. (1979) *IEEE J. Quant. Electron.,* **QE-13**, 763–769; *corrigendum: ibid,* **QE-15**, 54. [*113*]

Peyghambarian, N. and Gibbs, H.M. (1988) *Optical Nonlinearities and Instabilities in Semiconductors,* pp. 295–324, ed. H. Haug. Orlando: Academic Press. [*292, 293*]

Placzek, G. (1934) *The Rayleigh and Raman Scattering,* UCRL Transl. 256(L) (1962). Washington DC: US Dept. of Commerce. [*196*]

Pople, J.A. and Beveridge, D.L. (1970) *Approximate Molecular Orbital Theory.* New York: McGraw Hill. [*116, 119, 121*]

Puell, H. and Vidal, C.R. (1976) *Phys. Rev. A,* **14**, 2225–39. [*152*]

Pugh, D. and Morley, J.O. (1987) *Nonlinear Optical Properties of Organic Molecules and Crystals,* **1**: *Quadratic Nonlinear Optical Effects,* Chap. II–2, eds. D.S. Chemla and J. Zyss. Orlando: Academic Press. [*119*]

B Rabin, H. and Tang, C.L. eds. (1975) *Quantum Electronics: A Treatise,* **1**: *Nonlinear Optics.* New York: Academic Press.

B Reintjes, J.F. (1984) *Nonlinear Optical Parametric Processes in Liquids and Gases.* Orlando: Academic Press. [*10, 90, 224, 225*]

Robinson, F.N.H. (1967) *Bell Sys. Tech. J.,* **46**, 913–56. [*114, 116*]

Robinson, F.N.H. (1973) *Macroscopic Electromagnetism.* Oxford: Pergamon Press. [*87*]

Roothan, C.C.J. (1951) *Rev. Mod. Phys.,* **23**, 69–89. [*119*]

Schawlow, A.L. (1982) *Rev. Mod. Phys.,* **54**, 697–707.

Schiff, L.I. (1968) *Quantum Mechanics,* 3rd edn. New York: McGraw-Hill. [*37, 39, 103*]

Schmitt-Rink, S., Ell, C. and Hong, H. (1985) *Proc. 3rd Trieste Semicond. Symp.,* pp. 585–95, ed. M.H. Pilkuhn. Amsterdam: North-Holland. [*287*]

Schmitt-Rink, S., Miller, D.A.B. and Chemla, D.S. (1987) *Phys. Rev. B,* **35**, 8113–25. [*276, 288, 289–91*]

B Schubert, M. and Wilhelmi, B. (1986) *Nonlinear Optics and Quantum Electronics.* New York: Wiley-Interscience. [*144*]

Schultheis, L., Sturge, M.D. and Hegarty, J. (1985) *Appl. Phys. Lett.,* **47**, 995–7. [*210*]

Schultheis, L., Kuhl, J., Honold, A. and Tu, C.W. (1986) *Phys. Rev. Lett.,* **57**, 1797–800. [*210*]

Shah, J. (1985) *IEEE J. Quant. Electron.,* **QE-22**, 1728–43. [*263*]

Shang, C.C. and Hsu, H. (1987) *IEEE J. Quant. Electron.,* **QE-23**, 177–9. [*143, 300*]

Shank, C.V., Fork, R.L., Yen, R., Stolen, R.H. and Tomlinson, W.J. (1982) *Appl. Phys. Lett.,* **40**, 761–3. [*244*]

Shen, Y.R. (1975) *Prog. Quant. Electron.,* **4**, 1–34. [*236*]

B Shen, Y.R. ed. (1977) *Nonlinear Infrared Generation,* Topics in Applied Physics **16.** Berlin: Springer-Verlag.

Shen, Y.R. (1984a) *Phil. Trans. Roy. Soc. Lond. A,* **313**, 327–32. [*34*]

B Shen, Y.R. (1984b) *The Principles of Nonlinear Optics.* New York: Wiley-Interscience. [*xii, 216*]

Shoemaker, R.L. (1978) *Laser Coherence Spectroscopy,* ed. J.I. Steinfeld. New York: Plenum. [*210*]

Shore, D.W. and Menzel, B. (1968) *Principles of Atomic Spectra.* New York: Wiley. [*149*]

Slusher, R.E. (1974) *Progress in Optics,* **XII**, ed. E. Wolf. Amsterdam: North-Holland. [*210*]

Smith, R.A. (1959) *Semiconductors,* pp. 201–11. Cambridge Univ. Press. [*257, 268*]

Snyder, A.W. and Love, J.D. (1983) *Optical Waveguide Theory,* London: Chapman & Hall. [*239*]

Tang, C.L. and Erskine, D.J. (1983) *Phys. Rev. Lett.,* **51**, 840–3. [*263*]

Teng, C.-C. and Garito, A.F. (1983a) *Phys. Rev. Lett.,* **50**, 350–2. [*118*]

Teng, C.-C. and Garito, A.F. (1983b) *Phys. Rev. B,* **28**, 6766–73. [*118*]

Terhune, R.W., Maker, D. and Savage, C.M. (1962) *Phys. Rev. Lett.,* **8**, 404–6. [*316*]

Terhune, R.W. and Weinburger, D.A. (1987) *J. Opt. Soc. Am.,* **4**, 661–674. [*316*]

Tinkham, M. (1964) *Group Theory and Quantum Mechanics.* New York: McGraw-Hill. [*134*]

Tolman, R.C. (1938) *Principles of Quantum Statistics.* Oxford Univ. Press; paperback edn. (1980) New York: Dover. [*37*]

Tomlinson, W.J., Stolen, R.H. and Shank, C.V. (1984) *J. Opt. Soc. Am. B,* **1**, 139–49. [*244*]

Torrey, H.C. (1949) *Phys. Rev.,* **76**, 1059–75. [*171*]

Ward, J.F. (1965) *Rev. Mod. Phys.,* **37**, 1–18. [*75*]

Ward, J.F. and New, G.H.C. (1969) *Phys. Rev.,* **185**, 57–72. [*24, 225*]

Ward, J.F. and Miller, C.K. (1979) *Phys. Rev. A,* **19**, 826 – 33. [*116*]

Weiner, A.M., Heritage, J.P., Hawkins, R.J., Thurston, R.N., Kirschner, E.M., Leaird, D.E. and Tomlinson, W.J. (1988) *Phys. Rev. Lett.,* **61**, 2445 – 8. [*245*]

Weiner, J.S., Miller, D.A.B. and Chemla, D.S. (1987) *Appl. Phys. Lett.,* **50**, 842 – 4. [*315*]

B Wherrett, B.S. and Smith, S.D. eds. (1985) *Optical Bistability, Dynamical Nonlinearity and Photonic Logic.* London: Royal Society.

White, J.C. (1987) *Tunable Lasers,* Topics in Appl. Phys. **59**, pp. 115 – 207, eds. L.F. Mollenauer and J.C. White. Berlin: Springer-Verlag. [*232*]

Wigner, E.P. (1959) *Group Theory.* New York: Academic Press. [*131, 134, 149*]

B Williams, D.J. ed. (1983) *Nonlinear Optical Properties of Organic and Polymeric Materials.* Washington DC: Am. Chem. Soc. [*113*]

Winful, H.G. (1986) *Optical Fibre Transmission,* pp. 179 – 240, ed. E.E. Basch. Indianapolis: Sams. [*239, 245*]

Yao, S.S., Karaguleff, C., Gabel, A., Fortenberry, R., Seaton, C.T. and Stegeman, G.I. (1985) *Appl. Phys. Lett.,* **46**, 801 – 2. [*276*]

Yariv, A. (1978) *IEEE J. Quant. Electron.,* **QE-14**, 650 – 60. [*235*]

B Yariv, A. (1985) *Optical Electronics,* 3rd ed. New York: Holt Saunders.

B Yariv, A. and Yeh, P. (1984) *Optical Waves in Crystals.* New York: Wiley.

B Young, M. (1977) *Optics and Lasers: An Engineering Physics Approach,* Springer Series in Optical Sciences **5**. Berlin: Springer-Verlag.

Yuratich, M.A. (1976) *Nonlinear Optical Susceptibilities of Free Atoms and Molecules,* Ph.D. Thesis. University of Southampton, England. [*145*]

Yuratich, M.A. and Hanna, D.C. (1976) *J. Phys. B: Atom. Molec. Phys.,* **9**, 729 – 50. [*149*]

Yuratich, M.A. and Hanna, D.C. (1977) *Molec. Phys.,* **33**, 671 – 82. [*149*]

Zakharov, V.E. and Shabat, A.B. (1973) *Sov. Phys.-JETP,* **37**, 823 – 8. [*241*]

Zel'dovich, B.Y., Popovichev, V.I., Ragul'skii, V.V. and Faisullov, F.S. (1972) *Sov. Phys.-JETP Lett.,* **15**, 109 – 13. [*236*]

B Zernike, F. and Midwinter, J.E. (1973) *Applied Nonlinear Optics.* New York: Wiley-Interscience. [*305*]

Zyss, J. (1979) *J. Chem. Phys.,* **70**, 3333 – 49. [*121*]

Zyss, J. and Oudar, J.-L. (1982) *Phys. Rev. A.,* **26**, 2028 – 48. [*149*]

Zyss, J. and Chemla, D.S. (1987) *Nonlinear Optical Properties of Organic Molecules and Crystals,* **1**, pp. 23 – 191, eds. D.S. Chemla and J. Zyss. Orlando: Academic Press. [*121*]

Subject index

Further page references may be found in the Glossary of mathematical symbols.